Fourth IMA International Conference on Control Theory

The Institute of Mathematics and its Applications Conference Series

---

RECENT TITLES

Computational Methods and Problems in Aeronautical Fluid Dynamics, *edited by* B.L. Hewitt, C.R. Illingworth, R.C. Lock, K.W. Mangler, J.H. McDonnell, Catherine Richards and F. Walkden
Optimization in Action, *edited by* L.C.W. Dixon
The Mathematics of Finite Elements and Applications II, *edited by* J.R. Whiteman
The State of the Art in Numerical Analysis, *edited by* D.A.H. Jacobs
Fisheries Mathematics, *edited by* J.H. Steele
Numerical Software—Needs and Availability, *edited by* D.A.H. Jacobs
Recent Theoretical Developments in Control, *edited by* M.J. Gregson
The Mathematics of Hydrology and Water Resources, *edited by* E.H. Lloyd, T. O'Donnell and J.C. Wilkinson
Mathematical Aspects of Marine Traffic, *edited by* S.H. Hollingdale
Mathematical Modelling of Turbulent Diffusion in the Environment, *edited by* C.J. Harris
Mathematical Methods in Computer Graphics and Design, *edited by* K.W. Brodlie
Computational Techniques for Ordinary Differential Equations, *edited by* I. Gladwell and D.K. Sayers
Stochastic Programming, *edited by* M.A.H. Dempster
Analysis and Optimisation of Stochastic Systems, *edited by* O.L.R. Jacobs, M.H.A. Davis, M.A.H. Dempster, C.J. Harris and P.C. Parks
Numerical Methods in Applied Fluid Dynamics, *edited by* B. Hunt
Recent Developments in Markov Decision Processes, *edited by* R. Hartley, L.C. Thomas and D.J. White
Power from Sea Waves, *edited by* B. Count
Sparse Matrices and their Uses, *edited by* I.S. Duff
The Mathematical Theory of the Dynamics of Biological Populations II, *edited by* R.W. Hiorns and D. Cooke
Floods due to High Winds and Tides, *edited by* D.H. Peregrine
Third IMA Conference on Control Theory, *edited by* J.E. Marshall, W.D. Collins, C.J. Harris and D.H. Owens
Numerical Methods in Aeronautical Fluid Dynamics, *edited by* P.L. Roe
Numerical Methods for Fluid Dynamics, *edited by* K.W. Morton and M.J. Baines
Multi-Objective Decision Making, *edited by* S. French, R. Hartley, L.C. Thomas and D.J. White
Fourth IMA International Conference on Control Theory, *edited by* P.A. Cook

Fourth IMA International Conference on Control Theory

Based on the proceedings of a conference on Control Theory, organised by The Institute of Mathematics and its Applications and held at Robinson College, Cambridge 11-13 September, 1984

Edited by

P.A. COOK
University of Manchester
Manchester, UK

1985

ACADEMIC PRESS
Harcourt Brace Jovanovich, Publishers
London Orlando San Diego
New York Austin Montreal Sydney
Tokyo Toronto

ACADEMIC PRESS INC. (LONDON) LTD.
24/28 Oval Road
London NW1

United States Edition published by
ACADEMIC PRESS INC.
Orlando, Florida 32887

British Library Cataloguing in Publication Data
International IMA Conference on Control Theory,
(4th : 1984 : Robinson College, Cambridge)
Fourth IMA International Conference on Control Theory : based on the proceedings of a conference on Control Theory.
1. Control theory.
I. Title II. Institute of Mathematics and its Applications III. Cook, P.A.
629.8'312 QA402.3

ISBN 0-12-187260-2

Printed in Great Britain by
Whitstable Litho Ltd., Whitstable, Kent

CONTRIBUTORS

AMBROSINO, G. *Universita di Napoli, Dipartimento di Informatica e Sistemistica, Naples, Italy.*

BANKS, S.P. *University of Sheffield, Department of Control Engineering, Mappin Street, Sheffield, S1 3JD, UK.*

BELL, D.J. *UMIST, Department of Mathematics, P.O. Box 88, Manchester, M60 1QD, UK.*

BIRKETT, N.R.C. *University of Reading, Department of Mathematics, Whiteknights, Reading, RG6 2AX, UK.*

CELENTANO, G. *Universita di Napoli, Dipartimento di Informatica e Sistemistica, Naples, Italy.*

CHANDRASEKHAR, J. *I.I.T., Aeronautical Engineering Department, Bombay - 400 076, India.*

CHAPMAN, M.J. *Coventry (Lanchester) Polytechnic, Department of Mathematics, Priory Street, Coventry, CV1 5FB, UK.*

CHOTAI, A. *University of Lancaster, Department of Environmental Sciences, Lancaster, UK.*

CHRISTODOULAKIS, N.M. *University of Cambridge, Department of Applied Economics, Sidgwick Avenue, Cambridge, CB3 9DE, UK.*

CHU, E.W.K *University of Reading, Department of Mathematics, Whiteknights, Reading, RG6 2AX, UK.*

CORLESS, M. *University of California, Department of Mechanical Engineering, Berkeley, CA 94720, USA.*

CROUCH, P.E. *University of Warwick, Department of Engineering, Coventry, CV4 7AL, UK.*

DORLING, C.M. *University of Sheffield, Department of Applied and Computational Mathematics, Sheffield, S10 2TN, UK.*

FLETCHER, L.R. *University of Salford, Department of Mathematics and Computer Science, Salford, M5 4WT, UK.*

FOO, Y.K. *University of Oxford, Department of Engineering Science, Parks Road, Oxford, OX1 3PJ, UK.*

FRANKE, D. *Hochschule der Bundeswehr Hamburg, Fachbereich Elektrotechnik, Postfach 70 0822, D-2000 Hamburg 70, Germany.*

FRETWELL, P. *University of Hull, Department of Electronic Engineering, Hull, HU6 7RX, UK.*

GAROFALO, F. *Universita di Napoli, Dipartimento di Informatica e Sistemistica, Naples, Italy.*

GRAY, J.O. *University of Salford, Department of Electronic and Electrical Engineering, Salford, M5 4WT, UK.*

GRIMBLE, M.J. *University of Strathclyde, Industrial Control Unit, Department of Electronic and Electrical Engineering, Glasgow, G1 1XW, UK.*

HARRIS, C.J. *The Royal Military College of Science, School of Electrical Engineering and Science, Shrivenham, Swindon, Wilts, SN6 8LA, UK.*

HAYTON, G.E. *University of Hull, Department of Electronic Engineering, Hull, HU6 7RX, UK.*

HINRICHSEN, D. *University of Warwick, Department of Engineering, Coventry, CV4 7AL, UK.*

JOHNSON, M.A. *University of Strathclyde, Industrial Control Unit, Department of Electronic and Electrical Engineering, Glasgow, G1 1XW, UK.*

KALOGEROPOULOS, G. *The City University, Control Engineering Centre, School of Electrical Engineering and Applied Physics, Northampton Square, London, EC1V OHB, UK.*

KARCANIAS, N. *The City University, Control Engineering Centre, School of Electrical Engineering and Applied Physics, Northampton Square, London, EC1V OHB, UK.*

KAUTSKY, J. *The Flinders University of South Australia, School of Mathematical Sciences, Bedford Park, SA 5042, Australia.*

KOTSIOPOULOS, J.L. *Imperial College, Department of Electrical Engineering, London, SW7 2BT, UK.*

LEITMANN, G. *University of California, Department of Mechanical Engineering, Berkeley, CA 94720, USA.*

LJUNG, L. *Linkoping University, Department of Electrical Engineering, S-581 83 Linkoping, Sweden.*

LOCATELLI, A. *Politecnico di Milano, Dipartimento di Elettronica, 20133 Milano, Italy.*

MAGHSOODI, Y. *University of Oxford, Department of Engineering Science, Parks Road, Oxford, OX1 3PJ, UK.*

MARSHALL, J.E. *University of Bath, School of Mathematics, Claverton Down, Bath, BA2 7AY, UK.*

NICHOLS, N.K. *University of Reading, Department of Mathematics, Whiteknights, Reading, RG6 2AX, UK.*

NOLDUS, E.J. *University of Ghent, Automatic Control Laboratory, Grotesteenweg-Noord 2, B-9710 Ghent-Zwijnaarde, Belgium.*

OPITZ, H.P. *Furstenbergstr. 8, D-6450 Hanau 9, Federal Republic of Germany.*

O'REILLY, J. *University of Strathclyde, Industrial Control Unit, Department of Electronic and Electrical Engineering, Glasgow, G1 1XW, UK.*

OWENS, D.H. *University of Strathclyde, Department of Mathematics, 26 Richmond Street, Glasgow, G1 1XH, UK.*

OWENS, T.J. *University of Strathclyde, Industrial Control Unit, Department of Electronic and Electrical Engineering, Glasgow G1 1XW, UK.*

POSTLETHWAITE, I. *University of Oxford, Department of Engineering Science, Parks Road, Oxford, OX1 3PJ, UK.*

PRITCHARD, A.J. *University of Warwick, Control Theory Centre, Coventry, CV4 7AL, UK.*

PUGH, A.C. *Loughborough University of Technology, Department of Mathematics, Loughborough, Leics, UK.*

RICHARD, J.P. *Laboratoire d'Automatique et d'Informatique Industrielle, Institut Industriel du Nord, BP48, 59651 Villeneuve D'Ascq Cedex, France.*

ROTELLA, F. *Laboratoire de Systematique, U.E.R.d'I.E.E.A., Universite des Sciences et Techniques, 59655 Villeneuve d'Ascq, Cedex, France.*

RYAN, E.P. *University of Bath, School of Mathematics, Claverton Down, Bath, BA2 7AY, UK.*

SCATTOLINI, R. *Politecnico di Milano, Dipartimento di Elettronica, 20133 Milano, Italy.*

SCHIAVONI, N. *Politecnico di Milano, Dipartimento di Elettronica, 20133 Milano, Italy.*

TANTAWY, H.S. *Al-Azhar University, Department of Electrical Engineering, Cairo, Egypt.*

VALSAMIS, D. *University of Salford, Department of Electronic and Electrical Engineering, Salford, M5 4WT, UK.*

VERRALL, S. *Electrowatt Engineering Services (UK) Ltd., Horsham, West Sussex, RH12 1UP, UK.*

VIRK, G.S. *University of Southampton, Department of Aeronautics and Astronautics, Southampton, SO9 5NH, UK.*

VITTAL RAO, M.P.S. *University of Ulster at Jordanstown, Department of Electrical and Electronic Engineering, Shore Road, Newtownabbey, Co. Antrim, Northern Ireland, UK.*

WALTON, K. *University of Bath, School of Mathematics, Claverton Down, Bath, BA2 7AY, UK.*

WARWICK, K. *University of Newcastle-upon-Tyne, Department of Electrical and Electronic Engineering, Newcastle-upon-Tyne, UK.*

WILLEMS, K. *University of Ghent, Faculty of Engineering, Grotesteenweg-Noord 2, B-9710 Ghent-Zwijnaarde, Belgium.*

ZAMBETTAKIS, I. *Laboratoire de Systematique U.E.R.d'L.E.E.A., Universite des Sciences et Techniques, 59655 Villeneuve d'Ascq, Cedex, France.*

ZINOBER, A.S.I. *University of Sheffield, Department of Applied and Computational Mathematics, Sheffield, S10 2TN, UK.*

PREFACE

This is the fourth in a series of volumes containing the proceedings of conferences on Control Theory, organised under the auspices of the Institute of Mathematics and its Applications. The previous titles, with the locations and dates of the corresponding conferences, were "Recent Mathematical Developments in Control" (Bath, 1972), "Recent Theoretical Developments in Control" (Leicester, 1976) and "Third IMA Conference on Control Theory" (Sheffield, 1980), all published by Academic Press.

The Fourth IMA International Conference on Control Theory was held at Robinson College, Cambridge, on September 11th - 13th, 1984. It was organised by the IMA Control Group Committee, whose Chairman at that time was Dr. J.F. Marshall, with Dr. A.S.I. Zinober as Secretary. The thirty-four papers included here are all those which were presented at the conference, although in some cases the text has been subsequently revised. Three of the papers were invited articles from distinguished European academics, and these have been grouped together in Part I of the book. Part II comprises the remaining papers, which were accepted on the basis of extended abstracts, after stringent scrutiny by members of the organising committee. They have been classified into five chapters, each covering a particular subject area, inevitably somewhat loosely defined in view of the overlap between different topics. It will be seen that the largest group of papers is concerned with various aspects of Linear System Theory (mainly Algebraic), and that there are also a substantial number in the Optimisation and Optimal Control area. Perhaps the most notable development, however, by comparison with the earlier conferences, is the interest now shown in the control of uncertain systems, robustness against parameter variations, adaptive control and related topics, which are covered by the papers in the final chapter. Since these concepts are of manifest practical importance in Control Engineering, it is most encouraging to find that they are attracting such attention from theoreticians.

As editor of these proceedings, it is my pleasure to acknowledge the invaluable assistance of my colleagues on the Control Group Committee, and of Miss Catherine Richards and her secretarial staff at the IMA. I would also like, on behalf of the committee, to express our thanks to the authorities of Robinson College, whose help and cooperation made the conference possible.

Control Systems Centre,
UMIST,
Manchester.

P.A. Cook
February 1985

ACKNOWLEDGEMENTS

The Institute thanks the authors of the papers, the editor Dr. P.A. Cook (Control Systems Centre, UMIST), and also Miss P. Irving and Miss K. Jenkins for their contributions to the finished manuscript.

CONTENTS

PART I

DECENTRALIZED CONTROL SYSTEM STABILIZATION: TIME INVARIANT AND TIME VARYING FEEDBACK STRATEGIES

J.L. Willems*

(University of Ghent, Belgium)

ABSTRACT

This contribution deals with the problem of stabilizing a dynamic system with a decentralized structure by means of output feedback. The limitations of time-invariant feedback strategies are pointed out; criteria are given which enable one to check stabilizability. It is shown that time-varying periodic feedback strategies may be used to stabilize decentralized control systems which are not stabilizable by means of time-invariant feedback.

1. INTRODUCTION

Stabilization of dynamic systems is a fundamental objective in the design of controllers. Basic results on the stabilization of linear systems by state feedback or by dynamic output feedback are available in the control system literature[1,2]. However, for large-scale systems it is often not realistic to assume that all state or output measurements are transmitted to every control station. Indeed, for example if a system consists of geographically separated components, it is more reasonable to assume that at each control system only partial information on the outputs (or on the state variables) is available[3]. This information is called the local output; it may consist either of the local measurements only,

---

* The author gratefully acknowledges research support from the Fund for Joint Basic Research in Belgium and helpful discussions with Dr. R. Puystjens (University of Gent) concerning the appendices of the paper.

or of local measurements and some information transmitted from other control stations. In this way the cost and the reliability of communication links can be taken into account. Typical problems of this kind show up in electric power systems, traffic networks, communication networks, economic systems, to mention only a few. This leads to the consideration of systems with a decentralized control structure, where every control station can only use the local outputs to cooperate in the control of the global dynamic system. On the other hand, systems where all (state or output) measurements are available at every control station, are called systems with a centralized control structure. The constraints on the available information in decentralized systems complicate the stabilizability problem; in this contribution it is pointed out why control systems which are stabilizable by centralized control, may not be stabilizable by decentralized control. Criteria for analysing stabilizability by decentralized control are discussed. It is shown that the class of stabilizable systems can be considerably enlarged if time-varying decentralized feedback strategies are used.

In the present paper only discrete-time dynamic systems are considered. Similar results are valid for continuous-time dynamic systems.

2. PROBLEM STATEMENT

Consider the linear time-invariant control system described by the difference equation

$$x(t+1) = Ax(t) + Bu(t) \tag{1}$$

with output

$$y(t) = Cx(t) \tag{2}$$

In this model $t \in \mathbb{Z}$ denotes the discrete time variable, $x(t) \in \mathbb{R}^n$ represents the state of the system, $u(t) \in \mathbb{R}^m$ the input, and $y(t) \in \mathbb{R}^p$ the output. A, B and C are constant matrices of appropriate dimension. Dynamic output feedback uses a dynamic compensator

$$z(t+1) = Mz(t) + Py(t) \tag{3}$$

$$u(t) = Rz(t) + Sy(t) + v(t) \tag{4}$$

to influence the properties of the system. In these expressions $v(t) \in \mathbb{R}^m$ denotes the external input. The controlled system is governed by

$$\begin{bmatrix} x(t+1) \\ z(t+1) \end{bmatrix} = \begin{bmatrix} A+BSC & BR \\ PC & M \end{bmatrix} \begin{bmatrix} x(t) \\ z(t) \end{bmatrix} + \begin{bmatrix} B \\ 0 \end{bmatrix} v(t) \quad (5)$$

The question of pole assignment or stabilizability by means of centralized control is whether there exist matrices M, P, R, S, such that (5) has pre-assigned eigenvalues or is an asymptotically stable system. This model also includes the case of static feedback by considering $R = 0$, and the case of state feedback by letting C be the identity matrix. The solution of the pole assignment and the stabilizability problems for centralized control is well known[1,2]: pole assignment is possible iff the given system (1),(2) is controllable and observable, stabilizability by output feedback is equivalent to stabilizability and detectability of (1),(2), that is the property that all unobservable and/or uncontrollable eigenvalues of the matrix A are asymptotically stable.

If the system has a decentralized structure, then the control input is realized by a number of controllers (say N), such that u(t) consists of N local inputs:

$$u(t) = \begin{bmatrix} u_1(t) \\ u_2(t) \\ \cdot \\ \cdot \\ \cdot \\ u_N(t) \end{bmatrix}$$

where $u_i(t) \in \mathbb{R}^{m_i}$ $(i=1,\ldots,N)$ denotes the input realised by the i-th controller. Also only partial output information is available at each control station. Hence

$$y(t) = \begin{bmatrix} y_1(t) \\ y_2(t) \\ \cdot \\ \cdot \\ \cdot \\ y_N(t) \end{bmatrix}$$

where $y_i(t) \in \mathbb{R}^{p_i}$ $(i=1,\ldots,N)$ is the output available at the i-th control station; there a dynamic compensator is realized of the form

$$z_i(t+1) = M_i z_i(t) + P_i y_i(t) \tag{6}$$

$$u_i(t) = R_i z_i(t) + S_i y_i(t) + v_i(t) \tag{7}$$

The decentralized control system structure is shown in Fig. 1. Its model is a special case of the general model corresponding to (1)-(4), where now $z(t)$ and $v(t)$ consist of the subvectors $z_i(t)$ and $v_i(t)$ respectively, and where the matrices M, P, R, S are block diagonal:

$$M = \text{block diag}\,(M_1, M_2, \ldots, M_N)$$

$$P = \text{block diag}\,(P_1, P_2, \ldots, P_N)$$

$$R = \text{block diag}\,(R_1, R_2, \ldots, R_N)$$

$$S = \text{block diag}\,(S_1, S_2, \ldots, S_N)$$

Let B and C be partitioned as

$$B = [B_1 \; B_2 \; \ldots \; B_N]$$

and

$$C = \begin{bmatrix} C_1 \\ C_2 \\ \cdot \\ \cdot \\ \cdot \\ C_N \end{bmatrix}$$

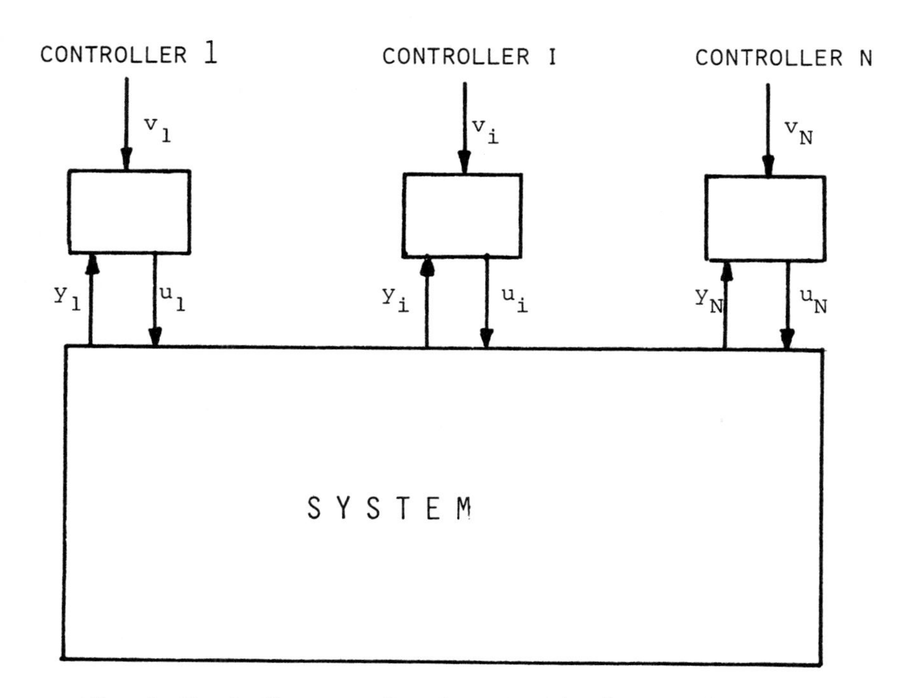

Fig. 1. Block diagram of a decentralized control system

such that

$$Bu(t) = \sum_{i=1}^{N} B_i u_i(t)$$

$$y_i(t) = C_i x(t)$$

If the matrix R_i is zero, the i-th control station uses a local static output feedback strategy. If all matrices R_i vanish, then the control corresponds to <u>decentralized static output feedback</u>.

The question of pole assignment or stabilizability by means of decentralized control is whether there exist matrices M, P, R and S, with the required block diagonal structure, such that the closed loop system (5) had preassigned eigenvalues or is asymptotically stable. It is obvious that pole assignment or stabilizability by means of centralized control is a necessary condition for the property to hold in the decentralized case; however the condition is not sufficient.

Fundamental results on these problems have been obtained by Wang and Davison[4] and by Corfmat and Morse[5]. However we will try to make the present contribution as self contained as possible; therefore most of the original results will be rederived. In addition some of the results are proved in a completely different and, in our opinion, simpler way.

3. STATIC OUTPUT FEEDBACK: THE CONCEPT OF FIXED MODES

We first consider the question which eigenvalues of the system matrix A can be relocated by static output feedback control. The motivation for this approach is given in Theorem 1 below, where a well known result for centralized control is reformulated. Let $\underline{S}$ be the set of block diagonal matrices

$$\underline{S} := \left\{ S \in \mathbb{R}^{m \times p} \mid S = \text{block diag}\,(S_1, S_2, \ldots, S_N),\; S_i \in \mathbb{R}^{m_i \times p_i} \right\}$$

The spectrum of a matrix M, that is the set of its eigenvalues, is denoted by $\sigma(M)$.

<u>Definition 1</u>

(a) The set of fixed modes of system (1),(2) is defined as

$$\Lambda(C,A,B) := \bigcap_{S \in \mathbb{R}^{m \times p}} \sigma(A+BSC)$$

(b) The set of fixed modes of system (1),(2) <u>with respect to the decentralized control</u> corresponding to the set $\underline{S}$ is defined as

$$\Lambda(C,A,B;\underline{S}) := \bigcap_{S \in \underline{S}} \sigma(A+BSC)$$

The sets $\Lambda(C,A,B)$ and $\Lambda(C,A,B;\underline{S})$ are clearly the sets of eigenvalues which cannot be moved by centralized or decentralized static output feedback, respectively. Obviously

$$\Lambda(C,A,B) \subset \Lambda(C,A,B;\underline{S})$$

It can readily be shown that the set of fixed modes $\Lambda(C,A,B)$ is the union of the set of uncontrollable modes and the set of unobservable modes of the system (1),(2). This also follows from Theorem A.1 in Appendix A. This observation yields the following reformulation of the classical pole assignment and stabilization result for centralized control systems:

Theorem 1 All eigenvalues of the closed loop control system (5) can arbitrarily be assigned by dynamic centralized output feedback (provided complex eigenvalues occur in complex conjugate pairs), except the fixed modes. These are fixed with respect to dynamic output feedback as well.

In the decentralized structure the static output feedback control

$$u_i(t) = S_i y_i(t) \tag{8}$$

yields the closed loop system matrix

$$A_{cl} = A + \sum_{i=1}^{N} B_i S_i C_i \tag{9}$$

which is the explicit expression of A+BSC for $S \in \underline{S}$. The fixed modes with respect to decentralized control can be identified by means of Theorem A.1 in Appendix A. Note that, without loss of generality, B_i and C_i may be assumed to have rank one, or B_i may be assumed to be a column vector and C_i a row vector ($m_i = p_i = 1$; $i=1,\ldots,N$); this follows from

$$\sum_{i=1}^{N} B_i S_i C_i = \sum_{i=1}^{N} \sum_{\alpha=1}^{m_i} \sum_{\beta=1}^{p_i} s_i^{\alpha\beta} b_i^{\alpha} c_i^{\beta}$$

where b_i^{α} is the α-th column of B_i, c_i^{β} the β-th row of C_i, and $s_i^{\alpha\beta}$ the entry on row α and column β and S_i.

Theorem 2 The eigenvalue λ^* of the matrix A is a fixed mode with respect to decentralized control if and only if there exist two disjoint subsets

$$\pi := \{i_1, i_2, \ldots, i_k\}$$

and

$$\rho := \{i_{k+1}, i_{k+2}, \ldots, i_N\}$$

of

$$\nu := \{1, 2, \ldots, N\}$$

with

$$\pi \cup \rho = \nu$$

such that the rank of the matrix

$$M(\pi,\rho) := \begin{bmatrix} \lambda^* I - A & B_{i_1} & B_{i_2} & \cdots & B_{i_k} \\ C_{i_{k+1}} & 0 & 0 & \cdots & 0 \\ C_{i_{k+2}} & 0 & 0 & \cdots & 0 \\ \vdots & & & \vdots & \\ C_{i_N} & 0 & 0 & \cdots & 0 \end{bmatrix}$$

is less than n, where I denotes the identity matrix in $\mathbb{R}^{n \times n}$.

Note that Theorem 2 clearly shows that any eigenvalue which is uncontrollable or unobservable (and hence a fixed mode with respect to centralized control) is also a fixed mode with respect to decentralized control; these cases correspond to $\pi = \phi$ or $\rho = \phi$. Theorem 2 also includes the case of a centralized control structure ($\nu = \{1\}$). The criterion of Theorem 2 is due to Anderson and Clements[6].

From Theorem 2 it follows that an eigenvalue λ^* such that rank $[M(\pi,\rho)] < n$, remains a fixed mode if all control stations corresponding to π are merged, as well as all control stations corresponding to ρ. Information exchange between control stations corresponding to π or between control stations corresponding to ρ is not helpful to move the mode λ^*.

If the rank of matrix (λ^*I-A) is equal to $(n-1)$, then the condition of Theorem 2

$$\text{rank } M(\pi,\rho) = n-1$$

is equivalent to the uncontrollability of the mode λ^* with respect to $B_{i_1}, B_{i_2},\ldots,B_{i_k}$, the unobservability of the mode λ^* with respect to $C_{i_{k+1}}, C_{i_{k+2}},\ldots,C_{i_N}$, and

$$\left[C_i(Is-A)^{-1}B_j\right]_{s=\lambda^*} = 0 \quad \text{for } i \in \rho,\ j \in \pi.$$

4. DYNAMIC OUTPUT FEEDBACK: STRONGLY CONNECTED SYSTEMS

It is well known for centralized control systems that eigenvalues which cannot be moved by static output feedback are also fixed modes for dynamic output feedback. It is readily seen that the same is true for decentralized control systems. Then the system equations are

$$x(t+1) = Ax(t) + \sum_{i=1}^{N} B_i u_i(t)$$

$$z_i(t+1) = M_i z_i(t) + w_i(t) \qquad (i=1,\ldots,N)$$

$$y_i(t) = C_i x(t) \qquad (i=1,\ldots,N)$$

$$r_i(t) = z_i(t) \qquad (i=1,\ldots,N)$$

where $u_i(t), w_i(t)$ is the input and $y_i(t)$, $r_i(t)$ the output at the i-th control station. Theorem 2 then immediately shows that fixed modes with respect to decentralized static output feedback are also fixed with respect to decentralized dynamic output feedback.

The next question is to what extent the non-fixed eigenvalues can be moved. It is very difficult to determine the eigenvalue locations that can be realized by static output feedback, even for centralized systems. For centralized control systems it is well known that arbitrary eigenvalue assignment can be achieved by dynamic output feedback. This question is now discussed for decentralized systems.

In classical (centralized) control system theory one particular approach to the problems of the pole assignment or stabilization by means of dynamic output feedback consists of using a static output feedback to make the system controllable from a single input channel and observable from a single output component; then the dynamic compensation can be considered for the system with scalar input and scalar output. In the present section a similar approach is used for the decentralized control problem; this idea is due to Corfmat and Morse[5]. For this analysis the concept of the graph of a decentralized control system is useful. The definitions are given in Appendix B.

As shown in Appendix B the decentralized control system can be decomposed into strongly connected subsystems which can be dealt with as completely decoupled for the analysis of stabilizability and pole assignment. A fixed mode of the global system may be, <u>with respect to the relevant subsystem</u>,

(i) either unobservable or uncontrollable, and hence a fixed mode with respect to centralized output feedback in the subsystem, or
(ii) only a fixed mode with respect to the decentralized control of the subsystem.

Case (i) is not analysed further since it is well known from classical linear control system theory. Only case (ii) is discussed further.

In the sequel it is therefore assumed that the decentralized control system under consideration is strongly connected and that it is controllable and observable as a centralized control system. The question now is whether or not there exists a decentralized <u>static</u> output feedback which makes the system controllable and observable with respect to a single control station. The results given in Appendix C show that this is indeed the case except if the system has fixed modes. This discussion leads to the conclusion that arbitrary eigenvalue assignment is possible except for the modes which are fixed with respect to static feedback. This is summarized in the following theorem.

<u>Theorem 3</u> In a decentralized control system arbitrary pole assignment can be achieved in every strongly connected

subsystem except for the fixed modes; in every subsystem the complex eigenvalues should occur in complex conjugate pairs. The pole assignment can be achieved by an (almost arbitrary) static output feedback and a dynamic output feedback from a single controller in every strongly connected subsystem.

Note that the result stated in Theorem 3 is slightly weaker than the property that all eigenvalues which can be moved by decentralized static output feedback can arbitrarily be assigned by decentralized dynamic output feedback provided complex eigenvalues occur in complex conjugate pairs; indeed the feature of complex conjugate pairs of eigenvalues must hold in every strongly connected component of the system. The dynamic compensator can be designed using the theory developed for centralized control systems; the design can be performed separately and independently in every strongly connected subsystem.

5. TIME- VARYING DYNAMIC OUTPUT FEEDBACK

One readily checks that in centralized feedback systems fixed modes (which are unobservable or uncontrollable modes) cannot be moved by time-varying static or dynamic output feedback. In the present section it is investigated whether or not the same is true for decentralized control systems. We only consider the case that all fixed modes are simple eigenvalues of the plant matrix, or can be made simple by a preliminary decentralized static output feedback. This implies that for every fixed mode the matrix $M(\pi,\rho)$ considered in section 3 has rank equal to $(n-1)$. It can then be assumed that the rank of the matrix (λ^*I-A) is $(n-1)$, which means that the preliminary feedback has been included if necessary.

The following periodic control strategy is used:

$$u_i(kr+j) = K_i(j)y_i(kr)$$

where $r \in \mathbb{Z}$ denotes the period, $j \in \{0,1,\ldots,r-1\}$ and $i \in \{1,2,\ldots,N\}$ denotes the control station. Then

$$x(kr+r) = A^r x(kr) + \sum_{j=0}^{r-1} \sum_{i=1}^{N} A^{r-j-1} B_i K_i(j) C_i x(kr) \tag{10}$$

This equation can be seen as the difference equation of a linear <u>time-invariant</u> discrete time system with sampling period r, with plant matrix A^r, and with rN control stations, whose pairs of input and output matrices are

$$(B_1,C_1),\ (AB_1,C_1)\ ,\ldots,\ (A^{r-1}B_1,C_1),$$
$$(B_2,C_2),\ (AB_2,C_2)\ ,\ldots,\ (A^{r-1}B_2,C_2),$$
$$\ldots$$
$$(B_N,C_N),\ (AB_N,C_N)\ ,\ldots,\ (A^{r-1}B_N,C_N).$$

Let r be chosen such that $A^r - \lambda^r I$ has rank (n-1) for every fixed mode λ of the original system, and also such that

$$\text{im}(B_i) + \text{im}(AB_i) + \ldots + \text{im}(A^{r-1}B_i)$$

is an A-invariant vector space for i=1,...,N. Then Appendix D shows that (10) has no fixed modes if the system is strongly connected and globally observable and controllable.

Thus the following algorithm is obtained for pole assignment in a strongly connected decentralized system which is observable and controllable as a centralized system: pick an (almost arbitrary) periodic feedback strategy $\{K_i(j)\}$, where r satisfies the conditions stated above; then a dynamic feedback in a single control station using the measurements $y_i(kr-r)$, $y_i(kr-2r)$,... to generate $u_i(kr)$ can be designed to achieve pole assignment. If a decentralized control system is not strongly connected, then for any strongly connected component the above procedure can be used. Hence all fixed modes can be assigned by means of dynamic time-varying output feedback except those which are uncontrollable or unobservable at the subsystem level.

Interesting features of this result are:

(i) the dynamic feedback element can be slow;
(ii) the dynamic feedback element can be designed using algorithms developed for linear time-invariant control systems.

The idea of using time-varying output feedback was suggested by Anderson and Moore[7] and applied to some particular examples by Wang[8]. It is an interesting fact that fixed eigenvalues can be relocated by time-varying output feedback for decentralized systems whereas this is not the case for centralized systems. An explanation for this phenomenon is that in the latter case the fixed modes are uncontrollable or unobservable ; for an uncontrollable eigenvalue the column eigenvector is independent of the feedback coefficients; for an unobservable eigenvalue the same is true for the row eigenvector. This is not the case for fixed modes of decentralized systems.

6. EXAMPLE

Consider the following decentralized control system with two control stations:

$$x(t+1) = \begin{bmatrix} 1 & 0 & 0 \\ 0 & 2 & 0 \\ 0 & 0 & 3 \end{bmatrix} x(t) + \begin{bmatrix} 0 \\ 1 \\ 1 \end{bmatrix} u_1(t) + \begin{bmatrix} 1 \\ 0 \\ 0 \end{bmatrix} u_2(t)$$

$$y_1(t) = [1 \quad 0 \quad 0]\, x(t)$$

$$y_2(t) = [0 \quad 1 \quad -2]\, x(t)$$

It is readily checked that the system is strongly connected and that it is observable and controllable as a centralized control system. The closed loop system matrix for decentralized static output feedback

$$A + k_1 b_1 c_1 + k_2 b_2 c_2 = \begin{bmatrix} 0 & k_2 & -2k_2 \\ k_1 & 2 & 0 \\ k_1 & 0 & 3 \end{bmatrix}$$

has a fixed mode $\lambda^* = 1$. Use the periodic feedback strategy ($k \in \mathbb{Z}$)

$$u_1(2k) = y_1(2k) + u_1'(2k)$$

$$u_2(2k) = y_2(2k) + u_2'(2k)$$

$$u_1(2k+1) = 0$$

$$u_2(2k+1) = 0$$

This yields the new decentralized control system

$$x(2k+2) = A_e\, x(2k) + b_{1e}u_1'(2k) + b_{2e}u_2'(2k)$$

$$y_1(2k) = [1 \quad 0 \quad 0]\, x(2k)$$

$$y_2(2k) = [0 \quad 1 \quad -2]\, x(2k)$$

where

$$A_e = \begin{bmatrix} 1 & 1 & -2 \\ 2 & 2 & 0 \\ 3 & 0 & 3 \end{bmatrix}$$

$$b_{1e} = \begin{bmatrix} 0 \\ 2 \\ 3 \end{bmatrix}, \quad b_{2e} = \begin{bmatrix} 1 \\ 0 \\ 0 \end{bmatrix}$$

This new decentralized control system is strongly connected, has no fixed modes and is observable and controllable with respect to the first control station. Eigenvalue assignment can hence be achieved by dynamic feedback, where the system relating $y_1(2k-2)$, $y_1(2k-4)$,... and $u_1'(2k)$ is time-invariant with respect to a sampling period equal to 2.

APPENDIX A

Lemma A.1

Let A and B be (m x n) - matrices with real or complex entries. Then

$$\text{rank}\begin{bmatrix} A \\ B \end{bmatrix} - \text{rank}\ |A\ B| = \dim\ (\text{im}\ A \cap \text{im}\ B) - \dim\ A^T \cap \text{im}\ B^T)$$

where dim denotes the dimension of a subspace and im the image or range space of a matrix.

Lemma A.2

Let A and B be (m x n)-matrices with real or complex entries, and let B have rank one. Then

$$\max\{\text{rank}(A+kB)\ |\ k \in \mathbb{C}\} = \min\{\text{rank}\begin{bmatrix} A \\ B \end{bmatrix},\ \text{rank}\ [A\ B]\}$$

Lemma A.1 is proved in a straightforward way; the latter lemma readily follows from the former one. Lemma A.2 can be generalized to yield the following theorem.

Theorem A.1

Let A, B_1, B_2,...,B_N be (m x n)-matrices with real or complex entries; let the matrices B_1, B_2,...,B_N have rank one. Then

$$\max\{\text{rank}(A+\sum_{i=1}^{N} k_i B_i)\ |\ k_i \in \mathbb{C}\}$$

$$= \min\{\text{rank}\ M(\pi,\rho)\ |\ \pi \subset \nu, \rho \subset \nu, \pi \cap \rho = \phi,\ \pi \cup \rho = \nu\}$$

with

$$\nu := \{1,2,...,N\}$$

$$M(\pi,\rho) := \begin{bmatrix} A & B_{i_1} & B_{i_2} & \cdots & B_{i_k} \\ B_{i_{k+1}} & 0 & 0 & \cdots & 0 \\ B_{i_{k+2}} & 0 & 0 & \cdots & 0 \\ \vdots & & & \vdots & \\ B_{i_N} & 0 & 0 & \cdots & 0 \end{bmatrix}$$

$$\pi := \{i_1, \ldots, i_k\}$$

$$\rho := \{i_{k+1}, \ldots, i_N\}$$

APPENDIX B

Consider the decentralized control system defined in Section 2. A directed graph is associated with this system in the following way: a node corresponds with a control station; a directed arc exists from node i to node j iff the transfer function matrix

$$H_{ji}(s) := C_j(Is-A)^{-1}B_i$$

does not vanish identically.

Definition B.1 A decentralized control system is strongly connected iff there exists a path between every pair of distinct nodes in the graph of the system.

If a decentralized control system is not strongly connected, then there exists a partition of the nodes (and thus of the control stations) such that $H_{ji}(s)$ is identically zero for $j \in \alpha$, $i \in \beta$, with

$$\alpha \cup \beta = \{1,2,\ldots,N\}$$

$$\alpha \cap \beta = \phi$$

$$\alpha \neq \phi \ , \ \beta \neq \phi$$

From geometric linear system theory it follows that there then exists a linear transformation of the state such that the system data have the following form:

$$A = \begin{bmatrix} A_{11} & A_{12} \\ 0 & A_{22} \end{bmatrix}$$

$$B_i = \begin{bmatrix} B_{i1} \\ 0 \end{bmatrix} \qquad (i \in \beta)$$

$$C_j = [0 \quad C_{j2}] \qquad (j \in \alpha)$$

with

$$A_{11} \in \mathbb{R}^{axa},\ A_{12} \in \mathbb{R}^{axb},\ A_{22} \in \mathbb{R}^{bxb},\ B_{i1} \in \mathbb{R}^{axm_i},\ C_{j2} \in \mathbb{R}^{p_i xb},$$

a+b=n. It is clear that, as far as the eigenvalues of the closed loop system matrix

$$A_{cl} = A + \sum_{i=1}^{N} B_i S_i C_i$$

are concerned, the decentralized control system splits up into two subsystems without interaction:

$$(S_1) \begin{cases} x_1(t+1) = A_{11}x_1(t) + \sum\limits_{i\in\beta} B_{i1}u_i(t) \\ y_i(t) = C_{i1}x_1(t) \qquad (i \in \beta) \end{cases}$$

and

$$(S_2) \begin{cases} x_2(t+1) = A_{22}x_2(t) + \sum\limits_{i\in\alpha} B_{i2}u_i(t) \\ y_i(t) = C_{i2}x_2(t) \qquad (i \in \alpha) \end{cases}$$

Subsystem (S_1) or subsystem (S_2) (or both) may have a centralized control structure; this corresponds to α or β being a singleton. Then the strong connectedness of the subsystem is checked. A subsystem which is not strongly connected, is decomposed into subsystems. In this way a decomposition of the system into strongly connected components is obtained; this corresponds to a block triangular structure of the transfer function matrix $C(Is-A)^{-1}B$. For the analysis of the pole assignment and stabilizability properties, the subsystems may be considered to be completely decoupled. The sum of the dimensions of the subsystems is equal to the dimension of the total system.

APPENDIX C

Consider the decentralized control system defined in Section 2. In the present section we discuss the question whether the system can be made observable from a single control station by means of decentralized static output feedback. First we consider the case of two control stations with input and output matrices of rank one, which we denote by b_1, b_2, and c_1, c_2 respectively. Observability of the pair $(A+s_1b_1c_1+s_2b_2c_2, c_1)$ is equivalent to

$$\text{rank}\begin{bmatrix} A+s_1b_1c_1+s_2b_2c_2-\lambda I \\ c_1 \end{bmatrix} = n \ \forall\lambda \in \mathbb{C} \qquad (C.1)$$

A mode λ^* is unobservable if (C.1) does not hold for $\lambda = \lambda^*$.

We first investigate whether there exist constant values of λ for which the above condition is not satisfied for whatever values of s_1 and s_2. Theorem A.1 of Appendix A requires

(i) either

$$\text{rank}\begin{bmatrix} A-\lambda I & b_2 \\ c_1 & 0 \end{bmatrix} < n \qquad (C.2)$$

(ii) or

$$\text{rank}\begin{bmatrix} A-\lambda I \\ c_1 \\ c_2 \end{bmatrix} < n \qquad (C.3)$$

If the system is observable as a centralized system, (C.3) cannot hold. Hence the only possible cases of constant modes which cannot be made observable by decentralized static output feedback are the fixed modes satisfying (C.2). Similarly the only possible cases of constant modes which cannot be made controllable with respect to the first control station by decentralized static output feedback are the fixed modes satisfying

$$\text{rank}\begin{bmatrix} A-\lambda I & b_1 \\ c_2 & 0 \end{bmatrix} < n$$

The second case to be considered is whether there exists a mode, dependent on s_1 and s_2, which remains unobservable for all s_1 and s_2. Then (C.1) requires the existence of an eigenvector $v(s_1,s_2)$ and an eigenvalue $\lambda(s_1,s_2)$ such that

$$(A+s_1b_1c_1 + s_2b_2c_2)v(s_1,s_2) = \lambda(s_1,s_2)v(s_1,s_2) \quad (C.4)$$

$$c_1v(s_1,s_2) = 0 \quad (C.5)$$

This yields

$$c_1Av(s_1,s_2) + s_2c_1b_2c_2v(s_1,s_2) = 0 \quad (C.6)$$

If $c_1b_2 \neq 0$, then

$$[A - (c_1b_2)^{-1}b_2c_1A]v(s_1,s_2) = \lambda(s_1,s_2)v(s_1,s_2)$$

Hence $\lambda(s_1,s_2)$ must be independent of s_1 and s_2. If c_1b_2 vanishes, but $c_1Ab_2 \neq 0$, then (C.6) yields

$$c_1Av(s_1,s_2) = 0$$

and (C.4) leads to

$$c_1A^2v(s_1,s_2) + s_2c_1Ab_2c_2v(s_1,s_2) = 0$$

and hence

$$[A - (c_1Ab_2)^{-1}b_2c_1A^2]v(s_1,s_2) = \lambda(s_1,s_2)v(s_1,s_2)$$

Again we conclude that $\lambda(s_1,s_2)$ must be independent of s_1 and s_2. This same conclusion can be obtained if $c_1A^kb_2 \neq 0$ for some $k \in \mathbb{N}$, that is if $c_1\exp(At)b_2$ does not vanish identically. This shows that if the system is strongly connected all modes except the fixed modes for which (C.2) holds, can be made observable from the first control station. This analysis can be generalized in a straightforward way to the case that there are more than two control stations and to the case that the input and output matrices are not of rank one. Also controllability can be dealt with in the same way. We obtain:

Theorem C.1 If a decentralized system is strongly connected, then all modes, except the fixed modes, can be made observable and controllable from a single control station by means of a decentralized static output feedback.

Note that the existence of at least one decentralized static output feedback law realizing observability and controllability of a mode implies that this property is achieved by almost any decentralized static output feedback strategy.

APPENDIX D

Let $A \in \mathbb{C}^{n\times n}$, $B \in \mathbb{C}^{n\times m}$, $C \in \mathbb{C}^{p\times n}$ and

$$M := \begin{bmatrix} A & B \\ C & O \end{bmatrix}$$

where O is the zero matrix in $\mathbb{C}^{p\times m}$.

Theorem D.1 If im(B) is A-invariant and

$$\operatorname{rank}(M) = \operatorname{rank}(A) \tag{D.1}$$

then

$$CB = O$$

and also

$$CA^kB = O \qquad \forall\, k \in \mathbb{N}$$

Proof From

$$\operatorname{rank}(A) \leq \operatorname{rank}\begin{bmatrix} A \\ B \end{bmatrix}, \ \operatorname{rank}\,[A\ B] \leq \operatorname{rank}(M)$$

and (D.1), we conclude the existence of a matrix $R \in \mathbb{C}^{p\times n}$ such that

$$[C \quad O] = R[A \quad B]$$

Since im(B) is A-invariant, there exists a matrix $Q \in \mathbb{C}^{m \times n}$ such that

$$AB = BQ$$

We readily obtain

$$CB = RAB = RBQ = O$$

Since im(B) is A-invariant, we also conclude

$$CA^{k}B = O \quad \forall\, k \in \mathbb{N}$$

7. REFERENCES

1. Kwakernaak, H., and Sivan, R. (1973) Linear Optimal Control Systems, Wiley-Interscience, New York.

2. Wonham, W.M. (1979) Linear Multivariable Control: A Geometric Approach, 2nd ed., Springer-Verlag, New York.

3. Aoki, M. (1972) On feedback stabilizability of decentralized dynamic systems, Automatica, Vol. 5, pp. 163-173.

4. Wang, S.H. and Davison, E.J. (1973) On the stabilization of decentralized control systems, IEEE Transactions on Automatic Control, Vol. AC-18, pp. 473-478.

5. Corfmat, J.P. and Morse, A.S. (1976) Decentralized control of linear multivariable systems, Automatica, Vol. 12, pp. 479-495.

6. Anderson, B.D.O. and Clements, D.J. (1981) Algebraic characterization of fixed modes in decentralized control, Automatica, Vol. 17, pp. 703-712.

7. Anderson, B.D.O. and Moore, J.B. (1981) Time-varying feedback laws for decentralized control, IEEE Transactions on Automatic Control Vol. AC-26, pp. 1133-1139.

8. Wang, S.H. (1982) Stabilization of decentralized control systems via time-varying controllers, IEEE Transactions on Automatic Control, Vol. AC-27, pp. 741-744.

FREQUENCY DOMAIN PROPERTIES OF IDENTIFIED TRANSFER FUNCTIONS

L. Ljung
*(Department of Electrical Engineering,
Linkoping University, Sweden)*

ABSTRACT

A black box view on the problem of estimating transfer functions of linear dynamical systems is taken. Parametric methods, such as the "least squares method", the "maximum likelihood method", the "output error method" etc are considered, but the properties of the resulting models are evaluated in terms of their frequency domain properties.

A general characterization of how the bias is distributed over the frequency domain is given and the effects of the bias of various design variables are displayed.

A simple asymptotic expression for the variance of the transfer function estimate is also given: The variance is the noise-to-signal ratio at the frequency in question multiplied by the "model order-to-number of data" ratio.

1. INTRODUCTION

Linear models of dynamical systems play an important role for many decisions and design issues in control, prediction, filtering and simulation. It is then important to keep in mind that the model never can give a true and exact description of the system in question. The model is always subject to errors and inaccuracies of different kinds. If it is constructed by modelling using basic physical laws, then certain effects must have been neglected or approximated. If the model has been identified from measured input-output data, then disturbances in these data cause errors and the limited amount of data makes it impossible to accurately describe every feature of the system.

In any case it is important to understand the properties of these errors and how to affect them, as well as to understand their effect on the design for which the model has been used. We shall in this paper undertake a study of how errors and disturbances in the measured input-output data affect the identified model's frequency domain properties.

The problem can be stated as follows.

Let a linear model of a discrete-time dynamical system be described by

$$y(t) = \sum_{k=1}^{\infty} g(k)u(t-k) + v(t) \quad t=1,2,... \tag{1.1}$$

Here $y(t)$, $u(t)$, $t=1,2,3...$ are the sequences of outputs and inputs, respectively, assuming the sampling interval to be one time unit. Moreover, $v(t)$ is an additive disturbance, whose character we shall discuss later. With (1.1) we associate the _transfer function or frequency function_

$$G(e^{i\omega}) = \sum_{k=1}^{\infty} g(k)e^{-ik\omega}, \quad -\pi \le \omega \le \pi \tag{1.2}$$

and it is well known how this function describes the properties of the model.

Suppose that the impulse response coefficients $g(k)$ somehow have been determined or estimated using the data set

$$Z^N: y(t), u(t) \quad t=1,2,...N \tag{1.3}$$

measured from the true system. Denote the corresponding transfer function by

$$\hat{G}_N(e^{i\omega}) \tag{1.4}$$

to emphasize that it is an estimate based on N measurements. Now, in practice the measurements (1.3) are always subject to errors and disturbances, and it is common to describe these as stochastic processes. The data set (1.3) is then a realization of a random variable (vector) which in turn means that the estimate (1.4) is a random variable. Let

$$G_N^*(e^{i\omega}) = E\,\hat{G}_N(e^{i\omega}) \tag{1.5}$$

If the expected value $G_N^*(e^{i\omega})$ does not give a correct description of the system we say that the estimate $\hat{G}_N(e^{i\omega})$ is _biased_. This is of course the typical situation in practice, since the true system in general is much more complex than our models can be. Often (1.5) will be (essentially) independent of N, and we shall drop the subscript N in the sequel. If we assume the true system itself to be linear (which of course may be unrealistic) with a true transfer function $G_O(e^{i\omega})$, then the bias can be quantified as

$$B(\omega) = G^*(e^{i\omega}) - G_O(e^{i\omega}) \tag{1.6}$$

The variance of (1.4) is

$$E|\hat{G}_N(e^{i\omega}) - G^*(e^{i\omega})|^2 \triangleq \overline{P}_N(\omega) \tag{1.7}$$

Remark. The transfer function is a complex valued variable. In (1.7) the variance is real valued, and defined through the absolute value of the difference. Sometimes it might be of interest to consider the matrix

$$E \begin{bmatrix} \mathrm{Re}[\hat{G}_N(e^{i\omega}) - G^*(e^{i\omega})] \\ \mathrm{Im}[\hat{G}_N(e^{i\omega}) - G^*(e^{i\omega})] \end{bmatrix} \begin{bmatrix} \mathrm{Re}[\hat{G}_N(e^{i\omega}) - G^*(e^{i\omega})] \\ \mathrm{Im}[\hat{G}_N(e^{i\omega}) - G^*(e^{i\omega})] \end{bmatrix}^T$$

$$\triangleq \overline{\Pi}_N(\omega). \tag{1.8}$$

Clearly $\overline{P}_N(\omega) = \mathrm{tr}\ \overline{\Pi}_N(\omega)$.

Combining (1.6) and (1.7) gives an expression for the mean square error of the estimate:

$$E|\hat{G}_N(e^{i\omega}) - G_O(e^{i\omega})|^2 = |B(\omega)|^2 + \overline{P}_N(\omega) \tag{1.9}$$

It is this error that is of interest when using the model $\hat{G}_N(e^{i\omega})$ but it is convenient to split its analysis into separate studies of the terms of the right hand side of (1.9).

We shall in this contribution study expressions for B and $\overline{P}_N$ for a common class of identification methods, and also discuss how these expressions are affected by design variables that may be at the user's disposal. A more thorough study of the bias term is given in [1], while the variance term is studied in detail in [2]. General aspects on the transfer function estimation problem are given in references[3-6].

In this paper we shall generally assume that the time system can be described by

$$y(t) = G_0(q)u(t) + v(t) \tag{1.10}$$

where $\{v(t)\}$ is a stationary stochastic process, independent of $\{u(t)\}$ and with spectrum

$$\Phi_v(\omega) \tag{1.11}$$

Moreover, q is the shift operator.

2. THE PREDICTION ERROR IDENTIFICATION METHOD FOR ESTIMATING TRANSFER FUNCTIONS

We shall in this section discuss prediction error methods for system identification. See [7] or [8], for a general discussion of this class of methods.

Let us postulate a set of models within which we shall look for the best description of the system:

$$y(t) = G(q,\Theta)u(t) + H(q,\Theta)e(t) \tag{2.1}$$

$$G(q,\Theta)u(t)=\sum_{n=1}^{\infty} g(n,\Theta)q^{-n}u(t)=\sum_{n=1}^{\infty} g(n,\Theta)u(t-n) \tag{2.2}$$

$$H(q,\Theta)e(t)=\left(1+\sum_{n=1}^{\infty} h(n,\Theta)q^{-n}\right)e(t)=e(t)+\sum_{n=1}^{\infty} h(n,\Theta)e(t-n) \tag{2.3}$$

In these expressions q^{-1} is the delay operator: $q^{-1}u(t)=u(t-1)$.

The vector Θ is a parameter vector that ranges over a subset D of R^d. This will define a set of candidate frequency functions;

$$\mathcal{G} = \{G(e^{i\omega},\Theta)\,|\,\Theta \in D\} \tag{2.4a}$$

$$\mathcal{H} = \{H(e^{i\omega},\Theta)\,|\,\Theta \in D\} \tag{2.4b}$$

The sequence $\{e(t)\}$ is assumed to be a sequence of independent random variables with zero mean values and variances λ. The model spectral density of the additive disturbance is thus

$$\Phi_v(\omega,\Theta) = \lambda|H(e^{i\omega},\Theta)|^2. \tag{2.5}$$

Behind the model set (2.1) hides essentially all linear, time-invariant models used in practice, including arbitrarily parametrized state space models, ARMAX models, the Box-Jenkins models, etc.

We shall give a few examples of such common models.

Example 2.1 "The least squares model"

Consider the difference equation

$$y(t)+a_1y(t-1)+\ldots+a_ny(t-n)=b_1u(t-1)+\ldots+b_mu(t-m)+e(t) \tag{2.6}$$

With q-notation it can be rewritten as

$$A(q^{-1})y(t) = B(q^{-1})u(t) + e(t), \tag{2.7}$$

where A and B are polynomials in q^{-1}. It is common to introduce the notation

$$\phi(t)=\left[-y(t-1)\ldots-y(t-n)\ u(t-1)\ldots u(t-m)\right]^T \tag{2.8}$$

$$\Theta = (a_1\ldots a_n\ b_1\ldots b_m)^T,$$

so that (2.6) can be represented as

$$y(t) = \Theta^T\phi(t) + e(t). \tag{2.9}$$

This expression is known as a linear regression for estimating Θ, for which the least squares method (see below) is a standard tool. For that reason, with abuse of terminology, the model (2.6) is sometimes known as "the least squares model" in the control literature.

We note that (2.6) is a special case of (2.1) with

$$G(q,\Theta) = \frac{B(q^{-1})}{A(q^{-1})} \quad H(q,\Theta) = \frac{1}{A(q^{-1})}. \tag{2.10}$$

□

Example 2.2 "Output error models"

When $H(q,\Theta)\equiv 1$ in (2.1) we have

$$y(t) = G(q^{-1},\Theta)u(t) + e(t) \tag{2.11}$$

where $G(q,\Theta)$ can be arbitrarily parametrized, e.g. as

$$G(q,\Theta) = \frac{b_1q^{-1}+\ldots+b_mq^{-m}}{1+f_1q^{-1}+\ldots+f_nq^{-n}} \tag{2.12}$$

$$\Theta = (b_1\ldots b_m\ f_1\ldots f_n)^T \tag{2.13}$$

For reasons that will become clear below, (2.11) is often called an "output error model".

Example 2.3 A state space model

Consider the state space model in innovations form

$$x(t+T)=A(\Theta)x(t)+B(\Theta)u(t)+K(\Theta)e(t)$$
$$y(t) = C(\Theta)x(t) + e(t) \tag{2.14}$$

where the matrices A,B, C and K depend on the parameter vector Θ in an arbitrary way. Clearly (2.14) is a special case of (2.1) with

$$G(q,\Theta) = C(\Theta)\left(qI-A(\Theta)\right)^{-1}B(\Theta) \tag{2.15}$$

$$H(q,\Theta) = 1+C(\Theta)\left(qI-A(\Theta)\right)^{-1}K(\Theta) \qquad \square$$

Example 2.4 The Box-Jenkins model

A common set of models is described by (2.1) where $G(q,\Theta)$ is parametrized as in (2.13) and

$$H(q,\Theta) = \frac{1+c_1q^{-1}+\ldots+c_nq^{-n}}{1+d_1q^{-1}+\ldots+d_nq^{-n}} \tag{2.16}$$

with

$$\Theta = (b_1\ldots b_m\ f_1\ldots f_n\ c_1\ldots c_n\ d_1\ldots d_n)^T \tag{2.17}$$

This is the model used in [5]. $\square$

To compare the model with the data we calculate the k-step ahead predictor of $y(t+k)$ based on $y(s)$, $s\leq t$ and $u(s)$, $s\leq t+k-1$. This is given by

$$\hat{y}(t+k|t,\Theta) = W_k(q,\Theta)G(q,\Theta)u(t+k)+\tilde{H}_k(q,\Theta)H^{-1}(q,\Theta)y(t) \tag{2.18}$$

where

$$H(q,\Theta)=\left(1+\sum_{r=1}^{k-1}h(r,\Theta)q^{-r}\right)+\sum_{r=k}^{\infty}h(r,\Theta)q^{-r}\triangleq\bar{H}_k(q,\Theta)+q^{-k}\tilde{H}_k(q,\Theta) \tag{2.19}$$

and

$$W_k(q,\Theta) = \bar{H}_k(q,\Theta)H^{-1}(q,\Theta). \tag{2.20}$$

The k-step prediction error is then given by

$$\varepsilon(t+k,t,\Theta)=y(t+k)-\hat{y}(t+k|t,\Theta)=W_k(q,\Theta)\{[G_o(q)-G(q,\Theta)]u(t+k)+v(t+k)\} \tag{2.21}$$

Now, a reasonable approach to select an estimate Θ is to choose that member in D that gives the "smallest" k-step ahead prediction error when applied to the observed data. That is, suppose y(t) and u(t) have been observed over the interval $0 \leq t \leq N$. Then calculate the sequence of k-step prediction errors corresponding to the nominal model Θ:

$$\varepsilon\big(r,(r-k),\Theta\big) \quad r=k,\ldots,N \tag{2.22}$$

To allow for more flexibility, we also filter this sequence through a linear filter L:

$$\varepsilon_F\big(r,(r-k),\Theta\big) = L(q)\,\varepsilon\big(r,(r-k),\Theta\big) = \sum_{i=0}^{\infty} \ell_i\,\varepsilon\big((r-i),(r-i-k),\Theta\big) \tag{2.23}$$

without loss of generality we may assume that $\ell_0=1$. We then form the criterion

$$V_N(\Theta) = \frac{1}{N}\sum_{r=k}^{N}\left[\varepsilon_F\big(r,(t-k),\Theta\big)\right]^2 \tag{2.24}$$

and select the estimate

$$\hat{\Theta}_N = \arg\min_{\Theta\in D} V_N(\Theta) \tag{2.25}$$

This estimate is very well-known for the case of k=1 and L(q)=1; it is then a "prediction error method" which coincides with the maximum likelihood method, when the noise e is Gaussian, see e.g. [7] and [8]. The possibilities using predictions further into the future as well as filtered prediction errors have been mentioned occassionally but they have not been extensively penetrated in the literature; see [8,9] for some aspects. As in [6], it is straightforward to show that the criterion (2.55) is approximately given by

$$\Theta_N \tilde{=} \arg\min_{\Theta\in D} \int_{\omega=-\pi}^{\pi} |\hat{G}_N(e^{i\omega})-G(e^{i\omega},\Theta)|^2 |L(e^{i\omega})|^2 \cdot |U_N(e^{i\omega})|^2 \cdot |W_k(e^{i\omega},\Theta)|^2 d\omega \tag{2.26}$$

where

$$\hat{G}_N(e^{i\omega}) = Y_N(e^{i\omega})/U_N(e^{i\omega}) \tag{2.27a}$$

with

$$Y_N(e^{i\omega}) = \frac{1}{\sqrt{N}}\sum_{r=1}^{N} y(r)e^{-ir\omega} \tag{2.27b}$$

$$U_N(e^{i\omega}) = \frac{1}{\sqrt{N}} \sum_{r=1}^{N} u(r) e^{-ir\omega} \tag{2.27c}$$

Notice that for k=1 we have

$$W_1(e^{i\omega},\Theta) = [H(e^{i\omega},\Theta)]^{-1} \tag{2.28}$$

so the weighting in (2.26)

$$|L(e^{i\omega})|^2 |U_N(e^{i\omega})|^2 / |H_N(e^{i\omega},\Theta)|^2 \tag{2.29}$$

becomes proportional to the model signal-to-noise-ratio at the frequency in question.

When the prediction horizon k, the filter L and the sets (2.4) have been decided upon, the method (2.55) uniquely determines the transfer function estimate

$$G(e^{i\omega},\hat{\Theta}_N) = \hat{G}_N(e^{i\omega})$$

3. THE DISTRIBUTION OF BIAS OVER FREQUENCIES

With the procedure described in Section 2, the transfer function estimate will depend on four quantities.

1. The model sets (2.4)

2. The input signal (input spectrum $\Phi_u(\omega)$)

3. The prefilter L

4. The prediction horizon k

The choice of these items will thus affect the properties and quality of the estimate. Generally speaking, these estimates can be shown to have a certain asymptotic distribution:

$$\hat{G}_N(e^{i\omega}) = G(e^{i\omega},\hat{\Theta}_N) \in \text{As } N\left(G(e^{i\omega},\Theta^*),\ \frac{1}{N} P(\omega)\right) \tag{3.1}$$

Here (3.1) means that the random variable

$$\sqrt{N}\left(G(e^{i\omega},\hat{\Theta}_N) - G(e^{i\omega},\Theta^*)\right)$$

converges in distribution to the normal distribution with zero mean and covariance $P(\omega)$ as N tends to infinity. In this section we shall focus our attention on the limit of the transfer function estimates, $G(e^{i\omega},\Theta^*)$. In Section 4 the variance P is discussed further.

We first assume that the input sequence is such that the following limits exist:

$$\lim_{N\to\infty} \frac{1}{N} \sum_{s=1}^{N} u(s)u(s+\tau) = r_u(\tau) \tag{3.2a}$$

We then define the input spectrum as

$$\Phi_u(\omega) = \sum_{s=-\infty}^{\infty} r_u(s)e^{-i\omega s}. \tag{3.2b}$$

The following result is not difficult to prove:

Theorem 3.1 Consider the estimate $\hat{\Theta}_N$ defined by (2.21)-(2.25). Suppose that the true system is given by (1.10), (1.11), and that the input is independent of $\{v(t)\}$. Then $\hat{\Theta}_N \to \Theta^*$ with probability 1, as $N\to\infty$ where

$$\Theta^* = \arg\min_{\Theta\in D} \bar{V}(\Theta)$$

where

$$\bar{V}(\Theta) = \int_{-\pi}^{\pi} \{|G_O(e^{i\omega})-G(e^{i\omega},\Theta)|^2\Phi_u(\omega) +$$

$$\Phi_v(\omega)\}|L(e^{i\omega})|^2|W_k(e^{i\omega},\Theta)|^2 d\omega \tag{3.3}$$

(cf (2.26)!)

Proof. We note that

$$\bar{V}(\Theta) = E[\varepsilon_F(t,t-k,\Theta)]^2 \tag{3.4}$$

It is thus sufficient to prove that $V_N(\Theta)$, defined by (2.24) converges to $\bar{V}(\Theta)$ with probability 1 and uniformly in $\Theta\in D$ as $N\to\infty$. That this is the case is proven, e.g. in [7].
We note that the function W in (3.3) depends only on the noise model:

$$W_k(e^{i\omega T},\Theta) = \frac{\bar{H}_k(e^{i\omega},\Theta)}{H(e^{i\omega},\Theta)} \tag{3.5}$$

(see (2.20)).

The theorem characterizes the limiting estimate Θ^* indirectly. A study of (3.3) (see [1] for details) reveals that $G(e^{i\omega},\Theta^*)$ can be characterized, in loose terms, as follows:

- "The limiting transfer function estimate $G(e^{i\omega},\Theta^*)$ is partly or entirely determined as the closest function to the true transfer function $G_0(e^{i\omega})$, measured in a quadratic norm over the frequencies $[-\pi, \pi]$ with a weighting function

$$Q(\omega,\Theta^*)=|L(e^{i\omega})|^2\Phi_u(\omega)\cdot\frac{|\bar{H}_k(e^{i\omega},\Theta^*)|^2}{|H(e^{i\omega},\Theta^*)|^2}$$ "

- "The resulting noise model

$$|H(e^{i\omega},\Theta^*)|^2$$

resembles the filtered error spectrum

$$\Phi_{er}(\omega,\Theta^*).\ |L(e^{i\omega})|^2$$

as much as possible", where

$$\Phi_{er}(\omega,\Theta) = |G_0(e^{i\omega})-G(e^{i\omega},\Theta)|^2\Phi_u(\omega)+\Phi_v(\omega)$$

This weighting function thus determines the distribution of bias over the frequencies. At frequencies, where $Q(\omega,\Theta^*)$ assumes high values, the bias $G_0(e^{i\omega})-G(e^{i\omega},\Theta^*)$ will be small and vice versa. With suitable values of Q, the bias distribution can be affected in a very flexible manner.

Now, the resulting weighting function $Q(\omega,\Theta^*)$ depends on a number of design variables:

- o The noise model $H(e^{i\omega},\Theta^*)$, which in turn depends on the a priori chosen set of noise models $\mathcal{H}$, (2.4b) and on the resulting filtered error spectrum.
- o The prediction horizon k, affecting the function $\bar{H}_k(e^{i\omega},\Theta^*)$.
- o The input spectrum $\Phi_u(\omega)$.
- o. The prefilter L.

A closer study of these effects is undertaken in [1], where also the role of the sampling interval is investigated.

Some particular points of this analysis may be stressed:

- First, suppose that we use a fixed noise model $H(q,\Theta)=H^*(q)$. Then the expression (3.3) specializes to

$$\Theta^*=\arg\min \int_{-\pi}^{\pi} |G_o(e^{i\omega})-G(e^{i\omega},\Theta)|^2 Q(\omega)\,d\omega \tag{3.6}$$

$$Q(\omega) = \Phi_u(\omega)\,|L(e^{i\omega})|^2\,|W_k^*(e^{i\omega})|^2 \tag{3.7}$$

  where W_k^* is computed from H^* as in (2.19)-(2.20). We thus notice that the design **variables** Φ_u, L, H^* and k affect the bias distribution only in the combination (3.7)! An unsuitable input spectrum can thus be compensated for by proper choices of H^*, k or L.

- Suppose that a goodness criterion for an identified model $\hat{G}_N$ can be stated as

$$J(\mathcal{D}) = \int_{-\pi}^{\pi} E|\hat{G}_N(e^{i\omega},\mathcal{D})-G_o(e^{i\omega})|^2 C(\omega)\,d\omega \tag{3.8}$$

  i.e. a weighted quadratic frequency domain norm. The symbol $\mathcal{D}$ here denotes the design variables that are available to the user, like choice of identification method, input spectra and other variables such as those listed earlier. The task of the user could be phrased as

$$\min_{\mathcal{D}\in D} J(\mathcal{D}) \tag{3.9}$$

  where D is a set of allowable designs. Suppose that we confine ourselves to the class of prediction error methods and that N is so large that $\hat{G}_N$ with good approximation can be replaced by G^*. Then it can be shown that with $\mathcal{D}=\{H^*,L,K,\Phi_u\}$

$$D_{opt} = \arg\min J(\mathcal{D}) \iff$$

$$\iff \Phi_u^{opt}(\omega)\,.\,|L^{opt}(e^{i\omega})|^2\,|W_k^{*opt}(e^{i\omega})|^2 = \alpha\cdot C(\omega) \quad \alpha>0 \tag{3.10}$$

  (see [11]).

We should thus choose the design variables so that the weighting function of the identification criterion matches that of the goodness criterion (3.8). Notice that there are many possibilities to achieve this:

- The model set of Example 2.1, which for k=1 gives the least squares (LS) method has

$$H(q,\Theta) = \frac{1}{A(q^{-1})}$$

which means that the corresponding weighting function for k=1 is

$$Q(\omega,\Theta) = \Phi_u(\omega) \cdot |L(e^{i\omega})|^2 \cdot |A(e^{i\omega})|^2 \quad (3.11)$$

Now, as often is the case, if the system is of low pass character, the function $|A(e^{i\omega})|^2$ will be high-frequency dominated. The LS-method thus has a tendency to overemphasize the fit at higher frequencies.

- The inclusion of a prefilter L gives rich possibilities to control the fit between model and system and to guide it to important and interesting frequency bands. The possibility to counteract the high frequency fit of the LS method by low pass or band pass filtering should be noticed in particular. In Figs. 1-3, it is illustrated how a second order model approximation is fitted to a fifth order system and how the frequency domain fit is affected by the prefilter. (The example is taken from [1].)

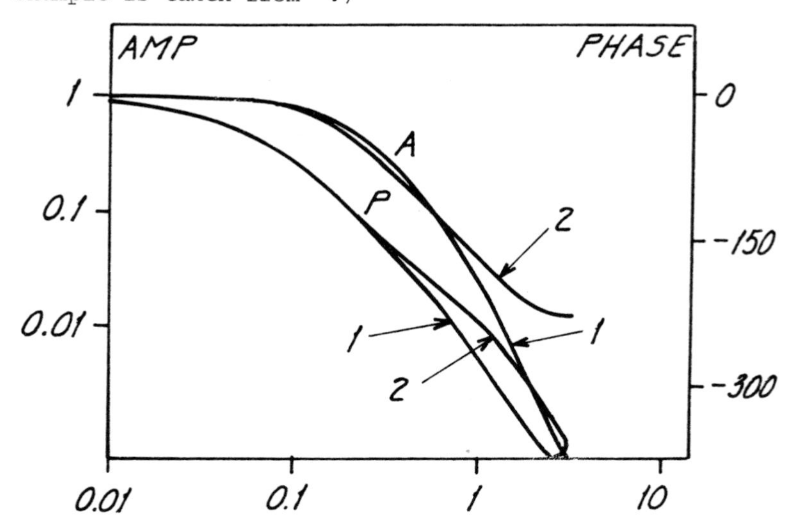

Fig. 1.
Bode plot of true 5th order system (1) and second order model (2) obtained by the output error method (H*=1, L=1).
A: Amplitude P: Phase

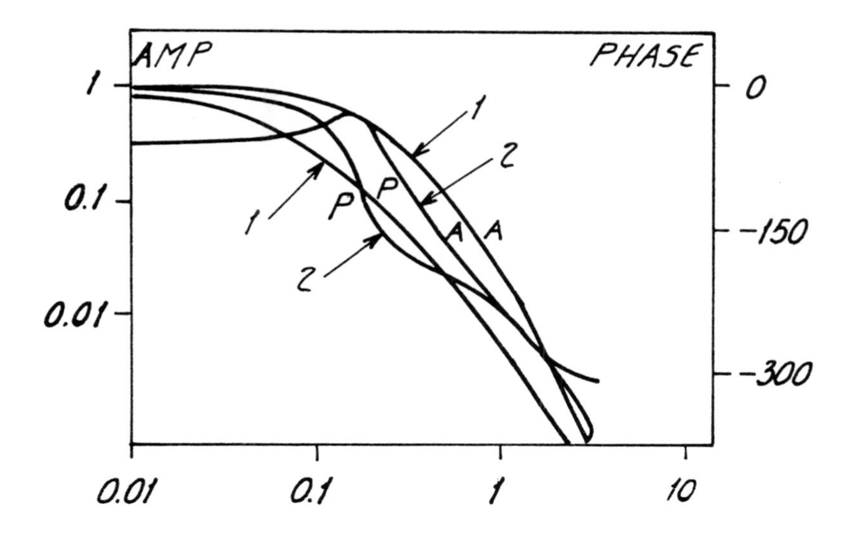

Fig. 2.
As Fig. 1, but the model is obtained by the LS method with L=1.

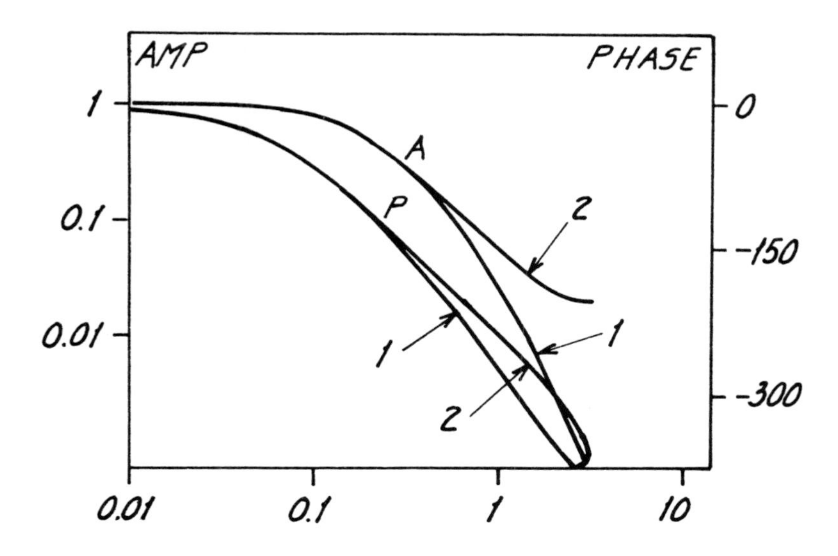

Fig. 3.
As Fig. 1 but L = Low pass filter with cutoff frequency = 0.1.

4. THE ASYMPTOTIC VARIANCE

The asymptotic variance $P(\omega)$ in (3.1) can be computed from corresponding results on the covariance matrix of the parameter estimate $\hat{\Theta}_N$ (see, e.g. [10]). Normally, this leads to very complicated expressions. It is therefore of interest that these expressions simplify considerably as the model order increases.

Suppose that the vector Θ in (3.1) can be decomposed so that

$$\Theta = \begin{pmatrix} \Theta_1 \\ \Theta_2 \\ \vdots \\ \Theta_n \end{pmatrix} \dim \Theta_k = s \quad d=n\cdot s \tag{4.1}$$

and

$$\frac{\partial}{\partial \Theta_k} G(q,\Theta) = q^{-k+1} \frac{\partial}{\partial \Theta_1} G(q,\Theta) \tag{4.2}$$

We shall call n the <u>order</u> of the model (3.1). In fact, the shift property (4.1)-(4.2) is typical for most black box models of linear systems, and applied, e.g., to all the models of the Examples 2.1, 2.2 and 2.4.

When the transfer function model is subject to (4.2), the following result can be proved [2]:

"Suppose that the true system is described by (1.10) - (1.11) and that it operates in open loop. Let $\hat{\Theta}_N(n)$ be the estimate obtained with an n:th order model, by minimization of (2.25). Let $\Theta^*(n)$ be defined as the limit $\hat{\Theta}_N(n)$ as N tends to infinity. Then

$$\sqrt{N}\,[G(e^{i\omega},\hat{\Theta}_N(n))-G(e^{i\omega},\Theta^*(n))]\in AsN(O,P_n(\omega)) \tag{4.3}$$

with

$$\lim_{n\to\infty} \frac{1}{n} P_n(\omega) = \frac{\Phi_v(\omega)}{\Phi_u(\omega)} \text{"} \tag{4.4}$$

The result (4.4) is remarkable in its generality. The limit variance depends only on the noise-to-signal ratio of the frequency in question, and is independent of the noise model set $\mathcal{H}$, of the transfer function model set $\mathcal{G}$ (as long as it is subject to (4.2)), of the prefilter $L(q)$ and of the prediction horizon k. The validity of the asymptotic expression (4.4) for low order models was tested in [2], and it was found to give reasonable approximations also for $n=2$ in the model set of Example 2.1. See Figs. 4-6. The result (4.3) can be written more suggestively as

$$E\left|G\left(e^{i\omega},\hat{\Theta}_N(n)\right)-G\left(e^{i\omega},\Theta^*(n)\right)\right|^2 \sim \frac{n}{N}\frac{\Phi_v(\omega)}{\Phi_u(\omega)} \qquad (4.5)$$

which shows that the variance is proportional to the "model order-to-number of data" ratio. Notice in particular that the variance increases with the order n, not with d, the dimension of Θ. This means, e.g. that the asymptotic variance in the model set (2.6) with $n=0$ is proportional to m, just as it is when $n=m$. The inclusion of a-parameters is thus "free", from the variance point of view. Most likely, the inclusion of these parameters will decrease the bias. For the least mean square error trade-off between bias and variance in (1.9) there is thus every reason to use a model structure (2.6) with $n=m$.

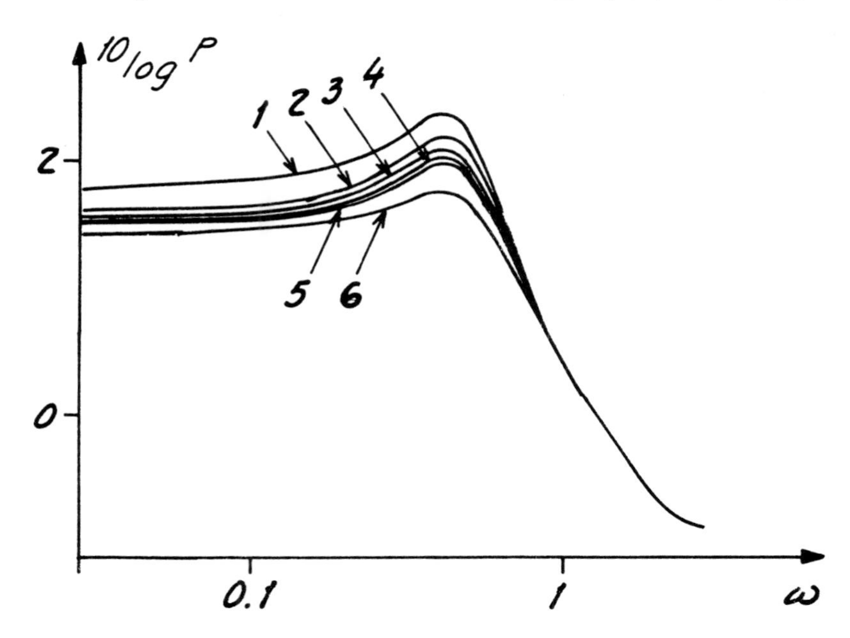

Fig. 4.
Bode plot of the variance of the transfer function estimate of the second order system $y(t)-1.5y(t-1)+0.7y(t-2)=u(t-1)+0.5u(t-2)$ when identified by the LS method. 1,-5: The exact variance expressions normalized by the model order n for $n=2,4,6,8$, and 10, respectively. 6: The asymptotic expression (4.5).

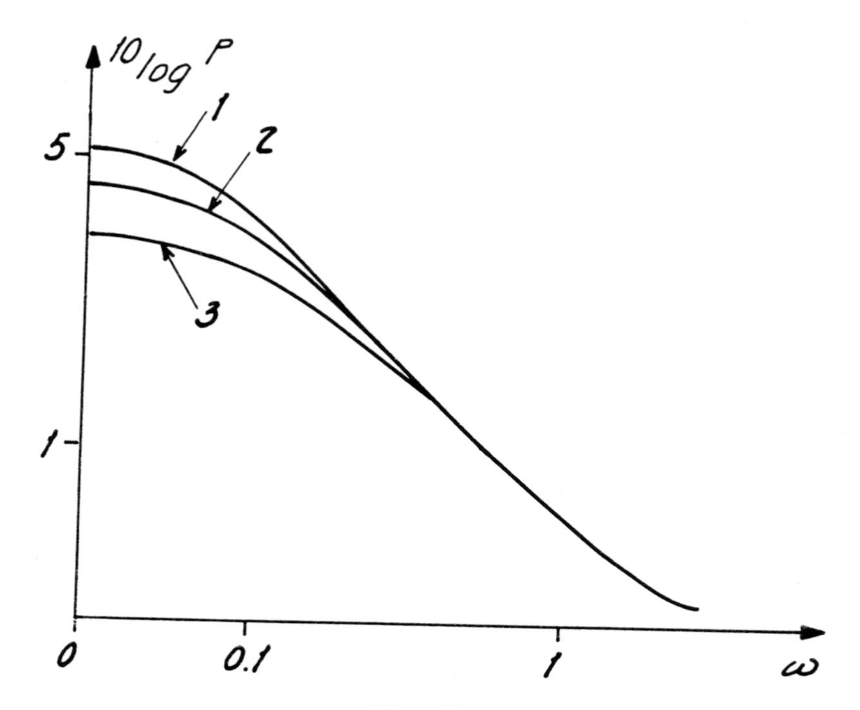

Fig. 5.
As Fig. 4 for the system y(t)-1.8y(t-1)+0.81y(t-2)=u(t-1).
1,2:Model orders n=2 and 4, 3:The asymptotic expression.

If we estimate both G and H and allow feedback in the input generation, the result (4.5) generalizes to

$$E\ \tilde{T}_N(e^{i\omega},n)\,\tilde{T}_N^T(e^{-i\omega},n) \sim \frac{n}{N}\,\Phi_v(\omega)\begin{bmatrix}\Phi_u(\omega) & \Phi_{ue}(\omega)\\ \Phi_{ue}(-\omega) & \lambda\end{bmatrix}^{-1} \tag{4.6}$$

Here

$$\tilde{T}_N(e^{i\omega},n) = \begin{pmatrix} G\left(e^{i\omega},\hat{\theta}_N(n)\right) - G\left(e^{i\omega},\theta^*(n)\right) \\ H\left(e^{i\omega},\hat{\theta}_N(n)\right) - H\left(e^{i\omega},\theta^*(n)\right)\end{pmatrix}$$

and

$\Phi_{ue}(\omega)$ = the cross spectrum between the input sequence u(t) and the white noise sequence e(t) in $v(t)=H_0(q)e(t)$, (see (1.10)).

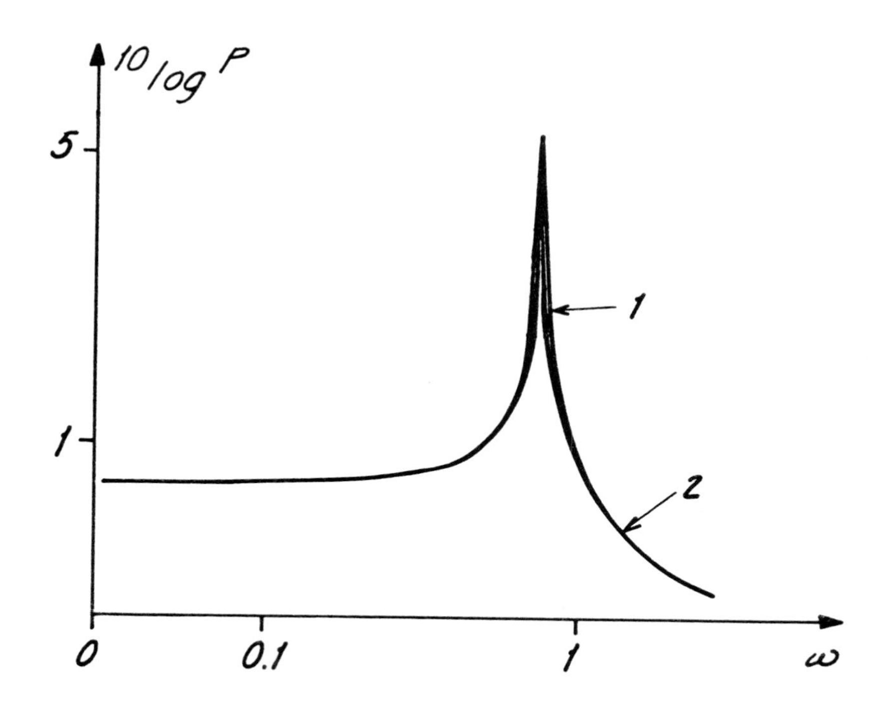

Fig. 6.

As Fig. 4 for the system $y(t)-1.4y(t-1)+0.98y(t-2)=u(t-1)+0.5u(t-2)$.
1: Model order 2 2: The asymptotic expression.

Moreover, λ is the variance of the true prediction errors $e(t)$.

The expressions (4.5) and (4.6) are quite explicit and can be used for concrete evaluation of the performance degradation, due to the incorrectness of the identified model, once it is used for some design purpose. See [2] and [12] for some explicit expressions. They can also be used for optimal input design as in [2] and [11]. We note in particular that if the design criterion is phrased as (3.8) and if the bias G^*-G_0 is negligible, then the variance $J(\mathcal{D})$ is minimized with respect to $\Phi_u(\omega)$, subject to the constraint

$$\int_{-\pi}^{\pi} \Phi_u(\omega)d\omega \leq \beta$$

by the optimal input

$$\Phi_u^{opt}(\omega) = \gamma \cdot \sqrt{\Phi_v(\omega) \cdot C(\omega)} \qquad (4.7)$$

where the constant γ is adjusted so that the input power constraint is met.

5. CONCLUSIONS

We have in this contribution taken a black-box view on the transfer function estimation problem. Even though the estimation is carried out in terms of an internal parametrization vector Θ, we have evaluated the model properties entirely through its "physical" behaviour in the frequency domain. In particular if the model is going to be used for control design it is very important to assess its properties in certain critical frequency bands.

With this philosophy we have derived expressions for the bias distribution and pointed to the effects of several design variables on this distribution. We have also given a very simple asymptotic expression for the variance of the transfer function estimate.

6. REFERENCES

1. Wahlberg, B., and Ljung, L., (1984) "Design variables for bias distribution in transfer function estimation." IEEE Conference on Decision and Control, Las Vegas, Dec 1984.

2. Ljung, L., (1985) "Asymptotic variance expressions for identified black-box transfer function models". IEE Trans. Automatic Control, Vol AC-30, Sept. 1985.

3. Goodwin, G.C., and Payne, R.L., (1977) "Dynamic Identification: Experiment Design and Data Analysis". Academic Press, New York.

4. Åström, K.J., and Eykhoff, P., (1971) "System identification-a survey. Automatica, vol 7, pp 123-167.

5. Box, G.E.P., and Jenkins, G.M., (1970) "Time Series Analysis, Forecasting and Control". Holden-Day, San-Francisco.

6. Ljung, L., (1984) "Estimation of transfer functions". Report LiTH-ISY-I-0656, Division of Automatic Control, Department of Electrical Engineering, Linköping University, Linköping, Sweden.

7. Ljung, L., (1978) "Convergence analysis of parametric identification methods". IEEE Transactions on Automatic Control, Vol AC-23, pp 770-783.

8. Åström, K.J., (1980) "Maximum likelihood and prediction error methods". Automatica, Vol. 16, pp 551-574.

9. van Zee, G.A., and Bosgra, O.H., (1982) "Validation of prediction error identification results by multivariable control implementation". Proc. 6th IFAC Symposium on Identification and System Estimation, Washington DC, pp 560-565.

10. Ljung, L., and Caines, P.E., (1979) "Asymptotic normality of prediction error estimation for approximate system models", Stochastics, vol. 3, pp 29-46.

11. Yuan, Z.D., and Ljung, L., (1985) "Unprejudiced open loop optimal input design for identification of transfer functions". Automatica, to appear 1985.

12. Ljung, L., (1984) "Performance evaluation of models, identified by the least squares method". INRIA, Nice.

CANONICAL FORMS AND PARAMETRIZATION PROBLEMS IN LINEAR SYSTEMS THEORY

Diederich Hinrichsen
*(Forschungsschwerpunkt Dynamische Systeme,
Universitat Bremen, 2800 Bremen 33, FRG)*

ABSTRACT

This paper gives a selective survey about canonical forms and their application to parametrization problems in linear systems theory. Topological properties of canonical forms are studied and it is shown how global canonical forms with certain topological properties can be used to obtain comprehensive information about topological invariants of orbit spaces of linear systems.

1. INTRODUCTION

As in linear algebra canonical forms play a fundamental role in linear systems theory. Areas of application are e.g. feedback design by pole placement[1], realisation[2], partial realisation[3] and identification[4,5,6].

In this paper we shall only deal with time-invariant linear models of the form

$$\dot{x}(t) = Ax(t) + Bu(t), \; y(t) = Cx(t), \; t \in [0,\infty) \qquad (1.1.a)$$

or

$$x(t+1) = Ax(t) + Bu(t), \; y(t) = Cx(t), \; t \in \mathbb{N} \qquad (1.1.b)$$

where

$$(A,B,C) \in K^{nxn} \times K^{nxm} \times K^{pxn}, \; K = \mathbb{R} \text{ or } \mathbb{C}.$$

Given a transformation group G acting on a space X of triples (A,B,C), a canonical form for this group action defines a parametrization of the associated orbit space X/G. This parametrization aspect is of particular interest in fields of system theory where state space systems are considered from an input output point of view. Since it is impossible to distinguish between two similar system models of the form (1.1) by input output measurements, the orbit spaces of reachable pairs (A,B) resp. minimal triples (A,B,C) modulo similarity are of special importance. Global problems where these orbit spaces have to be considered arise, for example, in convergence analysis of algorithms in identification, adaptive control and model reduction, when there is little a priori knowledge available except perhaps the system order or an upper bound to it (cf. Hannan[7], Deistler and Hannan[8], Byrnes[9]). This explains the growing interest in topological and geometrical properties of the above orbit spaces (cf. Glover[10], Hazewinkel and Kalman[11], Brockett[12], Clarke[13], Byrnes and Hurt[14], Hazewinkel[15], Delchamps[16], Helmke[17,18]).

Canonical forms (for the similarity action) provide a one-to-one correspondence between state space and input-output models, but in the multivariable case ($m \geq 2$) this correspondence is necessarily discontinuous[11]. Therefore several authors have argued that the application of canonical forms in identification algorithms may not be advisable due to numerical and statistical problems (see Glover and Willems[19], Ljung and Rissanen[20]). Instead, the use of local canonical forms (charts) is recommended. However, although local canonical forms, defined on euclidean subsets of the orbit manifold, are attractive from a numerical point of view, they cannot replace global ones as tools for the solution of global problems. One objective of this paper is to show that well constructed global canonical forms - in spite of possible discontinuities - yield very efficient tools for the global analysis of system spaces[21].

The paper is organised as follows. Section 2 introduces some general terminology concerning group actions, canonical forms and their topological properties.

In section 3 canonical forms for important group actions in linear systems theory are described and their smoothness properties are analysed. A general procedure for the construction of canonical forms is outlined which provides a common framework for most canonical forms in linear algebra and linear systems theory.

In section 4 it is shown how a global canonical form with nice topological properties can be used to obtain information about topological invariants of the corresponding orbit space.

In particular, the homology groups of the orbit space of reachable systems modulo similarity are completely determined and it is shown that the Kronecker canonical form is "of minimal discontinuity".

2. GROUP ACTIONS AND CANONICAL FORMS

In this section we introduce the general terminology concerning group actions and canonical forms which will be needed later.

Let X be a locally compact Hausdorff space and G a topological group with neutral element e. If

$$\alpha: G\times X \to X$$
$$(g,x) \to g.x$$

is a continuous map satisfying

$$(gh)\cdot x = g\cdot(h\cdot x), \quad g,h \in G,\ x \in X$$
$$e\cdot x = x, \qquad x \in X$$

then α is called a left action of the group G on X. Every such group action α induces an equivalence relation $\overset{\alpha}{\sim}$ on X by

$$x \overset{\alpha}{\sim} y \rightleftarrows \exists\ g \in G: \ y = g.x.$$

The equivalence classes of $\overset{\alpha}{\sim}$ are called the orbits of α and are denoted by

$$[x]_\alpha = G.x = \{y \in X;\ x \overset{\alpha}{\sim} y\}$$

The orbit space of α is the space of all equivalence classes of $\overset{\alpha}{\sim}$:

$$X/\alpha = \{[x]_\alpha;\ x \in X\}$$

provided with the quotient topology, i.e., the finest topology on X/α for which the canonical projection

$$\pi: X \to X/\alpha, \quad x \to [x]_\alpha$$

is continuous. With respect to this topology π is an open mapping. X/α is a Hausdorff space if and only if the graph of α

$$R_\alpha = \{(x,y) \in X \times X;\ x \overset{\alpha}{\sim} y\} \qquad (2.1)$$

is a closed subset of $X \times X$ (cf. Dieudonné[22]). A mapping $s: X/\alpha \to X$ is called a (global) section of π if $\pi(s(\bar{x})) = \bar{x}$ for all $\bar{x} \in X/\alpha$. A continuous section $s: X/\alpha \to X$ defines a homeomorphism of X/α onto $s(X/\alpha) \subset X$ with $\pi|s(X/\alpha)$ as its inverse.

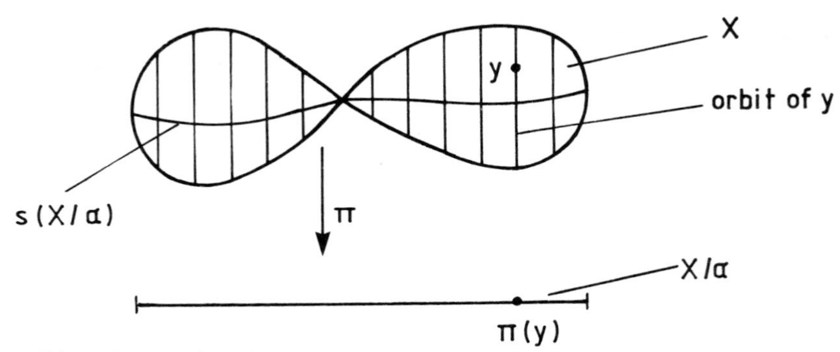

Fig. 1 Projection $\pi: X \to X/\alpha$ with continuous section s

The stabilizer of a point $x \in X$ is the closed subgroup

$$G_x = \{g \in G;\ g.x = x\}$$

of G. $g \to g.x$ induces a continuous bijection of G/G_x onto the orbit $[x]_\alpha$. G is said to act freely on X if all stabilizers are trivial, i.e. $G_x = \{e\}$ for all $x \in X$. Two orbits $[x]_\alpha$, $[y]_\alpha$ are said to be of the same type, if G_x and G_y are conjugate as subgroups of G, i.e., $G_y = g\,G_x g^{-1}$ for some $g \in G$.

If G acts freely on X, for each pair $(x,y) \in R_\alpha$ there exists a unique $\gamma(x,y) \in G$ such that $\gamma(x,y).x = y$. We say that G acts bicontinuously on X if $\gamma: R_\alpha \to G$ is continuous.

2.1 *Proposition*

Suppose that G acts freely and bicontinuously on X. If there exists a continuous section of $\pi: X \to X/\alpha$ on X/α then X is homeomorphic to the product $G \times (X/\alpha)$.

Proof

If $s: X/\alpha \to X$ is a continuous section of π, then $(g,\bar{x}) \to g.s(\bar{x})$ is a continuous bijection of $G \times (X/\alpha)$ onto X. Its inverse is $x \to (\gamma(s(\pi(x)),x),\pi(x))$.

□

In linear systems theory one may roughly distinguish between two different types of group actions. The first one describes linear coordinate transformations in the state, input and/or output spaces. In this case, systems lying in the same orbit may be considered to be "similar" or "isomorphic" (*similarity type group actions*). The second type describes the modification of systems by specific synthesis procedures (*feedback type group actions*).

2.2 *Example*

Let $L_{m,n,p}(K)$ be the space of linear models $(A,B,C) \in K^{nxn} \times K^{nxm} \times K^{pxn}$, $L^r_{m,p,n,}(K)$ (resp. $L^o_{m,n,p}(K)$) the open and dense subspaces of reachable (resp. observable) triples. The general linear group $Gl_n(K)$ acts on $L_{m,n,p}(K)$ (and, by restriction, on the σ-invariant subsets $L^r_{m,n,p}(K)$, $L^o_{m,n,p,}(K)$) via the similarity action

$$\begin{aligned} \sigma: \quad & Gl_n(K) \times L_{m,n,p}(K) \to L_{m,n,p}(K) \\ & (T, \ (A,B,C)\ ) \ \to (TAT^{-1}, \ TB, \ CT^{-1}) \end{aligned} \tag{2.2}$$

σ describes the effect of linear coordinate transformations $\tilde{x} = Tx$ on system equations of the form (1.1). If similar (i.e., σ-equivalent) triples are regarded as different representations of the same system, the "true" space of time-invariant linear systems with m input, n state and p output variables is not $L_{m,n,p}(K)$ but the orbit space

$$\hat{L}_{m,n,p}(K) = L_{m,n,p}(K)/\sigma. \tag{2.3}$$

Only few results are available concerning the similarity action on the whole space $L_{m,n,p}(K)$. Since σ has different orbit types, restrictions of σ to subspaces with constant orbit type have to be considered. Of particular importance are the open and dense subspaces $L^r_{m,n,p}(K)$, $L^o_{m,n,p}(K)$ and $L^{r,o}_{m,n,p}(K)$ (the space of minimal triples) on which σ is a free action. Triples (A,B,C) belonging to one of these spaces have orbits of principal type in the sense that the associated similarity

orbits are of highest dimension.
If outputs are neglected (i.e. $p = 0$), the principal orbits of the similarity action can be completely characterised in system theoretic terms. It is readily verified that the stabiliser of an input pair $(A,B) \in L_{m,n}(K) := K^{nxn} \times K^{nxm}$ under the similarity action is trivial if and only if (A,B) is reachable. □

2.3 *Example*

If linear coordinate transformations of the input, state and output are combined with static linear state feedback (see Fig. 2) we obtain the (right) state feedback action ϕ of the group

$$F_{m,n,p}(K) = Gl_p(K) \times \left\{ \begin{bmatrix} T & O \\ F & S \end{bmatrix}, \; T \in Gl_n(K),\; S \in Gl_m(K), F \in K^{mxn} \right\}$$

on the space $L_{m,n,p}(K)$

$$\phi \;:\; F_{m,n,p}(K) \times L_{m,n,p}(K) \to L_{m,n,p}(K) \tag{2.4}$$

$$((U, \begin{bmatrix} T & O \\ F & S \end{bmatrix}), \begin{bmatrix} A & B \\ C & O \end{bmatrix}) \mapsto \begin{bmatrix} T & O \\ O & U \end{bmatrix}^{-1} \begin{bmatrix} A & B \\ C & O \end{bmatrix} \begin{bmatrix} T & O \\ F & S \end{bmatrix}$$

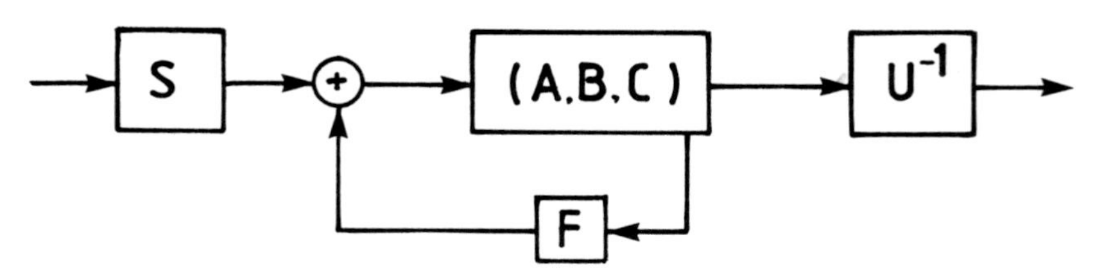

Fig. 2 State feedback action ϕ

The state feedback action has not as yet been analysed systematically on the whole space $L_{m,n,p}(K)$. However, the restriction of ϕ to the space of reachable input pairs has been extensively studied in the literature (see § 3). In particular, it is known that the orbit space $L^r_{m,n}(K)/\phi$ is finite □

The following definition taken from Birkhoff and McLane[23] is standard in the control theoretic literature[4,24].

2.4 *Definition*

A canonical form for a group action α on X is a map $\Gamma: X \to X$ which satisfies for all $x, y \in X$:

(C1) $$\Gamma(x) \overset{\alpha}{\sim} x$$

(C2) $$x \overset{\alpha}{\sim} y \iff \Gamma(x) = \Gamma(y)$$

Thus a canonical form is a complete invariant which associates with any $x \in X$ a uniquely determined "canonical" $\tilde{x}$ in the α-orbit of x. By the axiom of choice there always exists a canonical form for a given group action α. This indicates that - without additional requirements - the set theoretic concept defined above is not very useful. A natural first step is to ask whether there exists a continuous canonical form for a given action α. However, in most cases the answer will be "no" since a global continuous canonical form Γ would yield a global continuous section $[x]_\alpha \to \Gamma(x)$ of $\pi: X \to X/\alpha$ and hence would imply that X is homeomorphic to $G \times (X/\alpha)$ (Proposition 2.1). In many cases of interest $X \not\cong G \times (X/\alpha)$. Therefore we have to look for piecewise smooth versions. To define what "piecewise" means the following concept is needed.

2.5 *Definition*

Let X be a C^k-manifold ($k = 0, \infty, \omega$, i.e., a topological, smooth or analytical manifold, resp.). A partition $(X_i)_{i \in I}$ is said to be a C^k-stratification if

(i) Each "stratum" X_i is a C^k-submanifold of X.

(ii) $(X)_{i \in I}$ is locally finite (each point of X has a neighbourhood which meets only finitely many strata).

(iii) The boundary $\partial X_i = \overline{X}_i \setminus X_i$ of each stratum is contained in the union of all strata X_j of strictly lower dimension than X_i.

$(X_i)_{i \in I}$ is said to satisfy the frontier condition, if for all $i, j \in I$

(iv) $$X_i \cap \overline{X}_j \neq \phi \Rightarrow X_i \subset \overline{X}_j.$$

Fig. 3 Partitions of an open rectangle

A function f: X → K will be called <u>piecewise C^k</u> (continuous, smooth, analytical) if its restrictions $f|X_i$ to the strata of a suitable stratification $(X_i)_{i\in I}$ of X are C^k $(k = 0, \infty, \omega)$.

2.6 *Definition*

A C^k - stratification $(Y_i)_{i\in I}$ of a C^k - manifold Y $(k = 0, \infty, \omega)$ is said to be a C^k - cellular decomposition of Y if the submanifolds Y_i are C^k - diffeomorphic to euclidean spaces $\mathbb{R}^{n_i}$, $n_i \in \mathbb{N}$ (cells).

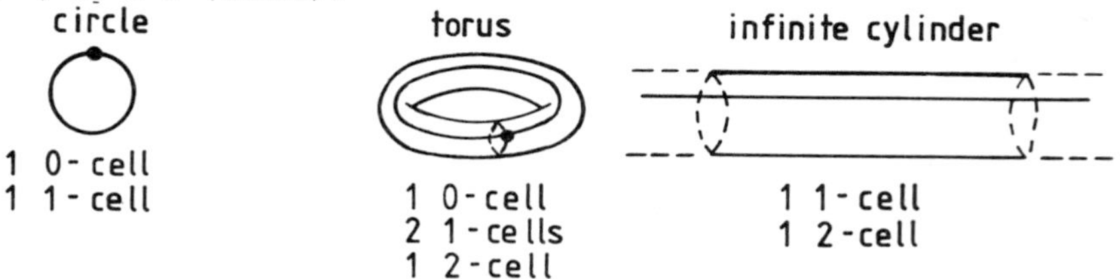

Fig. 4 Cellular decompositions

2.7 *Definition*

Let α: GxX → X be a group action on a C^k- manifold X $(k = 0, \infty, \omega)$.

(a) A canonical form Γ: X → X is said to be piecewise C^k if there exists an α-invariant C^k- stratification $(X_i)_{i\in I}$ such that $\Gamma|X_i$ is of class C^k for each $i \in I$.

(b) Γ is said to be C^k- cellular if the orbit space $Y = X/\alpha$ is a C^k- manifold and there exists a C^k- cellular decomposition $(Y_i)_{i\in I}$ of Y such that Γ is of class C^k on each stratum $X_i = \pi^{-1}(Y_i)$, $i \in I$.

If x moves continuously from one stratum X_i to another one, discontinuities may occur in $\Gamma(x)$. To obtain cellular canonical forms of "minimal discontinuity" it is natural to look for cellular decompositions of $Y = X/\alpha$ with a minimal number of cells. If Y is a C^k- manifold which admits a finite C^k- cellular decomposition then the smallest number of cells of a C^k- cellular decomposition of Y may be called the C^k-complexity of Y (denoted by $c_k(Y)$). Correspondingly, a C_k - cellular canonical form $\Gamma: X \to X$ will be called minimal if the associated cellular decomposition of $Y = X/\alpha$ has $c_k(Y)$ cells.

Some general results concerning the existence and properties of cellular and minimal cellular canonical forms for continuous, smooth and analytical group actions can be found in Helmke and Hinrichsen[21].

3. CONSTRUCTION OF CANONICAL FORMS

In this section we describe some of the available canonical forms for the system theoretic group actions mentioned in § 2 and discuss their smoothness properties. We conclude this section with an outline of a general procedure for the construction of these canonical forms.

A. The Kronecker Canonical Form for $\sigma: L^r_{m,n,p}(K) \to L^r_{m,n,p}(K)$

To construct the Kronecker canonical form for the similarity action on $L^r_{m,n,p}(K)$ one associates with every reachable system $(A,B,C) \in L^r_{m,n,p}(K)$ a special basis of K^n and expresses A,B,C with respect to this basis. Consider the following elimination procedure

> Eliminate in the reachability matrix $[B,AB,\ldots,A^{n-1}B]$ from the left to the right all the column vectors which are linearly dependent upon their predecessors. (3.1)

By reachability, the remaining vectors form a basis of K^n. They are ordered in the following way

$$S = S(A,B) := [b^1,\ldots,A^{\kappa_1-1} b^1,\ldots,b^m,Ab^m,\ldots,A^{\kappa_m-1} b^m] \quad (3.2)$$

where $\kappa_1,\ldots,\kappa_m$ are non-negative integers satisfying

$$\kappa_1 + \ldots + \kappa_m = n.$$

(If $\kappa_j = 0$, the corresponding block $[b^j, \ldots, A^{\kappa_j - 1} b^j]$ is absent in S(A,B).) S(A,B) is called the Kronecker basis and $\kappa = \kappa(A,B) = (\kappa_1, \ldots, \kappa_m)$ the list of Kronecker indices of (A,B). Note that $\kappa(A,B)$ is not a set but a family. It depends upon the ordering of the input channels i.e., on the ordering of the column vectors $b^1, \ldots b^m$ of B.

3.1 *Remark*

The set $\{\kappa_1, \ldots, \kappa_m\}$ of Kronecker indices of a reachable pair (A,B) coincides with the set of minimal indices of the singular pencil $[sI_n - A, B]$ as defined by Kronecker, see Gantmacher[25]. □

Any m-tuple $(\kappa_1, \ldots, \kappa_m) \in \mathbb{N}^m$ with sum equal to n is called a *combination* of n of length m. Let $K_{n,m}$ denote the set of these combinations. Then

$$\text{card } K_{n,m} = \binom{n+m-1}{n}$$

For $\kappa \in K_{n,m}$ define the associated Kronecker stratum by

$$\text{Kro } (\kappa) = \{(A,B,C) \in L^r_{m,n,p}(K) ; \kappa(A,B) = \kappa\} \tag{3.3}$$

Since $\kappa(A,B)$ is a similarity invariant, all the Kronecker strata are invariant with respect to σ. The Kronecker canonical form $\Gamma_K : L^r_{m,n,p}(K) \to L^r_{m,n,p}(K)$ is defined by

$$\Gamma_K(A,B,C) = (S^{-1}AS, S^{-1}B, CS) \tag{3.4}$$

where S = S(A,B) is defined by (3.2)

The structure of this canonical form is illustrated by the following example (for a complete specification, see Popov[24], Denham[4]).

3.2 *Example*

Suppose that n = 5, m = 3, p arbitrary. For the Kronecker list $\kappa = (2,0,3)$ we obtain the following structure of $\Gamma_K(A,B,C) = (A_K, B_K, C_K)$

$$A_K = \left[\begin{array}{cc|ccc} 0 & * & 0 & 0 & * \\ 1 & * & 0 & 0 & * \\ \hline 0 & * & 0 & 0 & * \\ 0 & * & 1 & 0 & * \\ 0 & 0 & 0 & 1 & * \end{array}\right], \quad B_K = \left[\begin{array}{ccc} 1 & * & 0 \\ 0 & 0 & 0 \\ \hline 0 & 0 & 1 \\ 0 & 0 & 0 \\ 0 & 0 & 0 \end{array}\right], \quad C_K \text{ arbitrary}$$

□

The results collected in the following theorem are due to Helmke[17].

3.3 *Theorem*

(a) $(\mathrm{Kro}(\kappa))_{\kappa \in K_{m,n}}$ is an analytical stratification of $L^r_{m,n,p}(K)$ satisfying the frontier condition.

(b) Let $\pi: L^r_{m,n,p}(K) \to L^r_{m,n,p}(K)/\sigma$ be the canonical projection. Then the projected Kronecker strata $\pi(\mathrm{Kro}(\kappa))$, $\kappa \in K_{m,n}$ form an analytical cellular decomposition of the orbit space (which is an analytical manifold).

(c) Γ_K, defined by (3.4), is a C^ω- cellular canonical form for the similarity action on $L^r_{m,n,p}(K)$ with underlying analytical stratification $(\mathrm{Kro}(\kappa))_{\kappa \in K_{m,n}}$.

It is noteworthy that the preceding theorem is valid over both fields $K = \mathbb{C}$ and $K = \mathbb{R}$.

B. Input output canonical form $\sigma: L^{r,o}_{m,n,p}(K) \to L^{r,o}_{m,n,p}(K)$

The space $L^{r,o}_{m,n,p}(K)/\sigma$ is of fundamental importance in linear systems theory, since it can be identified with the space $T^n_{m,p}(K)$ of p x m transfer matrices G(s) of McMillan degree $\delta(G(s)) = n$. The Kronecker canonical form for triples (A,B,C) $\in L^r_{m,n,p}(K)$ does not impose any special structure on the output matrices - quite adequately since the output matrices C of these triples are completely arbitrary. This changes in $L^{r,o}_{m,n,p}(K)$ and so one would require a "symmmetric" canonical form on this space, which imposes analogous constraints on B and on C. Such a canonical form has been constructed by Bosgra and Van der Weiden[26] and has been applied to the partial realisation problem by Bosgra[3]. To formulate their result we need the following notation. If $I \subset \{1,\ldots,n+p\}$, $J \subset \{1,..n+m\}$ are subsets of n numbers then $N(I) \in K^{(n+p)\times n}$ and $M(J) \in K^{n\times(n+m)}$ denote the submatrices consisting of the columns $i \in I$ in the identity matrix I_{n+p} and the rows $j \in J$ in the identity matrix I_{n+m}, resp.

3.4 *Theorem*

Let $(A,B,C) \in L^{r,o}_{m,n,p}(K)$ be given. There exist a unique $(\tilde{A},\tilde{B},\tilde{C}) \overset{\sigma}{\sim} (A,B,C)$, a unique permutation matrix P and unique sets $I \subset \underline{n+p}$, $J \subset \underline{n+m}$ of n numbers satisfying

(a) $[\tilde{B}\ \tilde{A}]M(I)^T$ is unipotent (i.e., upper triangular with ones on the diagonal);

(b) $N(I)^T \begin{vmatrix} \tilde{C} \\ P\tilde{A} \end{vmatrix} P^T$ and $P^T N(I)^T \begin{vmatrix} \tilde{C} \\ P\tilde{A} \end{vmatrix}$ are lower triangular with non-zero diagonal entries;

(c) rows $i \notin I$ of $\begin{vmatrix} \tilde{C} \\ P\tilde{A} \end{vmatrix}$ are linearly dependent of rows $1,2,\ldots,i-1$;

(d) columns $j \notin J$ of $[\tilde{B}\ \tilde{A}]$ are linearly dependent of columns $1,2,\ldots,j-1$.

Generically, P is the identity, $I = J = \underline{n}$ and $(\tilde{A},\tilde{B},\tilde{C})$ has the following structure

$$\tilde{C} = \left[\begin{array}{cccc|c} c_{11} & & & 0 & \\ \vdots & \ddots & & & 0 \\ & & \ddots & & \\ c_{p1} & \cdots & & c_{pp} & p \times (n-p) \end{array}\right] \quad c_{ii} \neq 0,$$

$$\tilde{B} = \begin{bmatrix} 1 & b_{12} & \cdots & b_{1m} \\ & \ddots & \ddots & \vdots \\ 0 & & \ddots & b_{m-1,m} \\ & & & 1 \\ \hline & & & \\ & 0_{(n-m)\times m} & & \end{bmatrix}; \quad \tilde{A} = \begin{bmatrix} a_{11} & \cdots & a_{1,p+1} & & \\ \vdots & & & \ddots & 0 \\ \vdots & & & & a_{n-p,n} \\ a_{m1} & & & & \vdots \\ 1 & \ddots & & & \vdots \\ & \ddots & \ddots & & \vdots \\ 0 & & 1 & a_{n,n-m+1} \cdots & a_{nn} \end{bmatrix}; a_{i,i+p} \neq 0$$

$\Gamma_{I/O}$: $(A,B,C) \to (\tilde{A},\tilde{B},\tilde{C})$ is called the I/O-canonical form on $L^{r,o}_{m,n,p}(K)$. Its construction is quite involved and its smoothness properties have not been investigated. However, there are some reasons to conjecture that $\Gamma_{I/O}$ is a C^{ω}-cellular form on $L^{r,o}_{m,n,p}(K)$. It would be interesting to confirm this conjecture since then the I/O-canonical form could eventually be used for a topological analysis of the orbit space $L^{r,o}_{m,n,p}(K)/\sigma$ in a similar way as the Kronecker canonical form has been used to study $L^{r}_{m,n,p}(K)/\sigma$ (see section 4).

C. The Brunovsky Canonical Form for ϕ: $L^{r}_{m,n}(K) \to L^{r}_{m,n}(K)$

The Brunovsky canonical form for the state feedback action ϕ on the space of reachable input pairs $L^{r}_{m,n}(K)$ is one of the first and best known canonical forms in linear systems theory[27].

It is easily derived from the Kronecker canonical form (which has, however, been constructed later by Popov[24]).

3.5 *Theorem*

(a) The ϕ-orbits in $L^{r}_{m,n}(K)$ are in bijective correspondence with $\{\kappa \in K_{m,n};\ \kappa_1 \geq \kappa_2 \geq \ldots \geq \kappa_m\}$, the set of m-partitions of n;

(b) Each orbit $[(A,B)]_{\phi}$ contains exactly one pair (A_{ϕ},B_{ϕ}) in Kronecker canonical form with decreasing Kronecker-list $\kappa_1 \geq \ldots \geq \kappa_m$ and all free parameters equal to zero.

In particular, the set of Kronecker indices is a complete invariant for ϕ whereas - due to input coordinate transformations - the Kronecker-list $\kappa(A,B)$ is not invariant under the action of the feedback group.

Although Brunovsky's canonical form has been known for many years it appears that until now no canonical form has been constructed for the feedback action ϕ on the space of triples $L^{r}_{m,n,p}(K)$.

D. A General Procedure for the Construction of Canonical Forms

The canonical forms in linear systems theory are mostly

the result of ad hoc constructions although behind them there is the common principle of reducing as many matrix entries as possible to 0 or 1. Prätzel-Wolters[28] developed a unifying framework for this type of canonical forms, interpreting the structural ones as result of a normalisation procedure and the structural zeros as result of a minimisation procedure within the subset of normalised elements of each orbit.

To define this minimisation procedure, a new relation of order is introduced on $\mathbb{R}$ (see Fig. 5) by taking 0 to be the smallest real number and taking on $\mathbb{R} \setminus \{0\}$ the usual order.

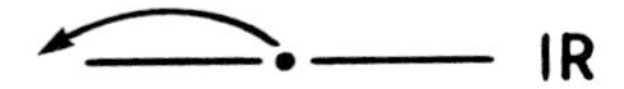

Fig. 5 The order relation < on $\mathbb{R}$

The vector spaces $\mathbb{R}^k$, $k \in \mathbb{N}$ are provided with the lexicographic order (from above) corresponding to <. Hence, for example,

$$\begin{bmatrix} 0 \\ 3 \\ 1 \\ 0 \end{bmatrix} < \begin{bmatrix} -1 \\ 0 \\ 0 \\ 0 \end{bmatrix}, \quad \begin{bmatrix} 0 \\ 2 \\ 0 \\ 1 \end{bmatrix} < \begin{bmatrix} 0 \\ 2 \\ -1 \\ 1 \end{bmatrix}$$

Identifying $a+bi \in \mathbb{C}$ with $\begin{bmatrix} a \\ b \end{bmatrix} \in \mathbb{R}^2$ a total order is defined on $\mathbb{C}^k$, $k \in \mathbb{N}$. Finally, in order to apply (<) to nxm matrices A these are identified in some way with nm-column vectors, e.g., with

$$\begin{bmatrix} a^1 \\ a^2 \\ \vdots \\ a^m \end{bmatrix} \text{(column order } <_c\text{) or } \begin{bmatrix} (a_1)^T \\ (a_2)^T \\ \vdots \\ (a_n)^T \end{bmatrix} \text{ (row order } <_r\text{)}.$$

Let α be any group action on a space X of (compound) matrices. In general there will not exist <-minimal elements of the orbits $[x]_\alpha$. However, normalising a sufficient number of parameters it is often possible to obtain a uniquely determined <-minimum among the normalised elements of each orbit. To determine those elements one proceeds in two steps. One firstly applies the screen function

$\Gamma_{I/O}: (A,B,C) \to (\tilde{A},\tilde{B},\tilde{C})$ is called the I/O-canonical form on $L^{r,o}_{m,n,p}(K)$. Its construction is quite involved and its smoothness properties have not been investigated. However, there are some reasons to conjecture that $\Gamma_{I/O}$ is a C^{ω}-cellular form on $L^{r,o}_{m,n,p}(K)$. It would be interesting to confirm this conjecture since then the I/O-canonical form could eventually be used for a topological analysis of the orbit space $L^{r,o}_{m,n,p}(K)/\sigma$ in a similar way as the Kronecker canonical form has been used to study $L^{r}_{m,n,p}(K)/\sigma$ (see section 4).

C. The Brunovsky Canonical Form for $\phi: L^{r}_{m,n}(K) \to L^{r}_{m,n}(K)$

The Brunovsky canonical form for the state feedback action ϕ on the space of reachable input pairs $L^{r}_{m,n}(K)$ is one of the first and best known canonical forms in linear systems theory[27].

It is easily derived from the Kronecker canonical form (which has, however, been constructed later by Popov[24]).

3.5 *Theorem*

(a) The ϕ -orbits in $L^{r}_{m,n}(K)$ are in bijective correspondence with $\{\kappa \in K_{m,n};\ \kappa_1 \geq \kappa_2 \geq \dots \geq \kappa_m\}$, the set of m-partitions of n;

(b) Each orbit $[(A,B)]_{\phi}$ contains exactly one pair (A_{ϕ},B_{ϕ}) in Kronecker canonical form with decreasing Kronecker-list $\kappa_1 \geq \dots \geq \kappa_m$ and all free parameters equal to zero.

In particular, the set of Kronecker indices is a complete invariant for ϕ whereas - due to input coordinate transformations - the Kronecker-list $\kappa(A,B)$ is not invariant under the action of the feedback group.

Although Brunovsky's canonical form has been known for many years it appears that until now no canonical form has been constructed for the feedback action ϕ on the space of triples $L^{r}_{m,n,p}(K)$.

D. A General Procedure for the Construction of Canonical Forms

The canonical forms in linear systems theory are mostly

the result of ad hoc constructions although behind them there is the common principle of reducing as many matrix entries as possible to 0 or 1. Prätzel-Wolters[28] developed a unifying framework for this type of canonical forms, interpreting the structural ones as result of a normalisation procedure and the structural zeros as result of a minimisation procedure within the subset of normalised elements of each orbit.

To define this minimisation procedure, a new relation of order is introduced on $\mathbb{R}$ (see Fig. 5) by taking 0 to be the smallest real number and taking on $\mathbb{R} \setminus \{0\}$ the usual order.

Fig. 5 The order relation $<$ on $\mathbb{R}$

The vector spaces $\mathbb{R}^k$, $k \in \mathbb{N}$ are provided with the lexicographic order (from above) corresponding to $<$. Hence, for example,

$$\begin{bmatrix} 0 \\ 3 \\ 1 \\ 0 \end{bmatrix} < \begin{bmatrix} -1 \\ 0 \\ 0 \\ 0 \end{bmatrix}, \quad \begin{bmatrix} 0 \\ 2 \\ 0 \\ 1 \end{bmatrix} < \begin{bmatrix} 0 \\ 2 \\ -1 \\ 1 \end{bmatrix}$$

Identifying $a+bi \in \mathbb{C}$ with $\begin{bmatrix} a \\ b \end{bmatrix} \in \mathbb{R}^2$ a total order is defined on $\mathbb{C}^k$, $k \in \mathbb{N}$. Finally, in order to apply ($<$) to $n \times m$ matrices A these are identified in some way with nm-column vectors, e.g., with

$$\begin{bmatrix} a^1 \\ a^2 \\ \vdots \\ a^m \end{bmatrix} \text{(column order } <_c\text{) or } \begin{bmatrix} (a_1)^T \\ (a_2)^T \\ \vdots \\ (a_n)^T \end{bmatrix} \text{ (row order } <_r\text{)}.$$

Let α be any group action on a space X of (compound) matrices. In general there will not exist $<$-minimal elements of the orbits $[x]_\alpha$. However, normalising a sufficient number of parameters it is often possible to obtain a uniquely determined $<$-minimum among the normalised elements of each orbit. To determine those elements one proceeds in two steps. One firstly applies the screen function

$$\omega: \mathbb{C}^k \to \{0,1\}^k, \quad \omega(x)_i = \begin{cases} 0 & \text{if } x_i = 0 \\ 1 & \text{if } x_i \neq 0 \end{cases}$$

to the elements of the orbit $[x]_\alpha$ to obtain the minimum

$$\omega^*_x = \min_{<_c} \{\omega(y) \; ; \; y \in [x]_\alpha\}.$$

Because $(<)$ is a total order and $\{0,1\}^k$ is finite, this minimum exists. The associated subset $\Omega_x = \{\hat{x} \in [x]_\alpha ; \omega(\hat{x}) = \omega^*_x\}$ contains in general many elements, but from the particular group action it is often clear how canonical elements in Ω_x can be obtained by normalising certain entries of elements in Ω_x.

Proceeding in this way, Prätzel-Wolters[28] has been able to characterise well-known canonical forms of linear algebra and linear system theory as normalised orbit minima with respect to the order $<$, e.g. the echelon canonical form, the Jordan canonical form, the Kronecker and the Brunovsky canonical forms.

Moreover, following this method, Prätzel-Wolters succeeded in constructing a canonical form for the similarity action on the full space $L_{m,n}(K)$ of (not necessarily reachable) input pairs. However, already in the single input case the resulting canonical form is too involved to be reported here (see Prätzel-Wolters[28] for an explicit description).

4. CELLULAR CANONICAL FORMS AND TOPOLOGICAL INVARIANTS OF THE ORBIT SPACE

This section gives a condensed survey about recent results concerning the orbit spaces $\hat{L}^r_{m,n,p}(K)$ and $\hat{L}^{r,o}_{m,n,p}(K)$. A general theorem which can be found in Dieudonné[29] implies that these spaces carry an analytical structure such that $\pi: L^r_{m,n,p}(K) \to \hat{L}^r_{m,n,p}(K)$ and $\pi: L^{r,o}_{m,n,p}(K) \to \hat{L}^{r,o}_{m,n,p}(K)$ are principal $Gl_n(K)$-bundles. By an easy connectivity argument it follows from Proposition 2.1 that for $m \geq 2$, these bundles are non-trivial and hence there does not exist a global continuous canonical form on these spaces. Then the cell decomposition induced by the Kronecker canonical form is used to determine the mod 2 Betti numbers of $\hat{L}^{r}_{m,n,p}(K)$. This is possible due to a beautiful theorem of Borel and Haefliger[30]. As a consequence one obtains the result that the Kronecker canonical form is of minimal discontinuity. Finally, some consequences are indicated concerning the critical point behaviour of objective functions on spaces of multivariable systems.

A. The Orbit Manifolds $\hat{L}^{r}_{m,n,p}(\mathbb{R})$ and $\hat{L}^{r,o}_{m,n,p}(\mathbb{R})$.

Let α: $G\times M \rightarrow M$ be a C^k-action of a Lie group G on an n-dimensional C^k-manifold M ($K = \infty, \omega$). One of the first questions about the orbit space M/α is whether it is again a C^k-manifold. In the system theoretic literature the existence of a compatible C^k-structure on an orbit space is usually derived by one of the following three methods.

(1) An atlas of C^k-compatible charts on M/α is described explicitly (cf. Hazewinkel and Kalman[11] for the similarity action on $L^{r}_{m,n}(K)$, Clark[13] for σ on $L^{r,o}_{m,n,p}(K)$).

(2) The orbit space M/α is embedded homeomorphically into some C^∞-manifold M'. If M/α can be represented as the fixed-point set for a compact Lie group acting on M' then, by a theorem of Bochner[31], M/α is a smooth manifold (cf. Byrnes and Duncan[32], Byrnes[9]).

(3) Whenever one can prove that the graph of α (2.1) is a closed C^k-manifold of $M\times M$ (by consequence, the orbits of α have to be closed submanifolds of M) the following theorem is applicable[29].

4.1 Theorem

There is a (unique) C^k-structure on the orbit space M/α compatible with the quotient topology such that π: $M \rightarrow M/\alpha$ is a C^k-submersion if and only if the graph of α is a closed C^k-submanifold of $M\times M$. In particular, if this condition is satisfied and α is free, then π: $M \rightarrow M/\alpha$ is a principal G-bundle.

The last method can easily be applied to the similarity action on $L^r_{m,n,p}(K)$. In fact, we know already that σ is free on this space and one can prove that the graph of σ on $L^r_{m,n,p}(K)$ is a closed analytical submanifold of $L^r_{m,n,p}(K) \times L^r_{m,n,p}(K)$ (Helmke[17]). As an immediate consequence one gets the following result (first obtained by Hazewinkel and Kalman[11], Byrnes and Hurt[14]).

4.2 *Proposition*

$\hat{L}^r_{m,n,p}(K) = L^r_{m,n,p}(K)/\sigma$ is an analytical manifold of dimension $n(m+p)$ and $\pi: L^r_{m,n,p}(K) \to \hat{L}^r_{m,n,p}(K)$ is a principal $Gl_n(K)$-bundle.

Since $L^{r,o}_{m,n,p}(K)$ is a σ-invariant open subset of $L^r_{m,n,p}(K)$ an analogous statement follows for the similarity action σ on $L^{r,o}_{m,n,p}(K)$ (cf. Clark[13]).

For the convergence analysis of algorithms defined on some manifold M it is of evident importance to know if M is connected. The following proposition collects some useful connectivity properties of the above orbit spaces in the case of principal interest $K = \mathbb{R}$

4.3 *Proposition*

Let $m,n,p \in \mathbb{N}$, $m,n \geq 1$. Then

(a) $L^r_{m,n,p}(\mathbb{R})$ is connected if and only if $m \geq 2$. For $m = 1$:

$$L^r_{1,n,p}(\mathbb{R}) \cong Gl_n(\mathbb{R}) \times \mathbb{R}^n \times \mathbb{R}^{p \times n} \quad ;$$

(b) $\hat{L}^r_{m,n,p}(\mathbb{R})$ is connected for all m,n,p. In particular,

$$\hat{L}^r_{1,n,p}(\mathbb{R}) \cong \mathbb{R}^n \times \mathbb{R}^{p \times n} \quad ;$$

(c) $L^{r,o}_{m,n,p}(\mathbb{R})$ is connected if and only if $\min(m,p) \geq 2$.

$L^{r,o}_{1,n,p}(\mathbb{R}) \cong Gl_n(\mathbb{R}) \times \hat{L}^{r,o}_{1,n,p}(\mathbb{R})$ and $L^{r,o}_{m,n,1}(\mathbb{R}) \cong Gl_n(\mathbb{R}) \times \hat{L}^{r,o}_{m,n,1}(\mathbb{R})$

(d) $\hat{L}^{r,o}_{m,n,p}(\mathbb{R})$ is connected if and only if max $(m,p) \geq 2$.

$\hat{L}^{r,o}_{1,n,1}(\mathbb{R})$ has n+1 connected components.

The analytical diffeomorphisms in (a), (b), (c) are constructed via the controllable (resp. observable) canonical forms for single input (resp. output) systems which are globally analytic. The equivalences in (a) and (c) follow from the fact that $L^{r}_{m,n,p}(\mathbb{R})$ and $L^{r,o}_{m,n,p}(\mathbb{R})$ are obtained from $\mathbb{R}^{n(m+n+p)}$ by cutting out submanifolds of codimension $\geq m$ (resp. $\geq m$ and $\geq p$). The last statement of (d) has to be proved separately (see Brockett[12]).

Since $Gl_n(\mathbb{R})$ is not connected, it follows from statement (a) that $L^{r}_{m,n,p}(\mathbb{R}) \not\cong Gl_n(\mathbb{R}) \times \hat{L}^{r}_{m,n,p}(\mathbb{R})$ if $m \geq 2$. Hence, Proposition 2.1 immediately yields the well-known result of Hazewinkel and Kalman[11].

4.4 *Corollary*

If $m \geq 2$, the $Gl_n(\mathbb{R})$-bundle π: $L^{r}_{m,n,p}(\mathbb{R}) \rightarrow \hat{L}^{r}_{m,n,p}(\mathbb{R})$ is non-trivial and hence there does not exist a continuous canonical form for the similarity action on $L^{r}_{m,n,p}(\mathbb{R})$.

By the same reasoning, applied to $L^{r,o}_{m,n,p}(\mathbb{R})$, one obtains from Prop. 2.1 and Prop. 4.3 (c) the analogous result for the similarity action on $L^{r,o}_{m,n,p}(\mathbb{R})$[33].

4.5 *Corollary*

If min $(m,p) \geq 2$, the $Gl_n(\mathbb{R})$-bundle π: $L^{r,o}_{m,n,p}(\mathbb{R}) \rightarrow \hat{L}^{r,o}_{m,n,p}(\mathbb{R})$ is non-trivial and there does not exist a continuous canonical form for $\sigma|L^{r,o}_{m,n,p}(\mathbb{R})$.

B. Betti Numbers of the Orbit-Manifolds $\hat{L}^{r}_{m,n,p}(\mathbb{R})$ and Applications

Let Z_2 denote the field $\mathbb{Z}/_{(2\mathbb{Z})}$ of two elements and, for any topological space Y, $H_q(Y,Z_2)$ the q-th singular homology group with coefficients in Z_2(vector space over Z_2). Then the mod 2

Betti numbers of Y are, by definition,

$$\beta_q(Y,Z_2) = \dim_{Z_2} H_q(Y,Z_2), \quad q \in \mathbb{N}.$$

Their sum will be denoted by $\beta(Y,Z_2)$. If Y is a topological n-manifold, the following version of the weak Morse inequalities can be proved (see Helmke and Hinrichsen[21]).

4.6 *Proposition*

If $(Y_i)_{i \in I}$ is a finite cellular decomposition of a topological n-manifold Y and $\gamma_q = \gamma_q(Y_i)_I$ the number of q-dimensional cells, $q = 0,1,\ldots,n$, then

$$\gamma_q \geq \beta_{n-q}(Y,Z_2), \quad q = 0,\ldots,n. \tag{4.1}$$

In particular, the C^0-complexity of Y is bounded below by the sum $\beta(Y,Z_2)$ of its mod 2 Betti numbers.

To see under which conditions equality holds in (4.1) let Y be an n-dimensional real analytical manifold. Recall that a closed subset $A \subset Y$ is said to be a locally analytical subvariety of Y if for every $x \in A$ there exists an open neighbourhood U of x in Y and finitely many analytical functions $f_j: U \to \mathbb{R}$, $j \in J$ such that

$$A \cap U = \{y \in U;\ \ f_j(y) = 0 \text{ for all } j \in J\}.$$

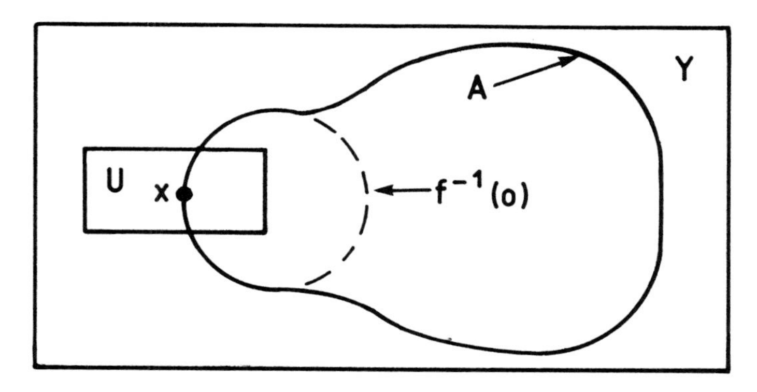

Fig. 6 Local analyticity of A at x

An analytical cellular decomposition $(Y_i)_{i\in I}$ of Y is said to have locally analytical closures, if the topological closures $\bar{Y}_i$ of the cells are locally analytical subvarieties of Y.

Fig. 7 C^ω-cellular decompositions of the circle

In Fig. 7 the cellular decomposition (a) has locally analytical closures whereas (b) has not. In fact, the following theorem of Borel and Haeflinger[30] shows that the number of k-dimensional cells of a C^ω-cellular decomposition of any real analytical manifold Y is completely determined by the homology of Y if it has locally analytical closures.

4.7 *Theorem*

Let Y be a real analytical manifold of dimension n and $(Y_i)_{i\in I}$ a finite C^ω-cellular decomposition of Y with locally analytical closures. Then

$$\gamma_q(Y_i)_I = \beta_{n-q}(Y,Z_2), \quad q = 0,\dots,n. \tag{4.2}$$

It follows immediately from (4.1) and Theorem 4.7 that every finite C^ω-cellular decomposition $(Y_i)_I$ with locally analytical closures is minimal (see § 2) and that the C^k-complexities of Y coincide: for $k = 0, \infty, \omega$:

$$c_k(Y) = \text{card } I, \quad k = 0,\infty,\omega$$

(compare Fig. 7).

Now let us apply these results to the cellular decomposition $(\hat{\text{Kro}}\,(\kappa))_{\kappa\in K_{n,m}}$ of $Y = \hat{L}^r_{m,n,p}(\mathbb{R})$, obtained by projecting the Kronecker strata $\text{Kro}(\kappa) \subset L^r_{m,n,p}(\mathbb{R})$ into the orbit space $\hat{L}^r_{m,n,p}(\mathbb{R})$ (see § 3):

$$\hat{Kro}(\kappa) = \pi(Kro(\kappa)), \quad \kappa \in K_{n,m}.$$

Γ_k induces analytical sections

$$s_\kappa : \hat{Kro}(\kappa) \to L^r_{m,n,p}(\mathbb{R}), \quad [(A,B,C)]_\sigma \to \Gamma_k(A,B,C)$$

on each of these cells - called Kronecker cells - and thus defines an analytical diffeomorphism between $\hat{Kro}(\kappa)$ and the affine space $s_k(\hat{Kro}(\kappa))$ of all triples $(\tilde{A},\tilde{B},\tilde{C})$ which are of Kronecker canonical form with index list κ. Counting parameters one obtains the following dimension formula for the analytical cells $\hat{Kro}(\kappa)$

$$N(\kappa) = \sum_{i,j=1}^{m} \min(\kappa_i,\kappa_j) + \sum_{i>j} \kappa_{ij} + np \tag{4.3}$$

where

$$K_{ij} = \begin{cases} 1 & \text{if } \kappa_i < \kappa_j \\ 0 & \text{if } \kappa_i \geq \kappa_j \end{cases}$$

For instance, in example 3.2 with $p = 0$, $N(2,0,3) = 10$.

To describe how these cells are pasted together in $\hat{L}^r_{m,n,p}(\mathbb{R})$ and to determine their closures, let $\leq$ denote the lexicographic order on $\bar{n}x\underline{m}$ where $\bar{n} = \{0,1,\ldots,n\}$. For any combination $\kappa \in K_{n,m}$ define

$$Y_\kappa = \{(i,j) \in \bar{n}x\underline{m};\ 0 \leq i \leq \kappa_j - 1\}$$

and

$$r_{ij}(\kappa) = \text{card } \{(k,\ell) \in Y_\kappa;\ (k,\ell) \leq (i,j)\}, \quad (i,j) \in \bar{n}x\underline{m}.$$

Introducing the "truncated reachability matrices"

$$R_{ij}(A,B) = [b^1 b^2 \ldots b^m Ab^1\ Ab^2 \ldots A^i b^j], \quad (i,j) \in \bar{n}x\underline{m}$$

one clearly has

$$r_{ij}(\kappa) = \text{rank } R_{ij}(A,B), \quad (A,B,C) \in Kro(\kappa).$$

The Kronecker order on $K_{n,m}$ is defined by

$$\lambda \preccurlyeq \kappa \Leftrightarrow r_{ij}(\lambda) \leq r_{ij}(\kappa) \text{ for all } (i,j) \in \underline{\bar{n}x\underline{m}}$$

$(\kappa,\lambda \in K_{n,m})$. The following characterisation of the adherence order $\text{Kro}(\lambda) \subset \overline{\text{Kro}(\kappa)}$ of the Kronecker cells has been derived by Helmke[17].

4.8 *Proposition*

The adherence order for the Kronecker cells in $\hat{L}^r_{m,n,p}(\kappa)$ is the Kronecker order on $K_{n,m}$:

$$\lambda \preccurlyeq \kappa \iff \text{Kro}(\lambda) \subset \overline{\text{Kro}(\kappa)} \iff \text{Kro}(\lambda) \cap \overline{\text{Kro}(\kappa)} \neq \phi \quad (4.4)$$

In particular, one obtains the following description of the closure of $\hat{\text{Kro}}(\kappa)$:

$\overline{\hat{\text{Kro}}(\kappa)} = \{[(A,B,C)]_\sigma \in \hat{L}^r_{m,n,p}(\mathbb{R});\ \text{rank } R_{ij}(A,B) \leq r_{ij}(\kappa)$ for all $(i,j) \in \bar{n}x\underline{m}\}$.

From this formula it is easily seen that the closures of the Kronecker cells are locally analytical submanifolds of $\hat{L}^r_{m,n,p}(\mathbb{R})$ Hence, by the theorem of Borel and Haefliger, we obtain the following result, due to Helmke[17].

4.9 *Theorem*

The decomposition $(\hat{\text{Kro}}(\kappa))_{\kappa \in K_{n,m}}$ of the orbit space $\hat{L}^r_{m,n,p}(\mathbb{R})$ into Kronecker cells is an analytical cellular decomposition with locally analytical closures. The mod 2 Betti numbers of $\hat{L}^r_{m,n,p}(\mathbb{R})$ are determined by

$$\beta_q(\hat{L}^r_{m,m,p}(\mathbb{R}),Z_2) = \text{card}\{\kappa \in K_{n,m};\ N(\kappa) = n(m+p)-q\},\ q \in \mathbb{N} \quad (4.5)$$

By means of the dimension formula (4.3) one can compute $\beta_q(\hat{L}^r_{m,n,p}(\mathbb{R}),Z_2)$ for all $m,n,p,q \in \mathbb{N}$. In particular, one observes that these Betti numbers do not depend upon p (in fact $\hat{L}^r_{m,n,p}(\mathbb{R})$ is a deformation retract of $\hat{L}^r_{m,n}(\mathbb{R})$).

As a corollary of Theorem 4.9 and Proposition 4.8 we obtain the announced minimality result for the Kronecker canonical form and a formula for the complexity of the orbit space $\hat{L}^r_{m,n,p}(\mathbb{R})$.[21]

4.10 *Corollary*

$\Gamma_K: L^r_{m,n,p}(\mathbb{R}) \to L^r_{m,n,p}(\mathbb{R})$ is a C^ω-cellular canonical form of minimal discontinuity. The C^k-complexity of the orbit space of reachable systems is

$$c_k(\hat{L}^r_{m,n,p}(\mathbb{R})) = \binom{m+n-1}{n}, \quad k = 0,\infty,\omega.$$

4.11 *Remark*

The problem to determine completely the Betti numbers of the space $T^n_{m,p}(\mathbb{R})$ of pxm-transfer matrices of McMillan degree n is still open. However, since $\hat{L}^{r,o}_{m,n,p}(\mathbb{R}) \cong T^n_{m,p}(\mathbb{R})$ is obtained from $\hat{L}^r_{m,n,p}(\mathbb{R})$ by removing an analytical subvariety of codimension p it can be proved (Helmke[17]) that the first p-1 mod 2 Betti numbers of $T^n_{m,p}(\mathbb{R})$ and $\hat{L}^r_{m,n,p}(\mathbb{R})$ coincide:

$$\beta_i(T^n_{m,p}(\mathbb{R}), Z_2) = \beta_i(\hat{L}^r_{m,n,p}(\mathbb{R}), Z_2), \quad i = 0,\dots,p-2$$

□

To indicate possible applications of the above results we conclude the paper with some remarks concerning their implications for the critical point behaviour of objective functions on $\hat{L}^r_{m,n}(\mathbb{R})$.

Let M be a C^∞-manifold. A function $f: M \to \mathbb{R}$ is said to have compact sublevel sets if $f^{-1}((-\infty,a])$ is compact in M for all $a \in R$. A critical point $x_o \in M$ of f is said to be nondegenerate if the Hessian matrix $H_f(x_o) = \left[\frac{\partial^2 f}{\partial x_i \partial x_j}(x_o)\right]$ (with respect to some coordinate system (x_i) about x_o) is non-singular. The index of a nondegenerate critical point x_o

is the number of negative eigenvalues of the (symmetric) Hessian $H_f(x_o)$. Nondegenerate critical points are isolated. The following Morse inequalities show that the Betti numbers of M yield lower bounds for the number of critical points of given index of f (cf. Milnor[34]).

4.12 *Theorem* (weak Morse inequalities)

Let f: M → R be a function with compact level sets and suppose that f has only nondegenerate critical points. If $c_i(f)$ denotes the number of critical points of index i of f then

$$c_i(f) \geq \beta_i(M, Z_2).$$

As an immediate consequence of Theorem 4.12 and 4.10 one obtains the following result.

4.13 *Proposition*

If f: $\hat{L}^r_{m,n,p}(\mathbb{R}) \to \mathbb{R}$ is a function with compact sublevel sets and only nondegenerate critical points then

$$c_q(f) \geq \text{card}\ \{\kappa \in K_{n,m};\ N(\kappa) = n(m+p)-q\},\ q \in \mathbb{N}.$$

In particular, f has at least $\binom{m+n-1}{n}$ critical points.

Results of this type are possibly of interest in the analysis of optimal model reduction procedures. Their importance for the convergence analysis of identification algorithms has been emphasised by Byrnes[9], especially with regard to asymptotically deterministic identification algorithms[35].

4.14 *Remark*

By duality the above results about the Betti numbers can directly be transferred to the orbit space of observable systems $\hat{L}^o_{m,n,p}(\mathbb{R})$ which is of more interest in identification than $\hat{L}^r_{m,n,p}(\mathbb{R})$. In fact, for given $p,n \geq 1$ the space $\hat{L}^o_{p,n,p}(\mathbb{R})$ may be interpreted as space of ARMA models corresponding to similarity classes of state space models of the form

$$x(t+1) = F\ x(t) + G\ \varepsilon\ (t)$$

$$y(t) = H\ x(t) + \varepsilon(t)$$

where (F,H) is observable and $\varepsilon(t)$ the driving and measurement noise (sequence of independent $\mathbb{R}^p$-valued random variables), see Ljung and Rissanen[20], Hannan[7].

REFERENCES

1. Ackermann, J. (1977) Entwurf durch Polvorgabe, Regelungstechnik, Heft 6, 173-179 and Heft 7, 209-215.

2. Kailath, T. (1979), A Course in Linear Systems Theory, Prentice-Hall, Englewood Cliffs, New York.

3. Bosgra, D.H. (1983) On parametrizations for the minimal partial realization problem, Systems and Control Letters 3, 181-187.

4. Denham, M.J. (1974), Canonical forms for the identification of multivariable linear systems, IEEE Trans. Autom. Control, Vol. AC-19, 646-656.

5. Guidorzi, R. (1975), Canonical structures in the identification of multivariable systems, Automatica 11, 361-374.

6. Hannan, E.J. (1981) System identification, Stochastic Systems: The Mathematics of Filtering and Identification and Applications (M. Hazewinkel and J.C. Willems, eds.), 221-246.

7. Hannan, E.J. (1979) The statistical theory of linear systems, in: Developments in Statistics (P. Krishnaiah, ed.), Vol. 2, 83-121, Academic Press, New York.

8. Deistler, M. and Hannan, E.J. (1981) Some properties of the parametrization of ARMA systems with unknown order, J. Multivariate Anal. 11, 474-484.

9. Byrnes, C.I. (1983) Geometric aspects of the convergence analysis of identification algorithms, in : Nonlinear Stochastic Problems (R.S. Bucy and J.F. Moura, eds.), D. Reidel, Dordrecht.

10. Glover, K. (1975) Some geometrical properties of linear systems with implications in identification, Proc. of the IFAC Congr. Boston 1975.

11. Hazewinkel, M. and Kalman, R. (1976) On invariants, canonical forms and moduli for linear, constant, finite-dimensional, dynamical systems, Lecture Notes in

Econ. Math. System Theory 131, Springer, New York, 48-60.

12. Brockett, R.W. (1976), Some geometric questions in the theory of linear systems, IEEE Trans. Autom. Control, Vol. AC-21, 449-455.

13. Clark, J.M.C. (1976), The consistent selection of local coordinates in linear system identification, Proc. JACC 576-580.

14. Byrnes, C.I. and Hurt, N. (1979), On the moduli of linear dynamical systems, Adv. in Math. Studies in Analysis, Vol. 4, 83-122.

15. Hazewinkel, M. (1980), (Fine) moduli (spaces) for linear systems: What are they and what are they good for? Geometrical Methods for the Theory of Linear Systems (C.I. Byrnes, C.F. Martin, eds.) D. Reidel, Dordrecht, 125-193.

16. Delchamps, D.F. (1982), The geometry of spaces of linear systems with an application to the identification problem, Ph.D Thesis Harvard University.

17. Helmke, U. (1982), Zur Topologie des Raumes linearer Kontrollsysteme, Ph.D Thesis, University of Bremen.

18. Helmke, U. (1984), The topology of the space of reachable linear systems I: the complex case, Universitat Bremen, FS Dynamische Systeme, Report 122, submitted: Math. Systems Theory, Springer.

19. Glover, K. and Willems, J.C. (1974), Parametrizations of linear dynamical systems: canonical forms and identifiability, IEEE Trans. Autom. Contr. AC-19, 640-645.

20. Ljung, L. and Rissanen, J. (1978), On canonical forms, parameter identifiability and the concept of complexity, Identification and System Parameter Estimation (Rajbman, ed.), North Holland.

21. Helmke, U. and Hinrichsen, H. (1983), Canonical forms and orbit spaces of linear systems, Universitat Bremen, FS Dynamische Systeme, Report Nr. 97, to appear in Proc. III. Working Conf. on System Modelling and Optimization, Hanoi 1983, Springer-Verlag.

22. Dieudonné, J. (1970), Foundations of Modern Analysis, Vol. 2, Academic Press.

23. Birkhoff, G. and McLane, S. (1977), A Survey of Modern

Algebra, Macmillan Publ. Co. Inc., New York, Fourth Edition.

24. Popov, V.M. (1972), Invariant description of linear time-invariant controllable systems, SIAM J. Control, Vol. 10, 252-264.

25. Gantmacher, E.R. (1959), The Theory of Matrices, Vol. 1 and 2, Chelsea, New York.

26. Bosgra, O.K. and Van der Weiden, A.J.J. (1980), Input-output invariants for linear multivariable systems, IEEE Trans. Autom. Contr., Vol. AC-25, 20-36.

27. Brunovsky, P. (1970), A classification of linear controllable systems, Kybernetics 3, 173-187.

28. Prätzel-Wolters, D. (1983), Canonical forms for linear systems, Linear Algebra and its Applications 50, 437-473.

29. Dieudonné, J. (1972), Foundations of Modern Analysis, Vol. 3, Academic Press.

30. Borel, A. and Haefliger, A. (1961), La classe d'homologie fondamentale d'un espace analytique, Bull. Soc. Math. France, 89, 461-513.

31. Bochner, S. (1945), Compact groups of differentiable transformations, Ann. Math. 46, 372-381.

32. Byrnes, C.I. and Duncan, T.E. (1982), On certain topological invariants arising in system theory, in: New Directions in Applied Mathematics, P.J. Hilton, G.S. Young (eds.), Springer-Verlag, 29-71.

33. Hazewinkel, M. (1977), Moduli and canonical forms of linear dynamical systems II: The topological case, Math. Systems Theory 10, 363-385.

34. Milnor, J.W. (1963), Morse Theory, Princeton University Press.

35. Ljung, L. (1981), Recursive identification, in: Stochastic Systems: The Mathematics of Filtering and Identification and Applications, D. Reidel, 221-246.

PART II

CHAPTER I

LINEAR SYSTEM THEORY

BILINEAR STRICT EQUIVALENCE OF MATRIX PENCILS

N. Karcanias and G. Kalogeropoulos

(*Control Engineering Centre, The City University, London*)

ABSTRACT

The notion of Bilinear-Strict Equivalence (BSE) of homogeneous matrix pencils $sF-\hat{s}G$ is introduced and a set of invariants is defined. It is shown that the sets of column, row minimal indices and the set of degrees of elementary divisors are invariant under BSE transformations. The search for a complete set of invariants is reduced to the study of invariants of homogeneous binary polynomials, $f(s,\hat{s})$, under Projective Equivalence (PE) transformations. It is shown that the definition of a complete set of invariants of $f(s,\hat{s})$ under PE transformations is reduced to a study of invariants of matrices under Extended Hermite type transformations. These results provide the means for the study of properties of linear state space system descriptions under space and frequency coordinate transformations.

1. INTRODUCTION

Linear, forced systems of the regular, or extended state space type, as well as a number of important problems of the geometric system theory[1,2] may be reduced to the study of properties of the differential system $S(F,G)$ [3,4], where

$$S(F,G):\quad F\dot{\underline{x}}(t) = G\underline{x}(t),\ F,G \in \mathbb{R}^{m\times n},\ \underline{x}(t) \in \mathbb{R}^{n} \tag{1.1}$$

The link of $S(F,G)$ description with the matrix pencil $sF-G$ indicates that matrix pencil theory[5] is the key for study of algebrogeometric and dynamic properties[3], as well as numerical aspects[6] of state space descriptions. The need for the simultaneous study of properties of $sF-G$ for finite and infinite frequencies, naturally leads to the definition of the homogeneous pencil $sF-\hat{s}G$. The classical theory of matrix pencils[5] deals with the study of invariants and canonical forms of $sF-\hat{s}G$ under

Strict Equivalence (SE). SE transformations may be interpreted as coordinate transformations in the domain and codomain of the ordered pair of operators (F,G). The aim of this paper is to extend the classical theory of SE, by studying the properties of $sF-\hat{s}G$ under the combined action of strict and bilinear (or projective type) transformations. The bilinear, or projective transformation on a straight line is defined by β: $(s,\hat{s}) \to (\lambda,\hat{\lambda})$, where

$$\beta: \begin{bmatrix} s \\ \hat{s} \end{bmatrix} = \begin{bmatrix} a & b \\ c & d \end{bmatrix} \begin{bmatrix} \lambda \\ \hat{\lambda} \end{bmatrix}, \quad \delta = ad-cb \neq 0, \ a,b,c,d \in \mathbb{R} \tag{1.2}$$

The pairs $(s,\hat{s}),(\lambda,\hat{\lambda})$ define the homogeneous coordinates of a point P of the straight line with respect to the reference points (A,B),(A',B') respectively[7]. Given that the projective parameters $s/\hat{s}$ and $\lambda/\hat{\lambda}$ may be interpreted as frequencies, we may say that (1.2) defines a coordinate transformation in the frequency domain. The aim of the paper is to introduce the study of invariants of $sF-\hat{s}G$ under the group of general coordinate transformations in the space (domain and codomain transformations i.e. strict equivalence[5]) and frequency domain (projective transformations on a straight line, i.e. bilinear equivalence). A number of preliminary nature results on the invariants of $sF-\hat{s}G$ under BSE transformations are given without a proof; the proof of these results, as well as the definition of a complete set of invariants, are given in reference 8).

2. DEFINITIONS AND STATEMENT OF THE PROBLEM

Let $L = \{L: L = (F,-G),\ F,G \in \mathbb{R}^{m\times n}\}$ be the set of ordered pairs of m×n matrices and let $\Theta = \{\theta: \theta = (\lambda,\hat{\lambda})\}$ be the set of ordered pairs of indeterminates. For $\forall\ L = (F,-G) \in L$ and $\theta = (s,\hat{s}) \in \Theta$, the matrix $\bar{L} = [F,-G] \in \mathbb{R}^{m\times 2n}$ will be called a *matrix representation* of L and the homogeneous polynomial matrix

$$L_\theta \triangleq L(s,\hat{s}) = [F,-G]\begin{bmatrix} sI_n \\ \hat{s}I_n \end{bmatrix} = sF-\hat{s}G \tag{2.1}$$

will be referred to as the θ-*matrix pencil* of L. By $L(\theta), L(\Theta)$ we denote the sets of all matrix pencils defined on L for a fixed $\theta = (s,\hat{s})$, all possible $\theta \in \Theta$ respectively. In the following, three types of equivalence are defined on L, or equivalently on $L(\Theta)$.

Consider first the set $K = \{k: k = (M,N), M \in \mathbb{R}^{m\times m}, N \in \mathbb{R}^{2n\times 2n}, |M|,|N| \neq 0\}$ and a composition rule (*) defined as follows: $*: K\times K \to K$: $\forall\ k_1 = (M_1,N_1),\ k_2 = (M_2,N_2) \in K$, then

$$k_1 * k_2 \triangleq (M_1,N_1)*(M_2,N_2) = (M_1M_2,N_2N_1) \tag{2.2}$$

It may be readily verified that $(K,*)$ is a group with identity element (I_m,I_{2n}). The action of K on L is defined by

$\circ$: $\mathcal{K}\times\mathcal{L}\to\mathcal{L}$: $\forall\ k=(M,N)\in\mathcal{K}$ and for an $L=(F,-G)\in\mathcal{L}$, then

$$k\circ L \triangleq k\circ(F,-G) = L' = (F',-G')\in\mathcal{L}, \text{ where } \bar{L}'=M\bar{L}N=M[F,-G]N \quad (2.3)$$

Such an action defines an equivalence relation on $\mathcal{L}$ and $\mathcal{K}(L)$ denotes the equivalence class of $L\in\mathcal{L}$ under $\mathcal{K}$. Three important subgroups of $\mathcal{K}$ are defined next:

(i) Strict Equivalence: The subgroup $\mathcal{H}$ of $\mathcal{K}$ defined by

$$\mathcal{H} \triangleq \{h:\ h=(P,Q), P\in\mathbb{R}^{m\times m}, Q=\mathrm{diag}(R,R), R\in\mathbb{R}^{n\times n}, |P|,|R|\neq 0\} \quad (2.4)$$

is called the *strict equivalence group* (SEG). The action of $\mathcal{H}$ on $\mathcal{L}$ is defined by $\circ$: $\mathcal{H}\times\mathcal{L}\to\mathcal{L}$: $\forall\ h\in\mathcal{H}$ and for $L=(F,-G)\in\mathcal{L}$, then

$$h\circ L \triangleq (P,Q)\circ(F,-G)=L'=(F',-G')\in\mathcal{L}, \text{ where } \bar{L}'=P[F,-G]\begin{bmatrix} R & O\\ O & R\end{bmatrix} \quad (2.5)$$

The equivalence relation $E_{\mathcal{H}}$ defined on $\mathcal{L}$ is called *strict equivalence* (SE)[5]. Two pencils $L^1_\theta = sF_1-\hat{s}G_1$, $L^2_\theta = sF_2-\hat{s}G_2$ are said to be strict equivalent, $L^1_\theta E_{\mathcal{H}} L^2_\theta$, if there exist $h\in\mathcal{H}$: $(F_2,-G_2)=h\circ(F_1,-G_1)$. By $\mathcal{H}(F,G)$ we denote the SE class of $L_\theta = sF-\hat{s}G$.

(ii) Bilinear Equivalence: The subgroup $\mathcal{B}$ of $\mathcal{K}$ defined by

$$\mathcal{B}=\{b:\ b=(I_m,\begin{bmatrix} aI_n & bI_n\\ cI_n & dI_n\end{bmatrix}) = (I_m,T_\beta), a,b,c,d\in\mathbb{R}, ad-bc\neq 0\} \quad (2.6)$$

is called the *bilinear equivalence group* (BEG). Every $b\in\mathcal{B}$ is generated by a projective transformation β of the type (1.2). The action of $\mathcal{B}$ on $\mathcal{L}$ is defined by $\circ$: $\mathcal{B}\times\mathcal{L}\to\mathcal{L}$: $\forall\ b\in\mathcal{B}$ and for an $L=(F,-G)\in\mathcal{L}$, then

$$b\circ L \triangleq b\circ(F,-G) = L' = (F',-G')\in\mathcal{L},$$

$$\text{where } \bar{L}' = I_m[F,-G]T_\beta = [aF-cG, bF-dG] \quad (2.7)$$

The equivalence relation $E_{\mathcal{B}}$, defined on $\mathcal{L}$, is called *bilinear equivalence* (BE). Two pencils $L^1_\theta = sF_1-\hat{s}G_1$, $L^2_{\theta'} = \lambda F_2-\hat{\lambda}G_2\in\mathcal{L}(\Theta)$, $\theta=(s,\hat{s})$, $\theta'=(\lambda,\hat{\lambda})$ are said to be bilinearly equivalent, $L^1_\theta E_{\mathcal{B}} L^2_{\theta'}$, if there exists a transformation β: $(s,\hat{s})\to(\lambda,\hat{\lambda})$ and thus a $b\in\mathcal{B}$: $(F_2,-G_2)=b\circ(F_1,-G_1)$. By $\mathcal{B}(F,G)$ we denote the BE class of $L_\theta = sF-\hat{s}G$.

The composition rule (*) is not commutative on $\mathcal{K}$; however the following property holds true:

Proposition (2.1): For $\forall\ b\in\mathcal{B}$ and $\forall\ h\in\mathcal{H}$, then $b*h=h*b$. □

(iii) Bilinear-Strict Equivalence: The subgroup $\mathcal{H}$-$\mathcal{B}$ of $\mathcal{K}$ defined by $\mathcal{H}\text{-}\mathcal{B}\triangleq\{r:\ r=h*b=b*h,\ \forall\ h\in\mathcal{H}, b\in\mathcal{B}\}$ is called the *bilinear-strict equivalence group* (BSEG) and its action on $(F,-G)\in\mathcal{L}$ is defined by $\circ$: $\mathcal{H}\text{-}\mathcal{B}\times\mathcal{L}\to\mathcal{L}$: $\forall\ r\in\mathcal{H}\text{-}\mathcal{B}$ and $L=(F,-G)\in\mathcal{L}$, then

$$r\circ L \triangleq h*b\circ(F,-G) = (aPFR-cPGR,\ bPFR-dPGR)\in\mathcal{L} \quad (2.8)$$

The equivalence relation E_{H-B} defined on L is called *bilinear-strict equivalence* (BSE). Two pencils $L^1_\theta = sF_1 - \hat{s}G_1$, $L^2_{\theta'} = \lambda F_2 - \hat{\lambda}G_2 \in L(\Theta)$, $\theta = (s,\hat{s})$, $\theta' = (\lambda,\hat{\lambda})$ are said to be bilinearly-strict equivalent, $L^1_\theta E_{H-B} L^2_{\theta'}$, if there exists a transformation β: $(s,\hat{s}) \to (\lambda,\hat{\lambda})$, and thus a $b \in B$, as well as an $h \in H$: $(F_2,-G_2) = h * b \circ (F_1,-G_1)$. By $H\text{-}B(F,G)$ we denote the BSE class of $L = sF - \hat{s}G$.

The characterisation of $H\text{-}B(F,G)$ by a set of invariants is considered next. The effect of the B group on the SE invariants of $H(F,G)$ is the starting point in our investigation.

3. PRELIMINARY RESULTS

Let $L = (F,-G) \in L$, $\theta = (s,\hat{s}) \in \Theta$ and $L_\theta = L(s,\hat{s}) = sF - \hat{s}G$. Let $S(s,\hat{s}) = \text{block-diag}\{f_1(s,\hat{s}),\ldots,f_\rho(s,\hat{s});0_{m-\rho,n-\rho}\}$ be the Smith form over $\mathbb{R}[s,\hat{s}]$ of $L(s,\hat{s})$, $F(F,G) = \{f_i(s,\hat{s}), i \in \underline{\rho}\}$ be the ordered set of homogeneous invariant polynomials (trivial non-zero elements included) of $S(s,\hat{s})$ and let $D(F,G)$ be the corresponding set of homogeneous elementary divisors over $\mathbb{R}$ of $L(s,\hat{s})$. Let us also denote by $I_c(F,G), I_r(F,G)$ the sets of column, row minimal indices respectively of $L(s,\hat{s})$[5,7]. Note that in the following the set of integers $\{1,2,\ldots,\rho\}$ shall be denoted by $\underline{\rho}$.

Theorem (3.1)[5,7]: $D(F,G)$, $I_c(F,G)$ and $I_r(F,G)$ form a complete set of invariants for the $H(F,G)$ equivalence class. □

The study of the effect of $b \in B$ on the SE invariants of $H(F,G)$ is intimately connected with the effect of the projective transformation β on homogeneous polynomials. Thus, let $\mathbb{R}_k\{\Theta\}$ be the set of homogeneous polynomials of degree k with coefficients from $\mathbb{R}$, in all possible indeterminates $\theta = (s,\hat{s}) \in \Theta$ and let β: $(s,\hat{s}) \to (\lambda,\hat{\lambda})$ be the projective transformation defined by (1.2) The action of β on $f(s,\hat{s}) \in \mathbb{R}_k\{\Theta\}$ is defined by

$$\beta \circ f(s,s) \triangleq \tilde{f}(\lambda,\hat{\lambda}) = f(a\lambda + b\hat{\lambda}, c\lambda + d\hat{\lambda}) \tag{3.1}$$

The polynomials $f_1(s,\hat{s}), f_2(\lambda,\hat{\lambda}) \in \mathbb{R}_k\{\Theta\}$ will be said to be *projectively equivalent* (PE), and shall be denoted by $f_1(s,\hat{s}) E_P f_2(\lambda,\hat{\lambda})$, if there exists a β: $(s,\hat{s}) \to (\lambda,\hat{\lambda})$ and a $c \in \mathbb{R}-\{0\}$ such that

$$\beta \circ f_1(s,\hat{s}) = c.f_2(\lambda,\hat{\lambda}) \tag{3.2}$$

Clearly, (3.2) defines an equivalence relation, E_P, on $\mathbb{R}_k\{\Theta\}$, which is called *projective equivalence* (PE). If $F_1 = \{f_i(s,\hat{s}) \in \mathbb{R}_{k_i}\{\Theta\}, i \in \underline{\rho}\}$, $F_2 = \{\tilde{f}_i(\lambda,\hat{\lambda}) \in \mathbb{R}_{k_i}\{\Theta\}, i \in \underline{\rho}\}$ are two ordered sets of homogeneous polynomials, then F_1, F_2 are *projectively equivalent*, $F_1 E_P F_2$, if $f_i(s,\hat{s}) E_P \tilde{f}_i(\lambda,\hat{\lambda})$ for $\forall\, i \in \underline{\rho}$. The projective equivalence class of $f(s,\hat{s})$ (F) shall be denoted by $E_P(f), (E_P(F))$.

Lemma (3.1): Let $f_1(s,\hat{s}) \in \mathbb{R}_{k_1}\{\Theta\}$, $f_2(s,\hat{s}) \in \mathbb{R}_{k_2}\{\Theta\}$ and let $g(s,\hat{s}) \in \mathbb{R}_p\{\Theta\}$ be their highest common factor (HCF). Let

β: $(s,\hat{s}) \to (\lambda,\hat{\lambda})$, $\tilde{f}_1(\lambda,\hat{\lambda}) = \beta \circ f_1(s,\hat{s})$, $\tilde{f}_2(\lambda,\hat{\lambda}) = \beta \circ f_2(s,\hat{s})$ and $\tilde{g}(\lambda,\hat{\lambda}) = \beta \circ g(s,\hat{s})$. Then $\tilde{g}(\lambda,\hat{\lambda})$ is a HCF of $\tilde{f}_1(\lambda,\hat{\lambda})$, $\tilde{f}_2(\lambda,\hat{\lambda})$. □

This property defines the covariance of the HCF under PE transformations.

Proposition (3.1): Let $L_1(s,\hat{s}) = sF_1-\hat{s}G_1$, $L_2(\lambda,\hat{\lambda}) = \lambda F_2-\hat{\lambda}G_2 \in \mathcal{L}(\Theta)$, $\mathcal{F}(F_1,G_1) = \{f_i(s,\hat{s}), i\in\underline{\rho_1}\}$, and $\mathcal{F}(F_2,G_2) = \{\tilde{f}_i(\lambda,\hat{\lambda}) i\in\underline{\rho_2}\}$. If $L_1(s,\hat{s})\mathcal{E}_{\mathcal{H}-\mathcal{B}}L_2(\lambda,\hat{\lambda})$ for some $b \in \mathcal{B}$, generated by β: $(s,\hat{s}) \to (\lambda,\hat{\lambda})$, and some $h \in \mathcal{H}$, then $\rho_1 = \rho_2 = \rho$ and

$$c_i\tilde{f}_i(\lambda,\hat{\lambda}) = \beta \circ f_i(s,\hat{s}),\ c_i \in \mathbb{R}-\{0\},\ i \in \underline{\rho} \tag{3.3}$$

□

Proposition (3.2)[7]: Let $L_1(s,\hat{s}) = sF_1-\hat{s}G_1$, $L_2(\lambda,\hat{\lambda}) = \lambda F_2-\hat{\lambda}G_2 \in \mathcal{L}(\Theta)$. If $L_1(s,\hat{s})\mathcal{E}_{\mathcal{H}-\mathcal{B}}L_2(\lambda,\hat{\lambda})$, then $\mathcal{I}_c(F_1,G_1) = \mathcal{I}_c(F_2,G_2)$ and $\mathcal{I}_r(F_1,G_1) = \mathcal{I}_r(F_2,G_2)$. □

Proposition (3.1) expresses the covariance property of the homogeneous invariant polynomials and Proposition (3.2) expresses the invariance property of column and row minimal indices of $L(s,\hat{s})$ under $\mathcal{E}_{\mathcal{H}-\mathcal{B}}$ equivalence. By using the above two propositions the following result is established.

Theorem (3.1): Let $L_1(s,\hat{s}) = sF_1-\hat{s}G_1$, $L_2(s,\hat{s}) = \lambda F_2-\hat{\lambda}G_2 \in \mathcal{L}(\Theta)$. $L_1(s,\hat{s})\mathcal{E}_{\mathcal{H}-\mathcal{B}}L_2(\lambda,\hat{\lambda})$ if and only if $\mathcal{I}_c(F_1,G_1) = \mathcal{I}_c(F_2,G_2)$, $\mathcal{I}_r(F_1,G_1) = \mathcal{I}_r(F_2,G_2)$ and $\mathcal{F}(F_1,G_1)\mathcal{E}_{\mathcal{P}}\mathcal{F}(F_2,G_2)$. □

From the above result it is clear that the search for a complete set of invariants is reduced to the study of invariants of $f(s,\hat{s}) \in \mathbb{R}_k\{\Theta\}$ under $\mathcal{E}_{\mathcal{P}}$ equivalence transformations.

4. PROJECTIVE EQUIVALENCE OF HOMOGENEOUS BINARY POLYNOMIALS

The aim of this section is the study of invariants of the PE class $\mathcal{E}_{\mathcal{P}}(f(s,\hat{s}))$ of a binary homogeneous polynomial $f(s,\hat{s}) \in \mathbb{R}_k\{\Theta\}$. The effect of PE transformations on the factorization of $f(s,\hat{s})$ into irreducible factors is considered first.

Lemma (4.1): Let $f(s,\hat{s}) = rs^2+ps\hat{s}+q\hat{s}^2$, $\tilde{f}(\lambda,\hat{\lambda}) = \tilde{r}\lambda^2+\tilde{p}\lambda\hat{\lambda}+\tilde{q}\hat{\lambda}^2 \in \mathbb{R}\{\Theta\}$ and let $\Delta = p^2-4rq$, $\tilde{\Delta} = \tilde{p}^2-4\tilde{r}\tilde{q}$ be their corresponding discriminants. If $f(s,\hat{s})\mathcal{E}_{\mathcal{P}}\tilde{f}(\lambda,\hat{\lambda})$, i.e. $\beta \circ f(s,\hat{s}) = c.\tilde{f}(\lambda,\hat{\lambda})$, then $\tilde{\Delta} = \delta^2/c^2.\Delta$, where $\delta = ad-bc \neq 0$ is the determinant of the projective transformation β. □

Remark (4.1): Lemma (4.1) implies that the discriminant Δ is an invariant of $f(s,\hat{s})$ of weight 2 under PE transformations. Furthermore, $\text{sign}\{\Delta\} = \text{sign}\{\tilde{\Delta}\}$ and the reducibility properties over $\mathbb{R}$ of $f(s,\hat{s}) \in \mathbb{R}_2\{\Theta\}$ are invariant under $\mathcal{E}_{\mathcal{P}}$ transformations.

By Lemma (4.1) and its remark we have:

Proposition (4.1): Let $p_i(s,\hat{s}) = (\gamma_i s-\delta_i\hat{s})$, $\gamma_i,\delta_i \in \mathbb{C}$, $(\gamma_i,\delta_i) \neq (0,0)$ be a prime over $\mathbb{C}$ of $f(s,\hat{s}) \in \mathbb{R}_k\{\Theta\}$. Under PE transformations, the following properties hold true:

(i) Any pair $p_i(s,s)$, $p_j(s,\hat{s})$ of distinct primes $((\gamma_i,\delta_i) \neq \zeta(\gamma_j,\delta_j),\ \zeta \in \mathbb{C})$ is mapped to a distinct pair of primes.

(ii) Any pair of complex conjugate primes $p(s,\hat{s}) = (\gamma s-\delta\hat{s})$, $\bar{p}(s,\hat{s}) = (\bar{\gamma}s-\bar{\delta}\hat{s})$ is mapped to a complex conjugate pair of primes.

(iii) Any pair of $p(s,\hat{s}), p'(s,\hat{s})$ of repeated primes ($(\gamma,\delta) \neq \zeta(\gamma',\delta')$, $\zeta \in \mathbb{C}$) is mapped to a pair of repeated primes. □

Let $f(s,\hat{s}) \in \mathbb{R}_k\{\theta\}$ and let us denote by $\mathcal{D}_{\mathbb{R}}(f) = \{(\alpha_i s-\beta_i\hat{s})^{\tau_i}, \alpha_i,\beta_i \in \mathbb{R}, (\alpha_i,\beta_i) \neq (0,0), i \in \underline{\tau}\}$, $\mathcal{D}_{\mathbb{C}}(f) = \{(\gamma_i s-\delta_i\hat{s})^{\rho_i}, \gamma_i,\delta_i \in \mathbb{C}, (\gamma_i,\delta_i) \neq (0,0), i \in \underline{\rho}\}$ the sets of real, complex elementary divisors (e.d.) of $f(s,\hat{s})$ respectively; note that $\mathcal{D}_{\mathbb{C}}(f)$ is symmetric, i.e. if $(\gamma_i s-\delta_i\hat{s})^{\rho_i} \in \mathcal{D}_{\mathbb{C}}(f)$, then $(\bar{\gamma}_i s-\bar{\delta}_i\hat{s})^{\rho_i} \in \mathcal{D}_{\mathbb{C}}(f)$. By Proposition (4.1) we have:

Proposition (4.2): Let $f(s,\hat{s}), \tilde{f}(\lambda,\hat{\lambda}) \in \mathbb{R}_k\{\theta\}$ and let $\mathcal{D}_{\mathbb{R}}(f), \mathcal{D}_{\mathbb{R}}(\tilde{f})$ and $\mathcal{D}_{\mathbb{C}}(f), \mathcal{D}_{\mathbb{C}}(\tilde{f})$ be the corresponding sets of real and complex e.d. $f(s,s) E_{\mathcal{P}} \tilde{f}(\lambda,\hat{\lambda})$, if and only if there exists a projective transformation β, such that the following conditions hold true:

(i) For $\forall\, e_i(s,\hat{s}) = (\alpha_i s-\beta_i\hat{s})^{\tau_i} \in \mathcal{D}_{\mathbb{R}}(f)$ there exists an $\tilde{e}_i(\lambda,\hat{\lambda}) = (\tilde{\alpha}_i\lambda-\tilde{\beta}_i\hat{\lambda})^{\tau_i}\mathcal{D}_{\mathbb{R}}(\tilde{f})$ such that $c_i.\tilde{e}_i(\lambda,\hat{\lambda}) = \beta \circ e_i(s,\hat{s})$, $c_i \in \mathbb{R}-\{0\}$, and vice-versa.

(ii) For $\forall\, e_i'(s,\hat{s}) = (\gamma_i s-\beta_i\hat{s})^{\rho_i} \in \mathcal{D}_{\mathbb{C}}(f)$ there exists an $\tilde{e}_i'(\lambda,\hat{\lambda}) = (\tilde{\gamma}_i\lambda-\tilde{\delta}_i\hat{\lambda})^{\rho_i} \in \mathcal{D}_{\mathbb{C}}(\tilde{f})$ such that $c_i'.\tilde{e}_i'(\lambda,\hat{\lambda}) = \beta \circ e_i'(s,\hat{s})$, $c_i' \in \mathbb{C}-\{0\}$ and vice-versa. □

The above result expresses the covariance property of the e.d. under PE and reduces the study of invariants of $E_{\mathcal{P}}(f)$ to the study of properties of the e.d. sets. Note that an elementary divisor $(\gamma s-\delta\hat{s})^{\rho}$ may be represented by an ordered triple (γ,δ,ρ), $\gamma,\delta \in \mathbb{C}$, $\rho \in \mathbb{Z}$; using this representation we may define the following sets:

Definition (4.1): (i) We define by $\mathcal{B}_i \triangleq \{(\alpha_j^i,\beta_j^i,\tau_i): \alpha_j^i,\beta_j^i \in \mathbb{R}, \tau_i \in \mathbb{Z}, j \in \underline{\nu}_i, (\alpha_j^i,\beta_j^i) \neq \xi(\alpha_k^i,\beta_k^i), \forall\, j \neq k, \xi \in \mathbb{R}-\{0\}\}$ the set of ordered triples corresponding to all elements of $\mathcal{D}_{\mathbb{R}}(f)$ with the same degree τ_i; τ_i will be referred to as the degree of $\mathcal{B}_i$. An ordering of the elements of $\mathcal{B}_i$ is defined by any permutation of its elements; such a permutation is denoted by $\pi(\mathcal{B}_i)$ and the set of all permutations by $\langle\mathcal{B}_i\rangle$.

(ii) $\mathcal{B}_{\mathbb{R}}(f) \triangleq \{\mathcal{B}_1,\ldots,\mathcal{B}_K: \tau_1 < \ldots < \tau_K\}$ corresponds to the set of all real e.d. and shall be referred to as the *real unique factorization set* ($\mathbb{R}$-UFS) of $f(s,\hat{s})$. The set $J_{\mathbb{R}}(f) \triangleq \{(\nu_1,\tau_1),\ldots,(\nu_K,\tau_K)\}$, where ν_i is the number of elements in $\mathcal{B}_i$, is defined as the *real list* of $f(s,\hat{s})$. Every permutation of $\mathcal{B}_{\mathbb{R}}(f)$ of the type $\pi(\mathcal{B}_{\mathbb{R}}(f) = \{\pi(\mathcal{B}_1),\ldots,\pi(\mathcal{B}_K): \pi(\mathcal{B}_i) \in \langle\mathcal{B}_i\rangle\}$ defines a *normal ordering* of $\mathcal{B}_{\mathbb{R}}(f)$ and the set of all such representation will be denoted by $\langle\mathcal{B}_{\mathbb{R}}(f)\rangle$.

(iii) Let $\pi(\mathcal{B}_{\mathbb{R}}(f)) = \{\pi(\mathcal{B}_1),\ldots,\pi(\mathcal{B}_K)\} \in \langle\mathcal{B}_{\mathbb{R}}(f)\rangle$, where $\pi(\mathcal{B}_i) = \{(\alpha_1^i,\beta_1^i,\tau_i),\ldots,(\alpha_{\nu_i}^i,\beta_{\nu_i}^i,\tau_i)\} \in \langle\mathcal{B}_i\rangle$. A *matrix representation* of $\pi(\mathcal{B}_{\mathbb{R}}(f))$ is defined by

$$[\mathcal{B}^{\pi}_{\mathbb{R}}(f)] = \begin{bmatrix} [\mathcal{B}^{\pi}_{1}] \\ \vdots \\ [\mathcal{B}^{\pi}_{k}] \end{bmatrix}, \text{ where } [\mathcal{B}^{\pi}_{i}] = \begin{bmatrix} \alpha^i_1 & \beta^i_1 \\ \alpha^i_2 & \beta^i_2 \\ \vdots & \vdots \\ \alpha^i_{\nu_i} & \beta^i_{\nu_i} \end{bmatrix} \in \mathbb{R}^{\nu_1 \times 2} \quad (4.1)$$

□

For the set $\mathcal{D}_{\mathbb{R}}(f)$ we may define in a similar manner the sets $\mathcal{B}'_i, \mathcal{B}_{\mathbb{C}}(f), J_{\mathbb{C}}(f)$ as well as the notions of normal ordering and matrix representation. We note however that from a pair of complex conjugate e.d., only one is now represented in $\mathcal{B}'_i$. $\mathcal{B}_{\mathbb{C}}(f)$ is referred to as the *complex unique factorization set* ($\mathbb{C}$-UFS) and $J_{\mathbb{C}}(f)$ as the *complex list* of $f(s,\hat{s})$. With this notation in mind we may state:

Theorem (4.1): Let $f(s,\hat{s}), \tilde{f}(\lambda,\hat{\lambda}) \in \mathbb{R}_k\{\Theta\}$ and let $(\mathcal{B}_{\mathbb{R}}(f), J_{\mathbb{R}}(f), \mathcal{B}_{\mathbb{C}}(f), J_{\mathbb{C}}(f))$, $(\mathcal{B}_{\mathbb{R}}(\tilde{f}), J_{\mathbb{R}}(\tilde{f}), \mathcal{B}_{\mathbb{C}}(\tilde{f}), J_{\mathbb{C}}(\tilde{f}))$ be the sets associated with e.d. of $f(s,\hat{s})$, $\tilde{f}(\lambda,\hat{\lambda})$ respectively. $f(s,\hat{s}) E_{\mathcal{P}} \tilde{f}(\lambda,\hat{\lambda})$ if and only if the following conditions hold true:

(i) $J_{\mathbb{R}}(f) = J_{\mathbb{R}}(\tilde{f})$ and $J_{\mathbb{C}}(f) = J_{\mathbb{C}}(\tilde{f})$.

(ii) There exist $\pi(\mathcal{B}_{\mathbb{R}}(f)) \in \langle\mathcal{B}_{\mathbb{R}}(f)\rangle$, $\pi'(\mathcal{B}_{\mathbb{C}}(f)) \in \langle\mathcal{B}_{\mathbb{C}}(f)\rangle$, $\tilde{\pi}(\mathcal{B}_{\mathbb{R}}(\tilde{f})) \in \langle\mathcal{B}_{\mathbb{R}}(\tilde{f})\rangle$, $\tilde{\pi}'(\mathcal{B}_{\mathbb{C}}(\tilde{f})) \in \langle\mathcal{B}_{\mathbb{C}}(\tilde{f})\rangle$, a projective transformation β, $\zeta_i \in \mathbb{R}-\{0\}$, and $\xi_i \in \mathbb{C}-\{0\}$ such that

$$[\mathcal{B}^{\tilde{\pi}}_{\mathbb{R}}(\tilde{f})] = \mathrm{diag}\{\zeta_i\}[\mathcal{B}^{\pi}_{\mathbb{R}}(f)][\beta]$$
$$[\mathcal{B}^{\tilde{\pi}'}_{\mathbb{C}}(\tilde{f})] = \mathrm{diag}\{\xi_i\}[\mathcal{B}^{\pi'}_{\mathbb{C}}(f)][\beta] \quad \text{where } [\beta] = \begin{bmatrix} a & b \\ c & d \end{bmatrix}, \ ad-cb \neq 0 \quad (4.2)$$

□

Remark (4.2): The real and complex list $J_{\mathbb{R}}(f), J_{\mathbb{C}}(\tilde{f})$ of $f(s,\hat{s})$ are invariants of the $E_{\mathcal{P}}(f)$ equivalence class.

Two pairs of sets $\mathcal{B}(f) = (\mathcal{B}_{\mathbb{R}}(f), \mathcal{B}_{\mathbb{C}}(f))$, $\mathcal{B}(\tilde{f}) = (\mathcal{B}_{\mathbb{R}}(\tilde{f}), \mathcal{B}_{\mathbb{C}}(\tilde{f}))$ for which $J_{\mathbb{R}}(f) = J_{\mathbb{R}}(\tilde{f})$, $J_{\mathbb{C}}(f) = J_{\mathbb{C}}(\tilde{f})$ and conditions (ii) of Theorem (4.1) hold true for some β and nonzero scalars ζ_i, ξ_i, will be called *normally projective equivalent* (NPE) and shall be denoted by $\mathcal{B}(f) E_{\tilde{\mathcal{P}}} \mathcal{B}(\tilde{f})$. It is clear that the problem of finding a complete set of invariants for $E_{\mathcal{P}}(f)$ is reduced to a study of invariants of sets $\mathcal{B}(f)$ under normal projective equivalence transformations $E_{\tilde{\mathcal{P}}}$.

5. BILINEAR-STRICT EQUIVALENCE INVARIANTS OF sF-ŝG

For every invariant polynomial $f_j(s,\hat{s}) \in \mathcal{F}(F,G) = \{f_i(s,\hat{s}), i \in \underline{\mu}\}$ the sets $J_{\mathbb{R}}(f_j), J_{\mathbb{C}}(f_j), \mathcal{B}_{\mathbb{R}}(f_j), \mathcal{B}_{\mathbb{C}}(f_j)$ may be defined. The sets defined by $J_{\mathbb{R}}(F,G) = \{\ldots; J_{\mathbb{R}}(f_j); \ldots\}$, $J_{\mathbb{C}}(F,G) = \{\ldots; J_{\mathbb{C}}(f_j); \ldots\}$, $\mathcal{B}_{\mathbb{R}}(F,G) = \{\ldots; \mathcal{B}_{\mathbb{R}}(f_j); \ldots\}$, and $\mathcal{B}_{\mathbb{C}}(F,G) = \{\ldots; \mathcal{B}_{\mathbb{C}}(f_j); \ldots\}$ are well defined and shall be called the *real list*, *complex list*, *real e.d. basis* and *complex e.d. basis* respectively of $sF-\hat{s}G$. By Theorems (3.2) and (4.1) we have:

Theorem (5.1): Let $L(s,\hat{s}) = sF-\hat{s}G$, $\tilde{L}(\lambda,\hat{\lambda}) = \lambda\tilde{F}-\hat{\lambda}\tilde{G} \in L\{\Theta\}$. $L(s,\hat{s}) E_{H-B} \tilde{L}(\lambda,\hat{\lambda})$ if and only if $I_c(\tilde{F},\tilde{G}) = I_c(\tilde{F},\tilde{G})$, $I_r(\tilde{F},\tilde{G}) = I_r(F,G)$, $J_{\mathbb{R}}(F,G) = J_{\mathbb{R}}(\tilde{F},\tilde{G})$, $J_{\mathbb{C}}(F,G) = J_{\mathbb{C}}(\tilde{F},\tilde{G})$, $B_{\mathbb{R}}(F,G) E_{\bar{P}} B_{\mathbb{R}}(\tilde{F},\tilde{G})$ and $B_{\mathbb{C}}(F,G) E_{\bar{P}} B_{\mathbb{C}}(\tilde{F},\tilde{G})$. □

Note that by $B_{\mathbb{R}}(F,G) E_{\bar{P}} B_{\mathbb{R}}(\tilde{F},\tilde{G})$ we mean that $\mu = \tilde{\mu}$ and $B_{\mathbb{R}}(f_j) E_{\bar{P}} B_{\mathbb{R}}(\tilde{f}_j)$ for $\forall\ j \in \underline{\mu}$ (a similar interpretation is given for $B_{\mathbb{C}}(F,G) E_{\bar{P}} B_{\mathbb{C}}(F,G)$).

Corollary (5.1.1): The sets $I_c(F,G)$, $I_r(F,G)$, $J_{\mathbb{R}}(F,G) J_{\mathbb{C}}(F,G)$ are invariants of the H-$B(F,G)$ equivalence class. □

By Theorem (5.1) it is clear that the search for a complete set of invariants is reduced to the study of the following type of equivalence defined on matrices.

Definition (5.1): Let $T_1, T_2 \in \mathbb{C}^{k\times 2}$. T_1, T_2 will be called *extended Hermite equivalent*, and shall be denoted by $T_1 E_{eh} T_2$, if there exist $\xi_i \in \mathbb{C}-\{0\}$, $i \in \underline{k}$, and $Q \in \mathbb{R}^{2\times 2}$, $|Q| \neq 0$, such that

$$T_2 = \text{diag}\{\xi_i\} T_1 Q \tag{5.1}$$

Before we proceed with the investigation of E_{eh} equivalence we give some useful notation[9]. $Q_{k,n}$ denotes the set of lexicographically ordered, strictly increasing sequences of k integers from $1,2,\ldots,n$. If $\underline{x}_{i_1},\ldots,\underline{x}_{i_k}$ is a set of vectors of $\mathbb{R}^n$, or $\mathbb{C}^n$, $k \leq n$, then by $\underline{x}_{i_1} \wedge \ldots \wedge \underline{x}_{i_k}$ we denote the exterior product of them and by a_ω, $\omega \in Q_{k,n}$ the ω-th coordinate characterised by the sequence $\omega = (i_1,\ldots,i_k)$. With this notation in mind, we may state the following.

Lemma (5.1): Necessary and sufficient conditions for the existence of a Q and of a set of ξ_i such that $T_1 E_{eh} T_2$ are:

(i) There exist $\xi_i \in \mathbb{C}$, $\xi_i \neq 0$, such that

$$\text{col-span}_{\mathbb{C}}\{T_2\} = \text{col-span}_{\mathbb{C}}\{\text{diag}\{\xi_i\}T_1\} \tag{5.2}$$

(ii) For the family of ξ_i solutions of (5.2) the following two conditions have to be satisfied:

$$\text{col-span}_{\mathbb{R}}\{\text{Re}(T_2)\} = \text{col-span}_{\mathbb{R}}\{\text{diag}\{\xi_i\}T_1\} \tag{5.3}$$

$$\text{col-span}_{\mathbb{R}}\{\text{Im}(T_2)\} = \text{col-span}_{\mathbb{R}}\{\text{diag}\{\xi_i\}T_1\} \tag{5.4}$$

□

Note, that if the matrices T_1, T_2 are both real, then condition (5.3) is necessary and sufficient for the existence of a real transformation Q. In the following attention is focussed on condition (i) of Lemma (5.1).

Proposition (5.1): Let $T_1 = [\underline{t}_{11}, \underline{t}_{12}]$, $T_2 = [\underline{t}_{21}, \underline{t}_{22}] \in \mathbb{C}^{k\times 2}$ and let $\underline{t}_{11}\wedge\underline{t}_{12}=(a^1_\omega)$, $\underline{t}_{21}\wedge\underline{t}_{22}=(a^2_\omega)$, $\omega\in Q_{2,k}$, be the exterior products of the columns of the two matrices. Necessary and sufficient conditions for the existence of a $Q\in\mathbb{C}^{2\times 2}$, $|Q|=\delta\neq 0$ and $\xi_i\in\mathbb{C}$, $\xi_i\neq 0$ such that equation (5.1) is satisfied are that:

$$a^2_\omega/a^1_\omega=\xi_\omega\delta,\quad \xi_\omega=\xi_{i_1}\xi_{i_2},\quad \omega=(i_1,i_2)\in Q_{2,k} \tag{5.5}$$

□

Conditions (5.5) guarantee the existence of a complex Q but not necessarily of a real one. In the case where both T_1,T_2 are complex conditions (5.3) and (5.4) have also to be used. For real matrices T_1,T_2 (5.5) guarantee a real Q.

Proposition (5.2): Let ξ_i, $i\in\underline{k}$ be a solution of equation (5.1).

(i) If $Q\in\mathbb{C}^{2\times 2}$, $|Q|=\delta=1$, then the ξ_i solution is uniquely defined.

(ii) If $Q\in\mathbb{R}^{2\times 2}$, $|Q|=\delta=\pm 1$, then if there exists another solution, this is defined by $\xi'_i=j\xi_i$, $i\in\underline{k}$.

□

The above preliminary results form the basis for the definition of a complete set of invariants and $\mathcal{E}_{eh}$ equivalence. In fact, it is shown in reference [8], that by eliminating the ξ_i parameters in (5.5) conditions, a new set of invariants is defined, namely the "Plücker vectors". For complex transformations Q, it is shown in reference[8], that the Plücker vectors together with the sets $J_{\mathbb{R}}(f)$,$J_{\mathbb{C}}(f)$ form a complete set of invariants for the $\mathcal{E}_{\mathcal{P}}(f)$ equivalence class. Some extra invariants are needed, however, to guarantee that Q is real. The study of $\mathcal{E}_{\mathcal{P}}$-equivalence on the set $\mathbb{R}_k\{\Theta\}$ forms the basis for the definition of a complete set of invariants of homogeneous matrix pencils under Bilinear-Strict Equivalence.

6. CONCLUSIONS

The notion of a Bilinear-Strict Equivalence of homogeneous matrix pencils has been introduced and a set of invariants for the $\mathcal{H}$-$\mathcal{B}$(F,G) has been defined. This set is made up from the column minimal indices, row minimal indices, and lists of the real and complex elementary divisors. The central notion in the analysis has been that of the Projective Equivalence, which has been defined on homogeneous binary polynomials. It has been shown, that the problem of defining a complete set of invariants for the $\mathcal{H}$-$\mathcal{B}$(F,G) class is equivalent to a study of invariants of matrices under Extended Hermite Equivalence transformations. The detailed study of this new type of equivalence defined on matrices is given in reference [8].

The study of $\mathcal{E}_{\mathcal{H}\text{-}\mathcal{B}}$ equivalence of matrix pencils provides the basis for the investigation of properties of the differential system S(F,G) under frequency and space type transformations. The algebraic, geometric and dynamic aspects of the subspaces of S(F,G) under $\mathcal{E}_{\mathcal{H}\text{-}\mathcal{B}}$ equivalence transformations is currently under investigation. Such a study may provide a classification of system properties such as controllability, observability, stability etc., into those which are frequency coordinate dependent or independent. The effect of $\mathcal{E}_{\mathcal{H}-\mathcal{B}}$ transformation on the numerical properties (such as condition number of the generalised eigenvalue-eigenvector problem) is also a topic under investigation.

7. REFERENCES

1. Wonham, W.M. (1974). Linear multivariable control: A geometric approach. Springer-Verlag, New York.
2. Willems, J.C. (1981). "Almost invariant subspaces: An approach to high gain feedback design, Part 1...", IEEE Trans. on Aut. Control, AC-26, 235-253.
3. Karcanias, N. and Hayton, G.E. (1981). "Generalized autonomous systems, algebraic duality, and geometric theory", Proc. 8th IFAC World Cong., Kyoto, Japan.
4. Jaffe, S. and Karcanias, N. (1981). "Matrix pencil characterisation of almost (A,B)-invariant subspaces...", Int.J. Control, 33, 51-93.
5. Gantmacher, G. (1959). Theory of Matrices, Vol.2, Chelsea, N.Y.
6. Van Dooren, P. (1981). "The generalized eigenstructure problem in linear system theory", IEEE Trans.Aut. Control, AC-26, 111-129.
7. Turnbull, H.W. and Aitken, A.C. (1932). An introduction to the theory of canonical matrices. Blackie & Son, London.
8. Karcanias, N. and Kalogeropoulos, G. (1984). "Bilinear-strict equivalence invariants of homogeneous matrix pencils". The City Univ., Control Eng.Centr.Res.Rep. CEC/NK-GK/10.
9. Marcus, M. and Minc, H. (1964). A survey of matrix theory and matrix inequalities. Allyn and Bacon, Boston.

ON TRANSFORMATIONS OF GENERALISED STATE SPACE SYSTEMS

G.E. Hayton and P. Fretwell
(Department of Electronic Engineering, University of Hull)

and

A.C. Pugh
(Department of Mathematics, Loughborough University of Technology)

ABSTRACT

The paper considers two equivalence relations between generalised state-space systems which leave their dynamic behaviour at both finite and infinite frequencies unchanged. The first of these, complete system equivalence, is defined in terms of transformations of the system matrices, the second, fundamental equivalence, by mappings of the solution sets of the describing differential equations together with mappings of the sets of restricted initial conditions. Finally these two concepts are unified.

1. INTRODUCTION

Recently much interest has been aroused by so-called generalised state-space systems[7,8,9] described by a set of first-order, linear, differential equations which, after Laplace transformation assuming non zero initial conditions, may be written

$$\begin{bmatrix} sE-A & B \\ -C & D \end{bmatrix} \begin{bmatrix} \chi(s) \\ -u(s) \end{bmatrix} = \begin{bmatrix} E\chi\ (0-) \\ -\ y(s) \end{bmatrix} \qquad (1.1)$$

E is possibly singular but $sE-A \not\equiv 0$. The corresponding system matrix is

$$P(s) = \begin{bmatrix} sE-A & B \\ -C & D \end{bmatrix} \qquad (1.2)$$

The classical state space form for proper systems is a special case of (1.1) with $E = I_n$ and clearly if E is nonsingular (1.1) may be reduced to the classical state space form. This is no longer so if E is singular and it should be noted that in this case the system equation (1.1) maps only a subset of $\mathbb{R}^n$. This subset $\{E\chi(0-) \mid \chi(0-) \in \mathbb{R}^n\}$ will be referred to as the restricted initial condition set. Such representations are of interest because they are the simplest form of system description which simultaneously display finite and infinite frequency behaviour. Accordingly we are interested in definitions of system equivalence which when applied to (1.1),(1.2) will leave invariant such behaviour. A number of definitions of equivalence of generalised state space systems have been proposed. The first of these, Rosenbrock's "restricted system equivalence" includes more invariants than those necessary to characterise the dynamic behaviour of the system. To overcome this problem Verghese[8] proposed the notion of "strong equivalence" describing it in terms of a set of permitted elementary operations which make it attractive from an algorithmic view. However no closed form description was given nor was one readily apparent. Subsequently Pugh et al.[5] provided such a description termed "complete system equivalence" based upon a study of the underlying transformations of matrix pencils. This paper first reviews complete system equivalence and then presents a number of results concerning what is felt to be a more natural approach to the problem by defining equivalence of generalised state-space systems in terms of mappings of the solution sets of the describing differential equations in the manner suggested by Pernebo[4] together with mappings of the sets of restricted initial conditions. This provides a conceptually pleasing definition of equivalence within which the invariance of the controllability and observability characteristics of the system is implicit. Further it is shown that systems equivalent under the new definition are, in fact, completely system equivalent.

2. COMPLETE SYSTEM EQUIVALENCE

The transformation of complete system equivalence arose from consideration of possible equivalence relations for the regular pencil sE-A. The classical relationship of "strict equivalence"[2] preserves both finite and infinite elementary divisors and hence preserves the structure of the pencil at both finite and infinite frequencies as required. However the set of infinite elementary divisors includes those of order one (giving rise to trivial Jordan blocks in the Kronecker canonical form) and these have no dynamic significance, hence strict equivalence is over restrictive and an attempt was made to generalise it without losing the desired properties. Two new transformations were proposed, "extended equivalence" and "complete equivalence" both of which differ from the equivalence relations covered in standard matrix theory literature in that they permit transformations between pencils of different dimensions and thus formally embody the operations of trivial expansion and deflation, useful in linear system theory, but previously handled on a rather ad hoc basis. Extended equivalence, despite apparent construction from constant transforming matrices, preserves only the finite elementary divisors of the pencil, infinite elementary divisors of any order and hence infinite frequency behaviour are not preserved. Of more interest in the context of generalised state space systems is complete equivalence which preserves both finite elementary divisors and infinite elementary divisors of order two or more, but does not preserve infinite elementary divisors of degree one. Thus complete equivalence retains both finite and infinite frequency behaviour whilst allowing change to the trivial Jordan blocks. Specialisation of complete equivalence to system matrices in generalised state-space form gives:-

Definition 1: Two generalised state space systems $\mathbb{P}_1$, $\mathbb{P}_2$ are said to be completely system equivalent if their system matrices are related by

$$\begin{bmatrix} M & O \\ X & I \end{bmatrix} \begin{bmatrix} sE_1-A_1 & B_1 \\ -C_1 & D_1 \end{bmatrix} = \begin{bmatrix} sE_2-A_2 & B_2 \\ -C_2 & D_2 \end{bmatrix} \begin{bmatrix} N & Y \\ O & I \end{bmatrix} \tag{2.1}$$

where M,X,N,Y are constant matrices of the appropriate dimensions and the composite matrices $[M \;\; sE_2-A_2]$, $[(sE_1-A_1)^t \;\; N^t]^t$ have neither finite nor infinite zeros[3].

The major results concerning complete equivalence are:-

Theorem 1: (i) Two generalised state-space systems are completely system equivalent if and only if they are strongly equivalent.

(ii) Complete system equivalence leaves invariant
 (a) the finite and infinite transmission zeros
 (b) the finite and infinite transfer function poles
 (c) the finite and infinite system poles and zeros
 (d) the finite and infinite decoupling zeros

Proof: The proof may be found in [5].

3. FUNDAMENTAL EQUIVALENCE

Complete system equivalence is defined as a transformation of system matrices; an alternative and intuitively more attractive way of defining equivalence of two systems is to so do in terms of maps between (i) the solution/input vector pairs $[\chi_i^t(s)\ u^t(s)]^t$ i=1,2 and (ii) the restricted initial condition/output vector pairs $[(E_i\chi_i(0-))^t\ y^t(s)]^t$ i=1,2. It is noted from the form of (1.1) that these pairs are fundamental to the system. The system equations themselves are mappings from an appropriate set of initial condition/output pairs to an appropriate set of solution/input pairs. The nature of these two sets characterises a system, for the solution/input pairs completely describe the controllability properties of the system, while the restricted initial condition/output pairs encapsulate the observability properties. These comments are probably best understood in the context of system equations in state space form for then the number of initial condition/output pairs which contain a given output describes the extent to which the system is unobservable. A system in state space form will, of course, be completely observable in case a given output occurs in one and only one initial condition/output pair. Dual statements may be made concerning the solution/input pairs and complete controllability. This leads to the following definition:-

Definition 2: Let $\mathbb{P}_1$, $\mathbb{P}_2$, be two generalised state space systems. They are said to be fundamentally equivalent if there exist

(i) a constant, injective map

$$\begin{bmatrix} \chi_2(s) \\ -u(s) \end{bmatrix} = \begin{bmatrix} N & Y \\ O & I \end{bmatrix} \begin{bmatrix} \chi_1(s) \\ -u(s) \end{bmatrix} \tag{3.1}$$

(ii) a constant, surjective map

$$\begin{bmatrix} E_2\chi_2(0-) \\ -y(s) \end{bmatrix} = \begin{bmatrix} M & O \\ X & I \end{bmatrix} \begin{bmatrix} E_1\chi_1(0-) \\ -y(s) \end{bmatrix}, \quad XE_1=O \tag{3.2}$$

Note that the special structure of these maps is dictated purely by the requirement that the input/output behaviour of $\mathbb{P}_1$, $\mathbb{P}_2$ be identical [6]. Also one might expect that the maps (3.1), (3.2) would need to be isomorphisms, but this is not so as the following result demonstrates:

Lemma 1: Let $\mathbb{P}_1$, $\mathbb{P}_2$ be two systems in generalised state space form which are fundamentally equivalent then the maps of (3.1),(3.2) are both bijective.

Proof: The restricted sets of initial conditions $\{E_1\chi_1(0-)\}$, $\{E_2\chi_2(0-)\}$ form vector spaces over $\mathbb{R}$. Let these spaces have dimensions t_1 (= rank E_1) and t_2 (= rank E_2) respectively. From (3.2) there exists a surjective map between the two spaces such that

$$E_2 \chi_2(0-) = ME_1 \chi_1(0-) \text{ so that } t_1 > t_2 \tag{3.3}$$

Additionally, from (1.1) and (3.1) the following injective map may be derived:-

$$\begin{aligned}&(sE_2-A_2)^{-1}E_2\chi_2(0-)-(sE_2-A_2)^{-1}B_2u(s)\\&= N(sE_1-A_1)^{-1}E_1\chi_1(0-)+N(sE_1-A_1)^{-1}B_1u(s)-Yu(s)\end{aligned} \tag{3.4}$$

Now (3.4) holds for all u(s) and, in particular, for u(s)=0, hence

$$E_2\chi_2(0-) = (sE_2-A_2)N(sE_1-A_1)^{-1}E_1\chi_1(0-) \tag{3.5}$$

is an injection between the restricted initial condition spaces and $t_1 < t_2$ and so $t_1=t_2=t$ and (3.3) and hence (3.2) is a bijection.

Now suppose (3.1) is not bijective, then clearly there exists an $E_2\chi_2(0-)$ which is independent of the image under (3.1) of any basis of $E_1\chi_1(0-)$ and this implies that $\{E_2\chi_2(0-)\}$ has dimension greater than t_1 which is a contradiction. Hence (3.1) is a bijection.

<u>Corollary 1</u>: With the matrices defined as in the Lemma:

$$(sE_2-A_2)^{-1}B_2-N(sE_1-A_1)^{-1}B_1+Y = 0 \tag{3.6}$$

<u>Proof</u>: This follows from the lemma since (3.4) holds for all u(s).

Corollary 2: With the matrices defined as in the Lemma there exists a bijective map M*, (not necessarily identical to M) between the restricted initial condition spaces such that:-

$$((sE_2-A_2)^{-1}M^*-N(sE_1-A_1)^{-1})E_1 \equiv 0 \tag{3.7}$$

<u>Proof</u>: From the Lemma, (3.5) is a rational bijection between the restricted initial condition spaces. I_n is a basis for $\chi_1(0-)$ and each basis element is mapped (not necessarily bijectively) to a basis vector $E_2\chi_2(0-)$

i.e. $$(sE_2-A_2)N(sE_1-A_1)^{-1}E_1 = Q \quad , Q \text{ constant} \tag{3.8}$$

Consideration of the constant, polynomial and strictly proper parts of (3.8) shows that it may be written as

$$(sE_2-A_2)N(sE_1-A_1)^{-1}E_1 = A_0E_1 \tag{3.9}$$

$M^* = A_0$ is a bijection between the restricted initial condition spaces and premultiplying both sides of (3.9) by $(sE_2-A_2)^{-1}$ gives (3.7).

To show that it is not necessary for $M^*=M$ consider two, identical, undriven, state space systems:-

$$P_i = \left[\begin{array}{ccc|c} s-1 & 0 & 0 & 0 \\ 0 & s & 0 & 0 \\ 0 & 0 & s & 0 \\ \hline 0 & -1 & -1 & 0 \end{array}\right] \begin{bmatrix} \chi_1(s) \\ \chi_2(s) \\ \chi_3(s) \\ \hline -u(s) \end{bmatrix} = \begin{bmatrix} \chi_1(0-) \\ \chi_2(0-) \\ \chi_3(0-) \\ \hline -y(s) \end{bmatrix} \quad i=1,2 \tag{3.10}$$

Choosing $N=I_3, M = \begin{bmatrix} 1 & 0 & 0 \\ 0 & 0 & 1 \\ 0 & 1 & 0 \end{bmatrix}$ fulfils the conditions of Definition (2.1)

but in this case $((sE_2-A_2)^{-1}M-N(sE_1-A_1)^{-1})E_1 \neq 0$

4. FUNDAMENTAL EQUIVALENCE AND THE KRONECKER STANDARD FORM

Any generalised state space system may be reduced to a particularly simple form $P_k(s)$, the Kronecker standard form, by a series of elementary row and column operations[7]. In matrix form:-

$$\begin{bmatrix} M & 0 \\ 0 & I \end{bmatrix} \begin{bmatrix} sE-A & B \\ -C & D \end{bmatrix} \begin{bmatrix} N & 0 \\ 0 & I \end{bmatrix} = \left[\begin{array}{cc|c} sI-\bar{A} & 0 & \bar{B} \\ 0 & I-sJ & \hat{B} \\ \hline -\bar{C} & -\hat{C} & \hat{D} \end{array}\right] \tag{4.1}$$

In this case the subsystem

$$\begin{bmatrix} sI-\bar{A} & \bar{B} \\ -\bar{C} & 0 \end{bmatrix} \begin{bmatrix} x(s) \\ -u(s) \end{bmatrix} = \begin{bmatrix} x(0-) \\ -y(s) \end{bmatrix}$$

will be referred to as the state space subsystem and the subsystem

$$\begin{bmatrix} I-sJ & \hat{B} \\ -\hat{C} & \hat{D} \end{bmatrix} \begin{bmatrix} \eta(s) \\ -u(s) \end{bmatrix} \begin{matrix} = \\ = \end{matrix} \begin{bmatrix} -J\eta(0-) \\ -y(s) \end{bmatrix}$$

will be referred to as the impulsive subsystem.

It should be noted that the above is a slight misuse of the term Kronecker standard form as it is only the block sE-A which is strictly of that form[2].

It is clear that reduction to Kronecker form is a special case of complete system equivalence. The following lemma shows that it may also be regarded as a transformation of fundamental equivalence.

Lemma 2: A generalised state space system is fundamentally equivalent to its Kronecker form and vice versa.

Proof: This follows readily from (1.1), (4.1) and definition 2.

Lemma 3: Let $\mathbb{P}_1$, $\mathbb{P}_2$ be fundamentally equivalent, generalised state space systems in standard form then

(i) the two state space sub-systems are fundamentally equivalent

(ii) the two impulsive sub-systems are fundamentally equivalent

Proof: The solution/input map of (3.1) may be partitioned and written

$$\begin{bmatrix} x_2(s) \\ \eta_2(s) \end{bmatrix} = \begin{bmatrix} N_1 & N_2 \\ N_3 & N_4 \end{bmatrix} \begin{bmatrix} x_1(s) \\ \eta_1(s) \end{bmatrix} - \begin{bmatrix} Y_1 \\ Y_2 \end{bmatrix} u(s) \tag{4.2}$$

Setting u(s)=P in (4.2) and comparing coefficients in the individual (block) elements shows that (4.2) may be reduced to the decoupled form

$$\begin{bmatrix} x_2(s) \\ \eta_2(s) \end{bmatrix} = \begin{bmatrix} N_1 & O \\ O & N_4 \end{bmatrix} \begin{bmatrix} x_1(s) \\ \eta_1(s) \end{bmatrix} - \begin{bmatrix} Y_1 \\ Y_2 \end{bmatrix} u(s) \tag{4.3}$$

(This is not unique, since N_2 can, in fact, be any matrix satisfying $N_2J_1=O$).

From Corollary 2 to Lemma 1 there exists a constant bijective map between initial condition/output pairs

$$\begin{bmatrix} x_2(O-) \\ J_2\eta_2(O-) \end{bmatrix} = \begin{bmatrix} M_1 & M_2 \\ M_3 & M_4 \end{bmatrix} \begin{bmatrix} x_1(O-) \\ J_1\eta_1(O-) \end{bmatrix}$$

such that

$$\left[\begin{bmatrix} (sI-A_2)^{-1} & O \\ O & (I-J_2)^{-1} \end{bmatrix} \begin{bmatrix} M_1 & M_2 \\ M_3 & M_4 \end{bmatrix} - \begin{bmatrix} N_1 & O \\ O & N_4 \end{bmatrix} \begin{bmatrix} (sI-A_1)^{-1} & O \\ O & (I-sJ_1)^{-1} \end{bmatrix}\right] \begin{bmatrix} I & O \\ O & -J_1 \end{bmatrix} = \begin{bmatrix} O & O \\ O & O \end{bmatrix} \tag{4.4}$$

Again by comparing coefficients in the individual (block) elements of (4.4) it may be shown that there exists a decoupled bijection between the initial condition pairs of the form

$$\begin{bmatrix} x_2(0-) \\ J_2\eta_2(0-) \end{bmatrix} = \begin{bmatrix} N_1 & 0 \\ 0 & N_4 \end{bmatrix} \begin{bmatrix} x_1(0-) \\ J_1\eta_1(0-) \end{bmatrix} \tag{4.5}$$

It should be noted that the structure of (4.5) is not unique since the element in (block) position (1,2) is any constant matrix in the kernel of J_1 and the element in (block) position (2,2) is any constant matrix satisfying $M_4J_1=N_4J_1$.

Finally comparison of output vectors gives

$$X_1 = 0 \text{ and } X_2E_1 = 0 \tag{4.6}$$

Summarising (4.3), (4.4) and (4.5)

$$\begin{bmatrix} x_2(s) \\ -u(s) \end{bmatrix} = \begin{bmatrix} N_1 & Y_1 \\ 0 & I \end{bmatrix} \begin{bmatrix} x_1(s) \\ -u(s) \end{bmatrix} \text{ is a bijection} \tag{4.7}$$

$$\begin{bmatrix} x_2(0-) \\ y(s) \end{bmatrix} = \begin{bmatrix} N_1 & 0 \\ 0 & I \end{bmatrix} \begin{bmatrix} x_1(0-) \\ y(s) \end{bmatrix} \text{ is a bijection} \tag{4.8}$$

that is the state space subsystems are fundamentally equivalent and

$$\begin{bmatrix} \eta_2(s) \\ -u(s) \end{bmatrix} = \begin{bmatrix} N_4 & Y_2 \\ 0 & I \end{bmatrix} \begin{bmatrix} \eta_1(s) \\ -u(s) \end{bmatrix} \text{ is a bijection} \tag{4.9}$$

$$\begin{bmatrix} J_2\eta_2(0-) \\ -y(s) \end{bmatrix} = \begin{bmatrix} N_4 & 0 \\ X_2 & I \end{bmatrix} \begin{bmatrix} J_1\eta_1(0-) \\ -y(s) \end{bmatrix} \text{ is a bijection} \tag{4.10}$$

that is the impulsive subsystems are fundamentally equivalent.

<u>Corollary 1</u>: If two generalised state space systems in standard form are fundamentally equivalent then their state space subsystems are system similar.

<u>Proof</u>: Recalling that N_1 in (4.7) of the Lemma is, in fact, M* of Corollary 2 to Lemma 1 it is clear that for this special case (3.7) may be simplified to

$$(sI-\bar{A}_2)N_1-N_1(sI-\bar{A}_1) = 0 \tag{4.11}$$

Substituting from (4.11) into (3.6) and equating coefficients gives

$$Y_1 = 0 \text{ and } \bar{B}_2 = N_1\bar{B}_1 \tag{4.12}$$

Further recalling that $x_2(0-) = N_1x_1(0-)$ is a bijection between initial condition spaces of the same dimension it is clear that the two state space subsystems have the same order and that N_1 is square and nonsingular. Finally the initial condition/output pair map of (4.8) ensures that any $x_1(0-)$ and its image $x_2(0-) = N_1x_1(0-)$ correspond to the same output $y(s)$ and, considering the undriven case, this leads to the equation

$$\bar{C}_1 = \bar{C}_2 N \tag{4.13}$$

so that

$$\begin{bmatrix} sI-\bar{A}_2 & \bar{B}_2 \\ -\bar{C}_2 & 0 \end{bmatrix} = \begin{bmatrix} N_1 & 0 \\ 0 & I \end{bmatrix} \begin{bmatrix} sI-\bar{A}_1 & \bar{B}_1 \\ -\bar{C}_1 & 0 \end{bmatrix} \begin{bmatrix} N_1^{-1} & 0 \\ 0 & I \end{bmatrix} \tag{4.14}$$

and hence the subsystems are system similar.

It should be noted that the result above actually holds for any pair of systems in the classical state space form which are fundamentally equivalent and is essentially that due to Pernebo[4]. When considering the impulsive subsystems of fundamentally equivalent generalised state space systems in standard form (3.7) becomes

$$((I-sJ_2)N_4-N_4(I-sJ_1))J_1 \equiv 0 \tag{4.15}$$

but it is not necessarily true that $(I-sJ_2)N_4-N_4(I-sJ_1) \equiv 0$, as may be seen from considering two identical, undriven systems described by

$$\begin{bmatrix} 1 & -s \\ 0 & 1 \end{bmatrix} \begin{bmatrix} \eta_1(s) \\ \eta_2(s) \end{bmatrix} = \begin{bmatrix} -\eta_1(0-) \\ 0 \end{bmatrix}$$

Solutions and initial conditions in both systems then take the form $\{(\eta_1(0-), 0)\}$ and an appropriate map is

$$N_4 = \begin{bmatrix} 1 & 0 \\ 0 & 0 \end{bmatrix}$$

It is readily ascertained that $J_2N_4 \neq N_4J_1$ although $(J_2N_4-N_4J_1)J_1 = 0$

This example highlights a basic difficulty in deriving a result analogous to Corollary 1 to Lemma 3 for the case of impulsive subsystems. The difficulty is avoided, however, by the following Lemma which shows that it is possible to replace N_4 by an alternative bijection N_4^* for which the required result holds.

Lemma 4: If two generalised state space systems in standard form are fundamentally equivalent there exists a constant map N_4^* such that

(i) $\begin{bmatrix} \eta_2(s) \\ -u(s) \end{bmatrix} = \begin{bmatrix} N_4^* & Y_2 \\ O & I \end{bmatrix} \begin{bmatrix} \eta_1(s) \\ -u(s) \end{bmatrix}$ is a bijection

(ii) $\begin{bmatrix} J_2\eta_2(s) \\ -u\ (s) \end{bmatrix} = \begin{bmatrix} N_4^* & O \\ X_2 & I \end{bmatrix} \begin{bmatrix} J_1\eta_1(O-) \\ -u(s) \end{bmatrix}$ is a bijection

(iii) $$(I-sJ_2)N_4^*-N_4^*(I-sJ_1) = O \qquad (4.16)$$

Proof: Rewriting (3.7) for the impulsive subsystems equating constants and substituting the unit matrix I for $\eta_1(O)$ gives $J_2Q = N_4J_1$ with Q constant.

Consider
$$N_4^* = N_4+(Q-N_4)-(Q-N_4)J_1J_1^t \qquad (4.17)$$
where J_1^t is a generalised inverse of J_1 so that $J_1J_1^tJ_1=J_1$. It is readily shown that N_4 in (4.9) and (4.10) can be replaced by N_4^*, (since the image of $J_1\eta_1(O-)$ is the same for both maps) and also that (4.16) holds.

The lemma allows a number of matrix relations to be derived. Particularly
$$\hat{B}_2 = N_4^*\hat{B}_1-Y_2 \qquad (4.18)$$
and
$$J_2Y_2 = O \qquad (4.19)$$

Finally note that the restricted initial condition/output pair map of (4.10) ensures that any $J_1\eta_1(O-)$ and its image $J_2\eta_2(O-)=N_4^*J_1\eta_1(O-)$ correspond to the same output $y(s)$ and this gives
$$\hat{C}_1 = X+\hat{C}_2N_4^* \qquad (4.20)$$
and
$$XJ_1 = O \qquad (4.21)$$
and
$$\hat{D}_1 = X\hat{B}_1+\hat{C}_2Y_2+\hat{D}_2 \qquad (4.22)$$

Lemma 4 with equations (4.18), (4.20) and (4.22) enable a matrix relationship between the two impulsive subsystems to be generated. It is shown in the next section that this relationship is one of complete system equivalence.

5. A RELATIONSHIP BETWEEN COMPLETE AND FUNDAMENTAL SYSTEM EQUIVALENCE

It is, in fact, possible to show that the intuitively attractive relationship of fundamental equivalence is an alternative characterisation for complete system equivalence - the next theorem gives part of this result.

Theorem 2: If two generalised state space systems, $\mathbb{P}_1$, $\mathbb{P}_2$, are fundamentally equivalent then they are completely system equivalent.

Proof: Assume that the systems are fundamentally equivalent then from Lemma 2 their standard forms are fundamentally equivalent and using the results of Corollary 1 to Lemma 3 the two state space subsystems are system similar. Since system similarity is a special case of complete system equivalence it remains only to consider the impulsive subsystems. Using the results of Lemma 4 together with (4.18), (4.19), (4.20), (4.22) there exists a matrix relationship

$$\begin{bmatrix} N_4^* & O \\ \hat{C}_1-\hat{C}_2N_4 & I \end{bmatrix} \begin{bmatrix} I-sJ_1 & \hat{B}_1 \\ -\hat{C}_1 & \hat{D}_1 \end{bmatrix} = \begin{bmatrix} I-sJ_2 & \hat{B}_2 \\ -\hat{C}_2 & \hat{D}_2 \end{bmatrix} \begin{bmatrix} N_4^* & Y_2 \\ O & I \end{bmatrix} \tag{5.1}$$

For this to be complete system equivalence it is necessary to show that neither $[N_4^* \quad I-sJ_2]$ nor $[N_4^{*t} \quad (I-sJ_1)^t]^t$ has infinite zeros (the special form of $I-sJ_i$ does not allow the presence of finite zeros). Consider $[N_4^* \; I-sJ_2]$ and replace s by 1/s to give a new matrix

$$G_\infty(1/s) = [N_4^* \quad I-J_2/s]$$

A matrix fraction description of this is

$$G_\infty(1/s) = [I-J_2J_2^t+sJ_2J_2^t]^{-1}[(I-J_2J_2^t+sJ_2J_2^t)N_4 \;\vdots\; I-J_2J_2^t+sJ_2J_2^t-J_2] \tag{5.2}$$

The (diagonal) denominator of $G_\infty(1/s)$ in (5.2) has full rank except at s=0. It may be shown that the numerator, N(s) has full rank at s=0 thus proving simultaneously that (5.2) is a relatively prime matrix fraction description and that $[N_4^* \ I-sJ_2]$ has no infinite zeros. A similar argument may be used to show that $[N_4^{*t} \ (I-sJ_1)^t]^t$ has no infinite zeros and hence the impulsive subsystems are complete system equivalent and, since any system matrix is complete system equivalent to its Kronecker form[5] it follows that systems which are fundamentally equivalent are also completely system equivalent.

In fact the converse of the above may also be shown to hold and so the results of Theorem 2 taken together with those of [5] unify the ideas of fundamental equivalence, complete system equivalence and operations of strong equivalence. Thus fundamental equivalence provides a conceptual framework of equivalence, while complete system equivalence gives a closed form matrix description of this. Operations of strong equivalence are useful from an algorithmic point of view since they describe the elementary operations allowed under this equivalence.

6. REFERENCES

1. Cobb, D (1982) "On the solutions of linear differential equations with singular coefficients", J. Differential Equations, 46, 310-323.

2. Gantmacher, F.R. (1959) "The theory of matrices", New York, Chelsea.

3. Kailath, T., (1980) "Linear Systems", Prentice Hall Inc. Englewood Cliffs, N.J.

4. Pernebo, L., (1977) "Notes on strict system equivalence", Int. J. Control Vol 25, No. 1, 21-38.

5. Pugh, A.C., Hayton,G.E., and Fretwell, P., (1983) "Some transformations of matrix equivalence arising from linear systems theory", Proc. of Automatic Control Conference, San Francisco, U.S.A.

6. Rosenbrock, H.H., (1970) "State space and multivariable theory", Nelson-Wiley.

7. Rosenbrock, H.H., (1974) "Structural properties of linear dynamical systems", Int. J. Control, Vol 20, 191-202.

8. Verghese, G., (1978) "Infinite frequency behaviour in generalised dynamical systems", Ph.D. Disst., Stanford Univ. California, U.S.A.

9. Verghese, G., Levy, B.C. and Kailath, T., (1981) "A generalised state-space for singular systems", I.E.E.E. Trans., Vol. AC-26, No. 4.

ON THE RELATIONSHIP BETWEEN THE ZEROS OF THE RATIONAL TRANSFER MATRICES OF G(s) AND $G^T(-s)G(s)$

M.A. Johnson and M.J. Grimble

(Industrial Control Unit, University of Strathclyde, Glasgow)

ABSTRACT

The optimal return difference relationship contains the matrix $G^T(-s)G(s)$ which plays an important role in the analysis of optimal root loci asymptotes. This paper reports the details of the relationships between the poles and zeros of G(s) and the composite matrix $G^T(-s)G(s)$. The use made of this theory, based on the Smith-McMillan form, in root loci analysis is reported.

1. INTRODUCTION

There is considerable interest in delineating the finite and infinite behaviour of the root loci of multivariable systems[1-9]. These approaches naturally encompass several different types of analysis, for example geometric techniques or dynamic transformations. The topic has now reached sufficient maturity to try and unify the known results. Analysis is used in the scientific method first to provide solutions, and second to permit the efficient computation of problem solutions. It is often the case that the techniques used to formulate solutions do not provide a useful computation route. The root loci analysis has certainly entered the phase of unification. A recent paper by Owens[9] has highlighted the need for more "uniformly based" proofs for various results in root locus theory. Very few researchers currently address the computational issues arising from the various root loci analyses and this is likely to be an area of interest in the near future.

The analysis reported in this paper is part of an attempt to unify root locus theory around the matrix/vector return difference expression and the use of the Smith-McMillan form. It was originally thought that this approach would form a useful framework for the presentation of root loci concepts since it depends largely on transfer function manipulation[10]. This paper reports some results in the continuing study of this approach.

The paper is divided into two parts, the first examines the finite zero relationship between the open loop transfer function matrix $G(s)$ and $G^T(-s)G(s)$. These results are then applied to the optimal return difference relationship. The second part presents similar results for the infinite pole-zero relationship of $G(s)$ and $G^T(-s)G(s)$. The use of such results in delineating the numbers and orders of infinite branches of the optimal root loci is demonstrated.

Proposition 1 (Verghese and Kailath, 1979[6]) For any $H(s) \varepsilon \mathbb{R}^{m \times \ell}(s)$ and $s_o \varepsilon \mathbb{C}_\infty \underline{\Delta}\ \mathbb{C}_U\{\infty\}$,

i) $\exists$ nonsingular rational matrices $L(s) \varepsilon \mathbb{R}^{m \times m}(s)$, $R(s) \varepsilon \mathbb{R}^{\ell \times \ell}(s)$, such that

(a) $L(s)$ and $R(s)$ have no poles or zeros at $s = s_o$

(b) $$L(s)H(s)R(s) = \begin{bmatrix} h_1(s) & & & & \\ & \ddots & & & \\ & & h_r(s) & & \\ & & & 0 & \\ & & & & \ddots \\ & & & & & 0 \end{bmatrix}$$

where $h_i(s) \varepsilon \mathbb{R}(s)$ and $r = \text{rank}\{H(s)\} \leqslant \min(m, \ell)$.

(ii) The pole and zero orders of $H(s)$ at $s = s_o$ are given by the pole and zero orders of $h_i(s)$, $i=1,\ldots,r$ at $s = s_o$.

The route to the optimal closed loop poles and their input direction vectors is the matrix optimal return difference relationship:

$$[I_m + \frac{1}{\rho} G^T(-s)G(s)]v = 0 \tag{1}$$

This particular relationship covers both the finite behaviour and the unbounded behaviour of the optimal closed loop poles (under the limiting condition $\rho \to 0$.) Proposition 1 describes the Smith-McMillan form in both these regions, and consequently provides the necessary tools for a Smith-McMillan transfer function analysis of equation(1). This requires some knowledge of the relationship between the zero structure of G(s) and that of $G^T(-s)G(s)$ as discussed in this paper.

2. THE FINITE ZEROS OF G(s) AND $G^T(-s)G(s)$

The method used is to rationalise the transfer function matrix writing it as $G(s) = \tilde{G}(s)/d(s)$, where $\tilde{G}(s)$ is a polynomial matrix, and d(s) the least common denominator polynomial. Subsequently, the operations to determine a Smith canonical form for $\tilde{G}(s)$ are executed, finally dividing by d(s), to yield the Smith-McMillan form of G(s).

2.1 A Basic Theorem

The main result for the connection between the Smith-McMillan forms of G(s) and $G^T(-s)G(s)$ has been derived previously:

Theorem 1 (Johnson and Grimble, 1982[8]). According to the numbers of input and output dimensions, the connections between the Smith-McMillan forms of G(s) and $G^T(-s)G(s)$ (normal rank of G(s) = min(r,m)) may be given as:

Case 1: $r < m$ If $G(s) = P(s)[\Gamma_r(s) \mid O_{rx(m-r)}]Q(s)$

then

$$G^T(-s)G(s) = P_1(s)\left[\begin{array}{c|c} \Gamma_r^T(-s)\Gamma_r(s) & O \\ \hline O & O_{(m-r)x(m-r)} \end{array}\right]Q_1(s) \tag{2}$$

where $P(s) \in \mathbb{R}^{rxr}(s)$; Q(s), $P_1(s)$ and $Q_1(s) \in \mathbb{R}^{mxm}(s)$ and P(s), Q(s), $P_1(s)$ and $Q_1(s)$ are unimodular matrices.

Case 2: $r = m$ If $G(s) = P(s)[\Gamma_m(s)]Q(s)$ then

$$G^T(-s)G(s) = P_1(s)[\Gamma_m^T(-s)\Gamma_m(s)]Q_1(s) \tag{3}$$

where P(s), Q(s), $P_1(s)$ and $Q_1(s)$ are all (mxm) unimodular matrices.

Case 3: $r > m$ If $G(s) = P(s)\begin{bmatrix}\Gamma_m(s)\\ O\end{bmatrix}Q(s)$ then

$$G^T(-s)G(s) = P_1(s)[\Gamma_m^T(-s)T_m^T(-s)T_m(s)\Gamma_m(s)]Q_1(s) \tag{4}$$

where $P(s) \varepsilon \mathbb{R}^{rxr}(s)$; $Q(s)$, $P_1(s)$ and $Q_1(s) \varepsilon \mathbb{R}^{mxm}(s)$, all being unimodular matrices. The diagonal matrix $T_m(s)$ compromises new finite zeros which modify the original finite pole/zero distribution of G(s).

It might be thought that the para-hermitian nature of $G^T(-s)G(s)$ would ensure that the reduction matrices in (2), (3) and (4) satisfy the relationship $P_1(s) = Q_1^T(-s)$. In such cases several interesting properties apply, for example, stable spectral factors follow directly as $Y^T(-s)Y(s) = G^T(-s)G(s) \Rightarrow Y(s) = T_m(s)\Gamma_m(s)Q_1(s)$ (case 3). Unfortunately, such a reduction is not always possible as can be shown by very simple counter examples. The following corollary gives sufficient conditions for this to occur.

Corollary 1 If the reduction of the polynomial matrix $\tilde{G}^T(-s)\tilde{G}(s)$ to its Smith canonical form can be achieved by manipulating the diagonal elements only, then the resulting reduction matrices will satisfy $P_1(s) = Q_1^T(-s)$.

2.2 Application in the optimal return difference

It is useful to examine how the forms of the theorem are used in the original return difference relationship (1). The three cases (r < m, r = m, r > m) have been treated individually previously[11] and only one case is given here:

Case 1: r < m Substituting (2) into (1) gives:

$$\left(\begin{bmatrix} I_r & O \\ O & I_{m-r} \end{bmatrix} + \frac{1}{\rho} P_1(s) \begin{bmatrix} \Gamma_r^T(-s)\Gamma_r(s) & O \\ O & O \end{bmatrix} Q_1(s) \right) v = O \tag{5}$$

Premultiply by $[P_1(s)]^{-1}$ and redefine the input direction vector, v, as $\tilde{v} = [P_1(s)]^{-1} v$, then (5) becomes:

$$\left(\begin{bmatrix} I_r & O \\ O & I_{m-r} \end{bmatrix} + \frac{1}{\rho} \begin{bmatrix} \Gamma_r^T(-s)\Gamma_r(s) & O \\ O & O \end{bmatrix} \begin{bmatrix} E_{11}(s) & E_{12}(s) \\ E_{21}(s) & E_{22}(s) \end{bmatrix} \right) \begin{bmatrix} \tilde{v}_1 \\ \tilde{v}_2 \end{bmatrix} = O \tag{6}$$

where $E(s) \triangleq Q_1(s)P_1(s) = \begin{bmatrix} E_{11}(s) & E_{12}(s) \\ E_{21}(s) & E_{22}(s) \end{bmatrix} \varepsilon \mathbb{R}^{mxm}(s)$ is unimodular, and $\tilde{v}$ is partitioned as $\tilde{v} = \begin{bmatrix} \tilde{v}_1 \\ \tilde{v}_2 \end{bmatrix}$. Equation (6) may be written as the pair of simultaneous equations:

$$[I_r + \frac{1}{\rho} \Gamma^T(-s)\Gamma(s)E_{11}(s)]\tilde{v}_1 + \Gamma_r^T(-s)\Gamma_r(s)E_{12}(s)\tilde{v}_2 = 0 \quad (7)$$

$$I_{r-m}\tilde{v}_2 = 0 \quad (8)$$

Clearly (7) and (8) reduce to

$$[I_r + \frac{1}{\rho} \Gamma_r^T(-s)\Gamma_r(s)E_{11}(s)]\tilde{v}_1 = 0 \text{ and } \tilde{v}_2 = 0 \quad (9)$$

There is one further step to be completed before (9) can be used unambiguously in a transfer function analysis of the finite behaviour of the optimal closed loop poles, as $\rho \to 0$. Although $E(s)$ is unimodular, $E_{11}(s)$ is not necessarily so, consequently a further Smith-McMillan reduction is needed, viz:

$$\Gamma_r^T(-s)\Gamma_r(s)E_{11}(s) = \tilde{P}_1(s)\Gamma_r^T(-s)T_r^T(-s)T_r(s)\Gamma_r(s)\tilde{Q}_1(s) \quad (10)$$

where $\tilde{P}_1(s)$ and $\tilde{Q}_1(s)$ are (rxr) unimodular matrices and the new finite zeros contributed by $E_{11}(s)$ are introduced via the diagonal matrix $T_r^T(-s)T_r(s)$.

Substituting (10) into (9) obtains:

$$[I_r + \frac{1}{\rho} \tilde{P}_1(s)\Gamma_r^T(-s)T_r^T(-s)T_r(s)\Gamma_r(s)\tilde{Q}_1(s)]\tilde{v}_1 = 0 \quad (11)$$

If (11) is premultiplied by $[\tilde{P}_1(s)]^{-1}$, and $\tilde{v}_1$ is redefined as:

$$\tilde{\tilde{v}}_1 = [\tilde{P}_1(s)]^{-1}\tilde{v}_1$$

a new but unambiguous return-difference relation results on which to base the root loci analysis, namely

$$[I_r + \frac{1}{\rho} \Gamma_r^T(-s)T_r^T(-s)T(s)\Gamma_r(s)\tilde{E}(s)]\tilde{\tilde{v}}_1 = 0 \quad (12)$$

where $E(\tilde{s})$ is (rxr) unimodular.

This method of reduction of the optimal return difference equation to obtain a relationship for optimal root loci analysis is similar in the remaining cases $r = m$, and $r > m$. In summary, the three reduced return difference expressions are:

(i) $r < m$ $[I_r + \frac{1}{\rho}\Gamma_r^T(-s)T_r^T(-s)T_r(s)\Gamma_r(s)\tilde{E}(s)]\tilde{v}_1 = 0$

$(v_2 = 0)$

(ii) $r = m$ $[I_m + \frac{1}{\rho}\Gamma_m^T(-s)\Gamma_m(s)E(s)]\tilde{v} = 0$

(iii) $r > m$ $[I_m + \frac{1}{\rho}\Gamma_m^T(-s)T_m^T(-s)T_m(s)\Gamma_m(s)E(s)]\tilde{v} = 0$

where the connection with the finite zeros and poles of G(s) is made via the appropriate $\Gamma(s)$. In the nonsquare cases, new finite zeros unrelated to the system transmission zeros are introduced via the terms $T^T(-s)T(s)$.

3. THE INFINITE ZEROS OF G(s) AND $G^T(-s)G(s)$

This section is concerned with providing a treatment analogous to that given for the finite zeros but for the infinite zeros. At a first glance it might be thought this would have few, if any, differences from the finite zero analysis. On the contrary there are several significant modifications which are of importance in the substitution and reduction of the optimal return difference equation, and in the topic of infinite zero invariance [10].

First a characterisation of the infinite zero structure of G(s) is given.

Proposition 2[1] Let $G(s) \varepsilon \mathbb{R}^{r\times m}(s)$ be a rational transfer function matrix, whose normal rank = min(r,m), then (I) There exists nonsingular $P(s) \varepsilon \mathbb{R}^{r\times r}(s)$ and $Q(s) \varepsilon \mathbb{R}^{m\times m}(s)$ such that:

(a) P(s) and R(s) have no poles or zeros at $s = \infty$.

(b) Case 1: $r < m$ $G(s) = P(s)[\Gamma_r(s)|0_{r\times(m-r)}]Q(s)$ where

$$\Gamma_r(s) = \text{block diag}\{1/s\, I_{d_0},\ 1/s^2 I_{d_1}, \ldots, 1/s^{i+1} I_{d_i}\} \text{ and}$$

$$\sum_{j=0}^{i} d_j = r.$$

(c) Case 2: $r = m$ $G(s) = P(s)\Gamma_m(s)Q(s)$ where

$$\Gamma_m(s) = \text{block diagonal } \{1/sI_{d_0}, 1/s^2 I_{d_i}, \ldots, 1/s^{i+1} I_{d_i}\}$$

$$\text{and} \quad \sum_{j=o}^{i} d_j = m.$$

(d) Case 3: $r > m$ $\quad G(s) = P(s)\begin{bmatrix}\Gamma_m(s)\\ O\end{bmatrix}Q(s)$

where $\Gamma_m(s)$ = block diagonal$\{1/sI_{d_o}, 1/s^2 I_{d_1}, \ldots, 1/s^{i+1} I_{d_i}\}$ and

$$\sum_{j=o}^{i} d_j = m.$$

(II) G(s) has no poles at $s=\infty$, and the number of (j+1)th order infinite zeros is d_j, $j=0,1,\ldots,i$. Note that if there are no (j+1)th order infinite zeros, then block I_{d_j} is deleted.

Proof This is a slight modification of the first two parts of Proposition 8.3 as given by Hung and MacFarlane[1]. Indeed, the example 7.2 of that paper discusses this very case and the results anticipate the proposition given here.

3.1 A Basic Theorem

With the details established, a theorem to describe the connection between the infinite zeros of G(s) and those of $G^T(-s)G(s)$ may now be given.

Theorem 2 According to the disparity of numbers of inputs and outputs, the relationship between the Smith-McMillan forms of G(s) and $G^T(-s)G(s)$ valid at $s=\infty$ may be given as:

Case 1: $r < m$ Using the characterisation of case 1, Proposition 2 for G(s) obtains:

$$G(s) = P(s)[\Gamma_r(s) \mid O_{rx(m-r)}]Q(s) \text{ and}$$

$$G^T(-s)G(s) = Q_1^T(-s)\left[\begin{array}{c|c}\Gamma_r^T(-s)\Gamma_r(s) & O\\ \hline O & O\end{array}\right]Q_1(s).$$

Case 2: $r = m$ Using the characterisation of case 2. Proposition 2 for G(s) obtains:

$$G(s) = P(s)\Gamma_m(s)Q(s) \text{ and}$$

$$G^T(-s)G(s) = Q_1^T(-s)\Gamma_m^T(-s)\Gamma_m(s)Q_1(s).$$

Case 3: r > m Using the characterisation of case 3, Proposition 2 for G(s) obtains:

$$G(s) = P(s) \begin{bmatrix} \Gamma_m(s) \\ 0 \end{bmatrix} Q(s) \text{ and}$$

$$G^T(-s)G(s) = Q_1^T(-s)[\Gamma_m^T(-s)\Gamma_m(s)]Q_1(s)$$

In each case $Q_1(s)$ is nonsingular with no poles or zeros at $s=\infty$, and $\Gamma_m(s)$ takes the structure given in the three cases of Proposition 2 as appropriate.

Remarks: (i) The infinite zero structure of $G^T(-s)G(s)$ is not modified by the introduction of new zeros as in the finite zero case. (ii) A reduction pair of the form $Q_1^T(-s), Q_1(s)$ can always be found so that non-symmetric pairs $P_1(s), Q_1(s)$ do not arise as in the finite zero case.

3.2 Applications in the Optimal Return Difference

As with the finite case, the use of Theorem 2 in the optimal return difference relationship is next considered. The three cases have been discussed individually elsewhere. Unlike the finite problems there are two ways of examining the optimal return difference relationship for the infinite zeros. Here the detail of the two methods is presented for just one case where the dimensions satisfy normal rank G(s) = r < m.

Method I, Case 1, r < m: This rearrangement follows that of the finite zero section. In the optimal return difference relationship, obtain from Theorem 2; case 1:

$$[I_m + \frac{1}{\rho}G^T(-s)G(s)]v = \left[I_m + \frac{1}{\rho}Q_1^T(-s)\left[\begin{array}{c|c} \Gamma_r^T(-s)\Gamma_r(s) & 0 \\ \hline 0 & 0_{m-r\times m-r} \end{array}\right]Q_1(s)\right]v = 0 \tag{13}$$

Premultiply (13) by $[Q_1^T(-s)]^{-1}$, and redefine $\tilde{v} = [Q_1^T(-s)]^{-1}v$ to obtain

$$\left[I_m + \frac{1}{\rho}\left[\begin{array}{c|c} \Gamma_r^T(-s)\Gamma_r(s) & 0 \\ \hline 0 & 0_{m-r\times m-r} \end{array}\right][Q_1(s)Q_1^T(-s)]\right]\tilde{v} = 0 \tag{14}$$

Define $D(s) = Q_1(s)Q^T(-s)$, partition both $D(s)$ and $\tilde{v}$, thus (14) becomes:

$$[I_r + \frac{1}{\rho}\Gamma_r^T(-s)\Gamma_r(s)D_{11}(s)]\tilde{v}_1 + \Gamma_r^T(-s)\Gamma_r(s)D_{12}(s)\tilde{v}_2 = 0,$$

$$I\tilde{v}_2 = 0 \qquad (15)$$

hence $\quad [I_r + \frac{1}{\rho}\Gamma_r^T(-s)\Gamma_r(s)D_{11}(s)]\tilde{v}_1 = 0$ and $\tilde{v}_2 = 0$

(16)

Recall that $Q_1(s)$ is nonsingular and regular at $s = \infty$ then (i) $D(s)$ is non-singular, and regular at $s = \infty$. (ii) Write $Q_1(s) = Q_1 + \frac{1}{s}\tilde{Q}_1(s)$, so that $D(s) = D + \frac{1}{s}\tilde{D}(s)$, where $D = Q_1^T Q_1$ and $D^T(-s) = D(s)$. (iii) $D = Q_1^T Q_1 \Rightarrow D > 0$.

Using Sylvester's theorem [12] yields D_{11}, the leading (rxr) block of $D(s)$ satisfies $D_{11} > 0$. Hence $D_{11}(s) = D_{11} + \frac{1}{s}\tilde{D}_{11}(s)$ is both nonsingular and regular at $s = \infty$.

In contrast to the similar treatment for the finite zero case, no further reduction is necessary (see equations (5) to (12)). Write for equation (16)

$$[I_r + \frac{1}{\rho}\Gamma_r^T(-s)\Gamma_r(s)E(s)]\tilde{v}_1 = 0 \quad \tilde{v}_2 = 0 \qquad (17)$$

In summary, the application of Theorem 2, when applied to the optimal return difference equation yields:

<u>Case 1: $r < m$</u> $\quad [I_r + \frac{1}{\rho}\Gamma_r^T(-s)\Gamma_r(s)E(s)]\tilde{v}_1 = 0 \quad \tilde{v}_2 = 0$:

<u>Cases 2 and 3: $r \geqslant m$</u>

$[I_m + \frac{1}{\rho}\Gamma_m^T(-s)\Gamma_m(s)E(s)]\tilde{v} = 0$, where in either case $E(s)$ of appropriate dimensions is nonsingular and regular at $s = \infty$ and furthermore satisfies $E^T(-s) = E(s)$.

<u>Method II, Case 1, $r < m$;</u> This is a slightly different rearrangement of the return difference analysis which has a useful connection with the analysis of Owens[3,4]. Recall equation (13), premultiply by $[Q_1^T(-s)]^{-1}$ and define $\hat{v} = Q_1(s)v$,

hence

$$\left[[Q_1^T(-s)]^{-1}[Q_1(s)] + \frac{1}{\rho}\left[\begin{array}{c|c} \Gamma_r^T(-s)\Gamma_r(s) & 0 \\ \hline 0 & 0 \end{array}\right]\right]\hat{v} = 0 \tag{18}$$

Define $D(s) = [Q_1(s)Q_1^T(-s)]^{-1}$, partition $D(s)$ and $\hat{v}$, then (18) becomes:

$$[D_{11}(s) + \frac{1}{\rho}\Gamma_r^T(-s)\Gamma_r(s)]\hat{v}_1 + D_{12}(s)\hat{v}_2 = 0$$

$$D_{22}^T(-s)\hat{v}_1 + D_{22}(s)\hat{v}_2 = 0 \tag{19}$$

The properties of $D(s)$ based on those of Q_1 imply $D_{22}(s)$ nonsingular, hence (19) becomes:

$$[D_{11}(s)-D_{12}^T(-s)D_{22}^{-1}(s)D_{12}(s) + \frac{1}{\rho}\Gamma_r(-s)\Gamma_r(s)]\hat{v}_1 = 0 \tag{20}$$

$$\hat{v}_2 = -D_{22}^{-1}(s)D_{12}^T(-s)\hat{v}_1 \tag{21}$$

Define $E(s) = D_{11}(s)-D_{12}^T(-s)D_{22}^{-1}(s)D_{12}(s)$, then (20) becomes

$$[E(s) + \frac{1}{\rho}\Gamma_r(-s)\Gamma_r(s)]\hat{v}_1 = 0 \tag{22}$$

where (i) $E^T(-s) = E(s)$ and (ii) $\det[D(\infty)] > 0$,

$$\det[D_{11}(\infty)] > 0, \quad \det[D_{22}(\infty)] > 0$$

$$\det[D(\infty)] = \det[D_{11}(\infty)-D_{12}^T(\infty)D_{22}^{-1}(\infty)D_{12}(\infty)].\det[D_{22}(\infty)] \Rightarrow$$

$$\det[D_{11}(\infty)-D_{12}^T(\infty)D_{22}^{-1}(\infty)D_{12}(\infty)] = \det[E(\infty)] > 0$$

In an expanded form, equation (22) is

$$\begin{bmatrix} E_{oo}(s)+1/\rho\gamma_o I_{d_o}, E_{o1}(s) & , E_{o2}(s) & ,\ldots\ldots & E_{oi}(s) \\ E_{o1}^T(-s) & , E_{11}(s)+1/\rho\gamma_1 I_{d_1}, & E_{12}(s) \quad ,\ldots\ldots & E_{1i}(s) \\ E_{oi}^T(-s),\ldots\ldots\ldots\ldots\ldots\ldots\ldots\ldots & & & E_{ii}(s)+1/\rho\gamma_i \; I_{d_i} \end{bmatrix}\begin{bmatrix}\hat{v}_o \\ \vdots \\ \hat{v}_i\end{bmatrix} = 0 \tag{23}$$

Apply the Theorem 1[4] then equation (23) will have the usual numbers of infinite zeros if matrices E_{oo}, $E_{11}, \ldots, E_{ii}$ are all nonsingular. In the optimal case a particularly simple structure applies to $E = \lim_{\rho \to o} E(s) = \Lambda^{-1}$ so that the conditions of Owens[4] are satisfied.

4. CONCLUSIONS

The basic finite zero relationship between the system transfers $G(s)$ and $G^T(-s)G(s)$ has been established for the cases of input/output dimensions $m > r$, $m=r$ and $m < r$. This analysis demonstrates (i) how new zeros are introduced for the nonsquare case, $r > m$, and (ii) that there were cases for which the unimodular reduction matrices are not para-hermitian.

The use of these results in the optimal return difference-input direction vector relationship was next examined. The significant result here was the similarity of form for the three cases, viz:

$$r < m, \ [I_r + \frac{1}{\rho}\Gamma^T(-s)T^T(-s)T(s)\Gamma(s)E(s)]v = 0$$

$$r = m, \ [I_m + \frac{1}{\rho}\Gamma^T(-s)\Gamma(s)E(s)]v = 0$$

$$r > m, \ [I_m + \frac{1}{\rho}\Gamma^T(-s)T^T(-s)T(s)\Gamma(s)E(s)]v = 0$$

where $T(s)$ introduced new finite zeros for the nonsquare cases and $E(s)$ was unimodular in each case.

Similar results were also demonstrated for the infinite zeros of $G(s)$ and $G^T(-s)G(s)$ based on the work of Hung and MacFarlane[1]. However, in contrast to the finite zero case the relationship differed in two aspects; (i) no new infinite zeros were introduced for the nonsquare cases, and (ii) parahermitian unimodular reduction matrices could always be found.

Applying these results to the reduction of the optimal return difference input direction vector relationship revealed (i) no new infinite zeros for either of the nonsquare cases and (ii) an arrangement of the relationship in the spirit of the Owens reduction[3] showed that no generic condition applied to the infinite branching. This confirms the results of Stevens[7], and more recently Owens[9].

5. REFERENCES

1. Hung, Y.S. and MacFarlane, A.G.J., (1981) "On the relationship between the unbounded asymptote behaviour of multivariable root loci, impulse response and infinite zeros". Int. J. Control, vol. 34, No. 1, pp 31-69.

2. Johnson, M.A. and Grimble, M.J., (1981) "On the asymptotic root loci of linear multivariable systems". Int. J. Control, vol 34, No. 2, pp 295-314.

3. Owens, D.H., (1978) "On structural invariants and the root loci of linear multivariable systems". Int. J. Control, vol. 28, No. 2, pp 187-196.

4. Owens, D.H., (1981) "On the orders of the optimal system infinite zeros". Systems and Control Letters, vol. 1, No. 2, pp 123-126.

5. Kouvaritakis, B., (1978) "The optimal root loci of linear multivariable systems". Int. J. Control, vol. 28, No. 1, pp 33-62.

6. Verghese, G. and Kailath, T., (1979) Comments on "On structural invariants and the root loci of linear multivariable systems". Int. J. Control, vol 29, No. 6, pp 1077-1080.

7. Stevens, P.K. (1982) "Optimal systems root loci: Relation to the McMillan structure of the open loop systems". IEEE Trans Automatic Control, Vol. AC-27, No. 6.

8. Johnson, M.A. and Grimble, M.J., (1983) "Asymptotic behaviour of the closed loop poles of linear optimal multivariable systems ". J. Dynamic Systems, Measurement and Control, Vol 105, pp 165-178.

9. Owens, D.H. , (1984) "On the generic structure of multivariable root loci". Int. J. Cont. vol 39, No. 2, pp 311-319.

10. Grimble, M.J. and Johnson, M.A., (1985) "Optimal Control and Stochastic Estimation Theory". John Wiley and Sons, Chichester, (to appear).

11. Johnson, M.A. and Grimble, M.J., (1982) "On the relationship between the zeros (finite and infinite) of the rational transfer matrices of G(s) and $G^T(-s)G(s)$", Report ICU/20, Industrial Control Unit, University of Strathclyde, Glasgow.

12. Markus, M., (1960) "Basic theorems in matrix theory", US National Bureau of Standards, Applied Mathematics Series, No 57.

POSSIBILITIES FOR THE LINEAR FEEDBACK CONTROL OF POWER SYSTEMS USING PARAMETRIC EIGENSTRUCTURE ASSIGNMENT

H.S. Tantawy
(Al-Azhar University)

and

J. O'Reilly
(Strathclyde University)

ABSTRACT

The possibilities, in general, for the linear state feedback control of a power system are investigated using parametric eigenstructure assignment.

1. INTRODUCTION

Over the last decade, power systems engineers have not been slow to realise the advantage of optimal linear state regulator control over conventional automatic voltage regulation (AVR) and turbine governor control. In its simplest form the power system under investigation usually consists of a single synchronous machine connected to an infinite bus through a transmission line. The basic design objective is to improve the highly oscillatory transient response of the power system using voltage regulator and speed governor controllers based upon linear quadratic[1-3] and pole placement[4,5] techniques.

The present article explores the possibilities, in general, for the linear state feedback control of a single machine power system, represented by an eighth-order linearised state-space model, using parametric eigenstructure assignment. The essence of the approach is the parameterisation of all the degrees of freedom available in the selection of the state feedback gain matrix K by the n eigenvalues $s_1, s_2, \ldots, s_n$ and n r-dimensional arbitrary parameter vectors $f_1, f_2, \ldots, f_n$. This explicit characterisation of design freedom in the free parameter vectors $f_1, \ldots, f_n$ is extremely useful in satisfying additional design criteria beyond eigenvalue assignment.[7-13] It will be used for direct tuning of the power system transient response.

2. MODEL AND STRUCTURAL CONTROLLABILITY OF POWER SYSTEM

2.1 The power system model

The power system under investigation consists of a synchronous machine unit connected to an infinite bus through a transmission line. It has both excitation voltage regulator and turbine speed governor controls. An eighth-order linearised state-space model[4] for the power system is

$$\dot{x}(t) = Ax(t) + Bu(t) \tag{2.1}$$

where the state vector

$$x(t) \triangleq [\delta,\ n,\ \psi_f,\ v_f,\ v_t,\ g_f,\ h]^T \tag{2.2}$$

and each state represents a deviation from its initial value. Nomenclature is as in [4]. The A and B matrices of the dynamical state-space model (2.1) are given by

$$A = \begin{bmatrix} 0.0 & 1.0 & 0.0 & 0.0 & 0.0 & 0.0 & 0.0 & 0.0 \\ -0.683 & -0.0346 & -0.816 & 0.0 & 0.0 & 0.882 & 0.0 & 1.32 \\ -0.0832 & 0.0 & -0.0254 & 0.137 & 0.0 & 0.0 & 0.0 & 0.0 \\ 0.130 & 0.0 & -0.107 & -0.137 & -1.370 & 0.0 & 0.0 & 0.0 \\ 0.0 & 0.0 & 0.0 & 0.0 & -0.274 & 0.0 & 0.0 & 0.0 \\ 0.0 & -0.0265 & 0.0 & 0.0 & 0.0 & -0.0616 & -1.370 & 0.0 \\ 0.0 & 0.0 & 0.0 & 0.0 & 0.0 & 0.0 & -13.7 & 0.0 \\ 0.0 & 0.0531 & 0.0 & 0.0 & 0.0 & 0.123 & 2.74 & -0.171 \end{bmatrix} \tag{2.3}$$

$$B = \begin{bmatrix} 0 & 0 & 0 & 0 & 1 & 0 & 0 & 0 \\ 0 & 0 & 0 & 0 & 0 & 0 & 1 & 0 \end{bmatrix}^T \tag{2.4}$$

The control vector u(t) is given by

$$u(t) = [u_v\ ,\ u_g]^T \tag{2.5}$$

2.2 Structural controllability of power system

Before embarking on any linear state feedback control of the power system (2.1)-(2.4), it is necessary to first examine the controllability properties of the system with regard to excitation and/or governor control. Since the non-zero entries of the model (2.1)-(2.4) inevitably contain inaccuracies on account of simplifying approximations and linearisation, it is most appropriate to consider its structural controllability.[6] For it is known[6] that if the system (2.1)-(2.4) is mapped into a diagraph (directed graph), then the properties of its physical structure will also appear in the diagraph itself and a visual inspection of this diagraph will reveal basic relationships between states and inputs.

Let us now test the power system model (2.1)-(2.4) for structural controllability. The diagraph of the system is shown in Fig. 1.

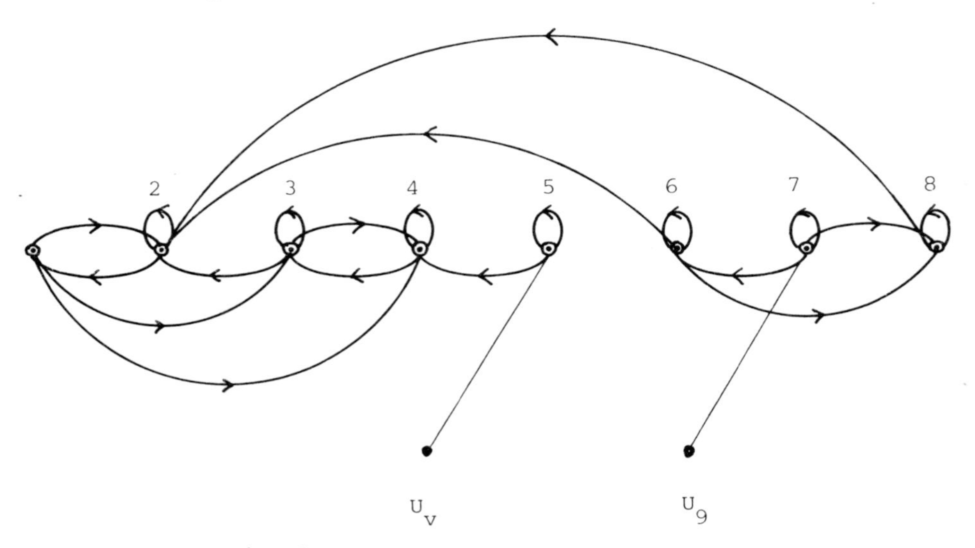

Fig. 1 Diagraph of the power system

By inspection of the diagraph, the states that are reachable by excitation control u_v and governor control u_g are respectively

$$u_v \rightarrow (5, 4, 3, 2, 1, 6, 8, 7)$$

$$u_g \rightarrow (7, 8, 2, 1, 3, 4, 6)$$

3. PARAMETRIC FORM OF LINEAR STATE FEEDBACK CONTROL

Having established that the power system (2.1)-(2.4) is structurally controllable with respect to excitation and governor control, we turn now to examine the design possibilities, in general, using linear state feedback control. It is well known that, apart from the case of single-input systems, specification of the closed-loop eigenvalues does not uniquely define a closed-loop system. The source of non-uniqueness can be identified as that coming from the freedom offered by state feedback, beyond eigenvalue assignment, in selecting the associated eigenvectors and generalised eigenvectors from an admissible class.[7] A major step towards the development of systematic design procedures for eigen-structure (eigenvalue and eigenvector) assignment has been recently taken in a series of papers.[7-14]

Consider the linear, time invariant, completely controllable system

$$\dot{x}(t) = Ax(t) + Bu(t) \tag{3.1}$$

where $x(t) \in R^n$ is the state vector, $u(t) \in R^r$ is the control vector, and A and B are real matrices of compatible dimensions. If the linear state feedback control law

$$u(t) = Kx(t) \tag{3.2}$$

is applied to the open-loop system (3.1), the resulting closed-loop system takes the form

$$\dot{x}(t) = A_c x(t), \quad A_c \triangleq A + BK \tag{3.3}$$

The eigenstructure assignment problem is one of determining a real feedback gain matrix K such that the closed-loop system (3.3) is assigned an arbitrary self-conjugate set of n eigenvalues $\{s_i\}$, together with any associated set of eigenvectors and generalised eigenvectors. In essence the parametric eigenstructure assignment approach[7,8] consists of parameterising <u>all</u> the degrees of freedom available in the selection of the state feedback gain matrix K by the n eigenvalues $s_1, s_2, \ldots, s_n$ and n r-dimensional arbitrary parameter vectors $f_1, f_2, \ldots, f_n$.

Let us now make the simplifying though inessential assumption that the desired closed-loop eigenvalues $\{s_i\}$ of A_c are distinct and no closed-loop eigenvalue is equal to an open-loop eigenvalue. Then, specialising the results of the articles[7,8] we have the following proposition.

Proposition 1 The (non-unique) state feedback gain matrix K which assigns the eigenvalue spectrum $\{s_i\}$ to the closed-loop system (3.3) is given by

$$K = FV^{-1} \tag{3.4}$$

where F is an r x n parameter matrix defined as

$$F = [f_1 \; f_2 \; \dots \; f_n] \tag{3.5}$$

and V = V(F) is an n x n modal matrix of the closed-loop system matrix A_c defined by

$$V = [v_1 \; v_2 \; \dots \; v_n] \tag{3.6}$$

where v_i is the i-th closed-loop eigenvector given by

$$v_i = (s_i I - A)^{-1} B f_i \; , \quad i = 1,2,\dots,n \tag{3.7}$$

Moreover, the parameter vectors f_i, i= 1,2,...,n are arbitrarily chosen under the conditions:

(i) $\det[V] \neq 0$

(ii) $f_i \in R^r$ for a real eigenvalue s_i, whereas $f_j = f_i \in R^r$ or $f_j = f_i^* \in C^r$ for a complex conjugate pair of eigenvalues s_i, $s_j = s_i^*$.

It is observed in Proposition 1 that the state feedback gain matrix K takes the explicitly parameterised form $K = K(s_1, s_2 \dots, s_n; f_1, f_2, \dots, f_n)$. The freedom in choosing the parameter vectors $\{f_i\}$ reflects the freedom offered by state feedback in assigning the eigenvectors $\{v_i\}$, beyond the assignment of the closed-loop eigenvalues $\{s_i\}$. Note, however, that for single-input systems (r=1) such as the power system (2.1), (2.2) with excitation control only, the vectors $\{f_i\}$ reduce to scalars and, accordingly, the gain matrix K of (3.4) is unique.

4. POWER SYSTEM CONTROL

Consider now the application of the parametric approach for the linear state feedback control of the power system (2.1)-(2.4). The open-loop eigenvalues listed in Table 1 show that while the uncontrolled system is stable its transient performance is highly oscillatory (see also Fig. 3). Accordingly, since the overall speed of response of the system will be dominated by the closed-loop eigenvalues $\{s_i\}$ it is necessary to choose appropriate positions for $\{s_i\}$ in the open left-half of the complex plane. A reasonable choice for the closed-loop eigenvalues, based on a sector criterion[4], is given in Table I.

Table I

Eigenvalues of the power system

Eigenvalue s_i	Open-loop value	Closed-loop value
s_1	-0.0114	-0.1641
s_2	±j0.7986	±j0.2432
s_3	-0.0572	-0.2750
s_4	-0.0772	-0.3761
s_5	±j0.1146	±j0.11458
s_6	-0.1952	-0.9719
s_7	-0.2740	±j0.7056
s_8	-13.7	-14.7

Having selected values for the closed-loop eigenvalues $\{s_i\}$, the task remains of choosing the free parameter vectors $\{f_i\}$ to further improve the transient performance of the system. Also, the free parameter vectors $\{f_i\}$ may be chosen to satisfy additional design criteria such as minimum gain control.[12] Of more immediate interest, in the context of the present power system application, is to illustrate the ease with which the transient response of the closed-loop system may be quickly tuned through random selection of the free parameters.

As first guess, let us arbitrarily choose the parameter matrix F as

$$F = F_1 = \begin{bmatrix} 0.01 & 0.01 & 0.01 & 0.05 & 0.05 & 0.01 & 0.01 & 0.03 \\ 0.0 & 0.0 & 0.01 & 0.0 & 0.0 & 0.02 & 0.02 & 0.02 \end{bmatrix} \tag{4.1}$$

in accordance with condition (ii) of Proposition 1. The modal matrix V_1, associated with F_1, when calculated from (3.6) and (3.7), satisfies condition (i) of Proposition 1. The feedback

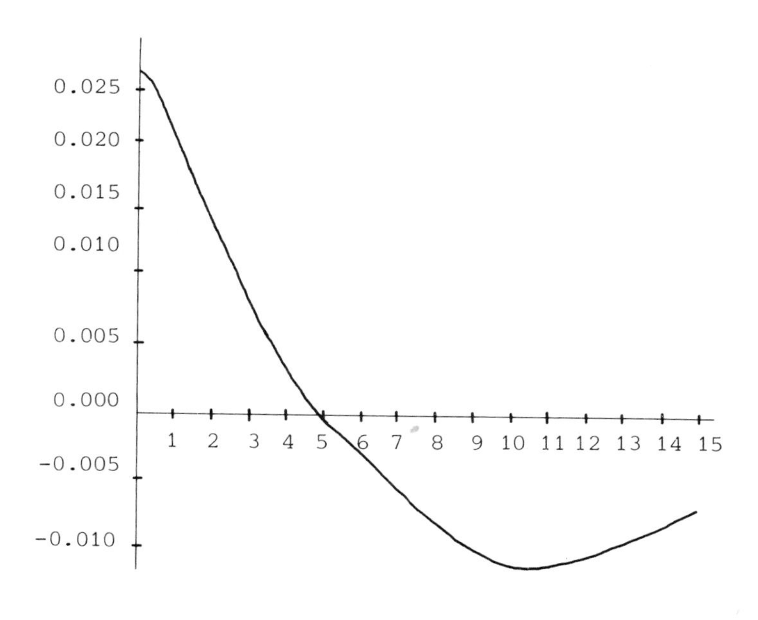

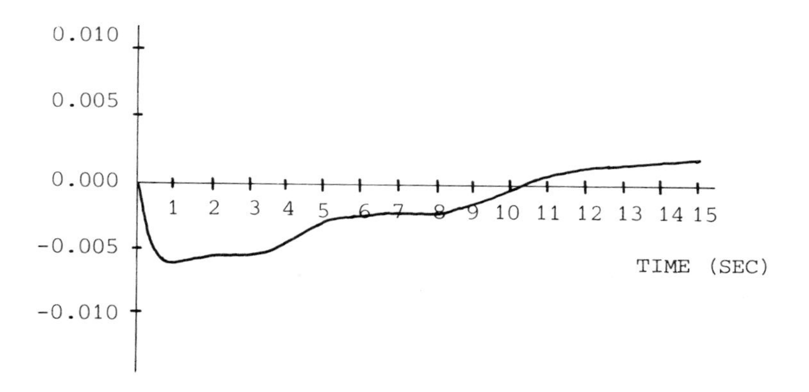

Fig. 2. System Response

gain matrix K_1 is

$$K_1 = \left[\begin{array}{c|c|c|c|c|c|c|c} 30.5599 & 44.2219 & 4.740 & 1.3715 & -1.5762 & -109.0367 & 2.5654 & -30.6868 \\ 10.5380 & 1.3590 & 1.5580 & 0.1810 & -0.2540 & -13.6751 & -2.0194 & -12.2318 \end{array}\right] \quad (4.2)$$

Integration of (3.3) with gain matrix (4.2) and the initial state $x(0) = [0.025, 0.0, 0.009, 0.0155, 0.0, 0.0, 0.0, 0.0]^T$ yields the time histories of the eight state variables of which $x_1 = \delta$ and $x_2 = n$, are depicted in Fig. 2.

A second choice of parameter matrix

$$F_2 = \begin{bmatrix} 3.0 & 3.0 & 3.0 & 4.0 & 4.0 & 5.0 & 5.0 & 8.0 \\ 6.0 & 6.0 & 4.0 & 6.0 & 6.0 & 3.0 & 3.0 & 7.0 \end{bmatrix} \quad (4.3)$$

results in the state feedback gain matrix

$$K_2 = \left[\begin{array}{c|c|c|c|c|c|c|c} -3.2352 & -5.2222 & 1.7467 & 1.5606 & -2.3065 & -48.1492 & -0.6252 & -22.2662 \\ -0.955 & -6.6674 & 0.8665 & 0.6304 & -1.1329 & -13.4388 & -1.2891 & -9.0136 \end{array}\right] \quad (4.4)$$

for which the transient response of the closed-loop system, with the same initial condition as before, is shown in Fig. 3. It is noted that the elements of the state feedback gain matrix (4.4), associated with the system response of Fig. 3, are not unduly large.

Fig. 3 and Fig. 2 illustrate the considerable effect the choice of free parameters, after eigenvalue assignment, has on the "shape" of the closed-loop system response. The parameterization of design freedom lends itself to on-line tuning of the system response so as to compensate for unmodelled non-linearities, dynamics, disturbances and other effects which might degrade the system response. This is in contrast with optimal linear regulator design where the selection of the cost weighting matrices is usually carried out in an off-line solution of a matrix Riccati equation.

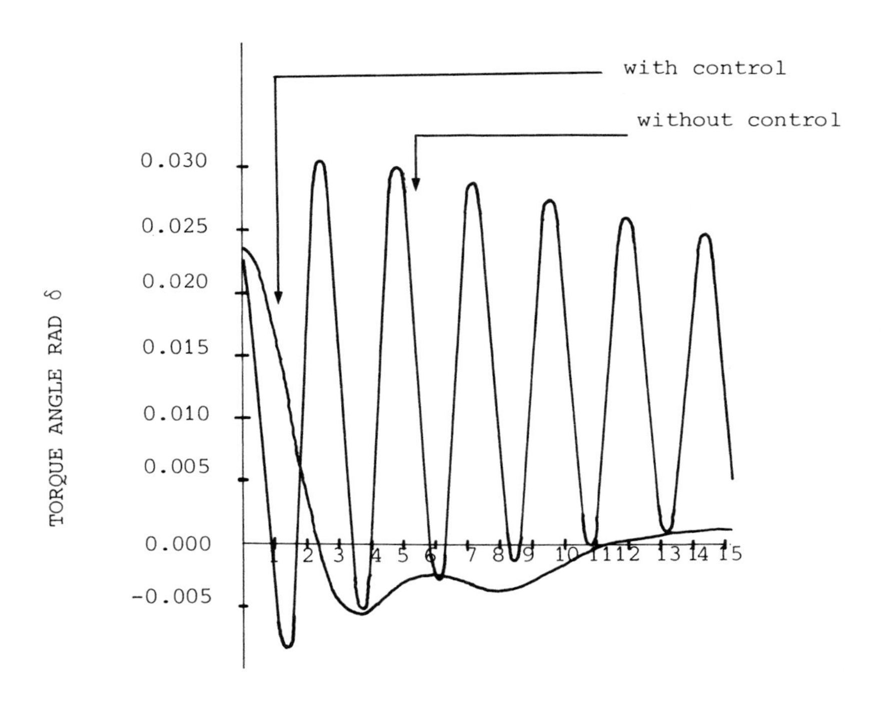

Fig. 3

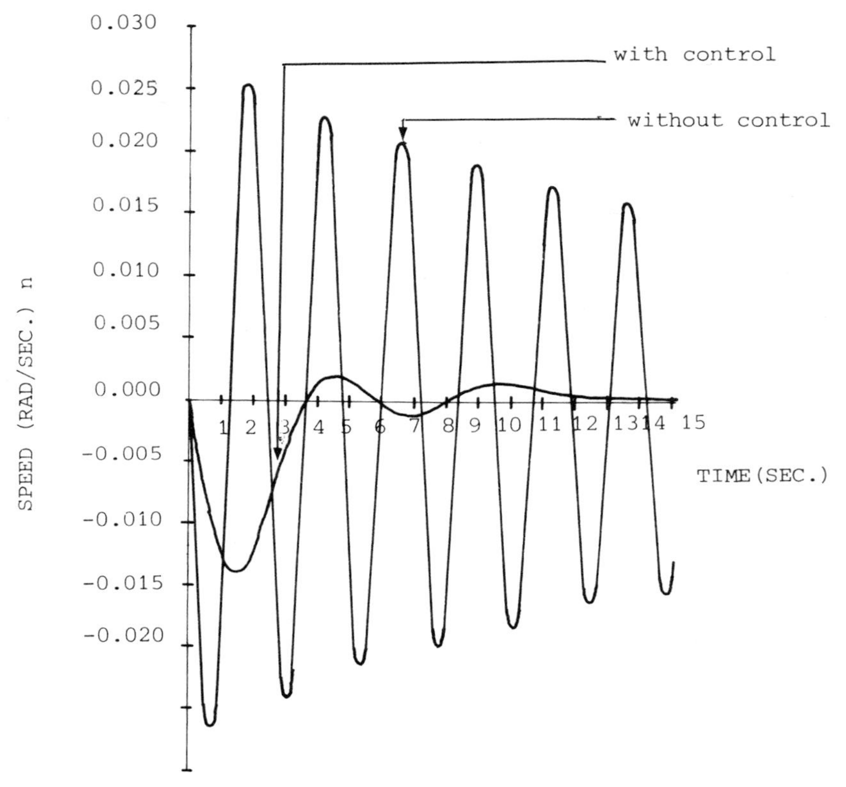

Fig. 3

5. CONCLUSION

For ease of exposition, the single machine power system model and state feedback control investigated are somewhat idealised. It is emphasised that the methods of parametric eigenstructure assignment described herein have direct application to augmented state-space models incorporating integral action[3] and to sampled-data or discrete identified state-space models of power systems.[3]

Developments of the parametric approach to linear state feedback design are presented in [9-13]. A noteworthy feature of the feedback gain matrix is that it possesses redundancy. The minimum number of free parameters (degrees of freedom) is determined for different classes of multivariable systems in[9-11]. This specification of the minimum number of free parameters has obvious advantages when we come to systematically choose them to satisfy design criteria, other than closed-loop eigenvalue assignment, such as minimum gain control.[12] Also, preliminary examples indicate that the selection of the free parameters can be used to improve the robustness of the system in the face of model parameter variations. These and other analytical issues, such as output feedback control, are currently under investigation for general multivariable systems.

6. REFERENCES

1. Mousa, H.A.M. and Yu, Y-N., (1972) "Optimal power system stabilization through excitation and/or governor control", IEEE Trans. Power Apparatus and Systems, vol. PAS-91, pp. 1164-1174.

2. Newton, M.E. and Hogg, B.W., (1976) "Optimal control of a micro-alternator system", IEEE Trans. Power Apparatus and Systems, vol. PAS-95, pp. 1822-1833.

3. Pullman, R.T. and Hogg, B.W., (1979) "Discrete state-space controller for a turbogenerator," Proc. IEE, vol. 126, pp. 87-92.

4. Pai, M.A., Prabhu, S.S. and Ramana, I.V., (1976) "Modal control of a power system", Int. J. Systems Sci., vol. 6, pp. 87-100.

5. Kumar, A.B.R. and Richards, E.F., (1982) "An optimal control law by eigenvalue assignment for improved dynamic stability in power systems", IEEE Trans. Power Apparatus and Systems, vol. PAS-101, pp. 1570-1577.

6. Franksen, O.I., Falster, P. and Evans, F.J., (1979) "Structural aspects of controllability and observability, Parts I and II," J. Franklin Inst., vol. 308, pp. 79-124.

7. Fahmy, M.M. and O'Reilly, J., (1982) "On eigenstructure assignment in linear multivariable systems', IEEE Trans. Automat. Control, vol. AC-27, pp. 690-693.

8. Fahmy, M.M. and O'Reilly, J., (1983) "Eigenstructure assignment in linear multivariable systems--a parametric solution", Proc. 21st IEEE Conf. on Decision and Control, Orlando, Florida, 8-10 Dec. 1982, pp. 1308-1311; also, IEEE Trans. Automat. Control, vol. AC-28, pp. 990-994.

9. Fahmy, M.M. and O'Reilly, J., (1984) "Organisation of the non-uniqueness of a canonical structure", Int. J. Systems Sci., vol. 14, pp. 585-601.

10. Fahmy, M.M. and O'Reilly, J., (1983) "Dead-beat control of linear discrete-time systems", Int. J. Control, vol. 37, pp. 685-705.

11. Fahmy, M.M. and O'Reilly, J., (1985) "The minimum number of degrees of freedom in state feedback control", Int. J. Control, 41, to appear.

12. Fahmy, M.M. and O'Reilly, J., (1983) "Use of the design freedom in time-optimal control", System and Control Letters, vol. 3, pp. 23-30.

13. Fahmy, M.M. and Tantawy, H., (1984) "Eigenstructure assignment via linear state feedback control", Int. J. Control, 40, 161-178.

14. Roppenecker, G., (1981) "Polevorgable durch zustandsruck - fuhrung", Regelungstechnik, vol. 29, pp. 228-233.

RECENT RESULTS ON POLE ASSIGNMENT IN DESCRIPTOR SYSTEMS

L.R. Fletcher
(Department of Mathematics and Computer Science, University of Salford)

ABSTRACT

The existence and uniqueness of solutions of the initial value problem for the implicit system $J\dot{x} = Ax$ of linear time-invariant ordinary differential equations cannot be taken for granted when J is singular. Moreover the number of linearly independent solutions depends on rather subtle features of the matrix pair (A,J); it need not be equal to the rank of J for example. For control systems in descriptor form these issues are still more complicated as they can be very strongly affected by feedback. In this paper we discuss the pole assignment problem with special attention being paid to these issues.

1. INTRODUCTION

In this paper we consider the continuous time descriptor system

$$J\dot{x} = Ax + Bu, \quad x(0) = x_o \tag{1}$$

$$y = Cx \tag{2}$$

Here x, u, y are functions of t with values in $\mathbb{R}^n$, $\mathbb{R}^m$, $\mathbb{R}^p$, respectively, and J, A, B, C are real matrices of the appropriate sizes. Many examples of sytems where descriptor models are useful were described by Luenberger[3]; moreover there are intriguing, but unexplored, similarities with the work of Manitius[4] on retarded differential systems.

To introduce the main concerns of this paper it will be convenient to review the issues of existence and uniqueness of solutions of the initial value problem

$$J\dot{x} = Ax, \quad x(0) = x_o. \tag{3}$$

Wilkinson[6] discusses these issues in detail and we highlight the following points from his discussion:

- if (3) has a solution then it is unique provided that

$$\det(A-\lambda J) \not\equiv 0 \tag{4}$$

  regarded as a polynomial in λ; if (4) is satisfied we shall say that the pair (A,J) is regular.

- if (3) has a solution and (4) is not satisfied then the solution set of (3) contains a number of arbitrary functions;

- the initial value problem (3) has a solution provided x_o lies in a certain subspace of $\mathbb{R}^n$;

- the dimension of this subspace is less than or equal to rank J, but equality is <u>not</u> necessary.

Suppose we now add a feedback control law

$$u = -K\underline{y} \tag{5}$$

in the system (1), (2). We can refer to the discussion above with regard to the existence and uniqueness of solutions of the closed loop initial value problem

$$J\dot{x} = (A-BKC)x, \quad x(0) = x_o. \tag{6}$$

First we observe that the word "control" can only be used in a meaningful way when we take care to ensure that

$$\det(A-BKC-\lambda J) \not\equiv 0 \tag{7}$$

(regarded as a polynomial in λ). It is not necessarily the case that provided (4) is satisfied then (7) is satisfied independently of K. Conversely, even when (4) is not satisfied it is possible that a feedback K can be found to fulfil some useful control objective such as pole assignment and to satisfy (7). With regard to the space of admissible initial conditions for (6), its dimension is <u>not</u>, in general, independent of K.

In this paper we discuss (finite) pole assignment in descriptor systems paying careful attention to the points we have just made. We shall investigate in particular the questions of controllability and observability for the system (1), (2); this we do in Section 2 by reference to the work of Yip and Sincovec[7]. In Section 3 we discuss regularity and point out that a controllable and observable system (1), (2) can be made regular by means of a suitable feedback (5). We give a simple characterisation of the reachable and observable subspaces of the system (1), (2) in Section 4. This enables us to give an upper bound on the number of finite closed loop poles and we state a condition which ensures that this number is equal to rank J. In Section 5 we state a basic theorem concerning pole assignment by output feedback and briefly indicate how it might be proved.

It is a pleasure to acknowledge many stimulating conversations on the topic of descriptor systems with Dr. E.K.W. Chu and Dr. N.K. Nichols of Reading University, Mr. S.N. Farrington of Marconi plc., Dr. J. Kautsky of the Flinders University of South Australia, Dr. V.A. Nye of Strathclyde University and Dr. P. van Dooren of Phillips Research Laboratories, Brussels. Many of these conversations would not have been possible without financial support from the Science and Engineering Research Council.

2. CONTROLLABILITY AND OBSERVABILITY

In this section we give our definitions of controllability and observability for the system (1), (2) and relate them to those given by other workers. Since we are taking pole assignment as our model control problem it is appropriate to consider only systems which have no uncontrollable or unobservable modes so we have

Definition 2.1

We shall say that the system (1) is controllable if for any $\mu \in \mathbb{C}$, $v \in \mathbb{C}^n$

$$v^T A = \mu v^T J, \quad v^T B = 0 \Rightarrow v = 0 \tag{8}$$

Definition 2.2

We shall say that the system (1), (2) is observable if for any $\mu \in \mathbb{C}$, $v \in \mathbb{C}^n$

$$Av = \mu Jv, \; Cv = 0 \Rightarrow v = 0. \tag{9}$$

Definition 2.1 is equivalent to the concept of (R-D) reachable introduced by Pandolfi[5] and to R - controllability in the paper of Yip and Sincovec[7]. Definition 2.2 is equivalent to Yip and Sincovec's definition of observability[7].

3. REGULARITY

In this section we discuss regularity of the closed loop system obtained from (1), (2) with feedback (5). We first note a fundamental result:

Theorem 3.1

Suppose that the system (1), (2) is controllable and observable and that $\lambda \in \mathbb{R}$. Then there is a real matrix K such that

$$\det(A-BKC-\lambda J) \neq 0.$$

Thus we can, and will, assume that if the system (1), (2) is controllable and observable then (4) is satisfied, a preliminary feedback having been applied if necessary. It is important now to have some criterion which guarantees that (7) is satisfied as it is not easy to check. Suppose $\lambda_1,\ldots,\lambda_q$ are distinct complex numbers such that

$$\det(A-BKC-\lambda_i J) = 0 \quad i=1,\ldots,q \tag{10}$$

Then there exist vectors $s_1,\ldots,s_q$; $t_1,\ldots,t_q$ such that

$$(A-BKC-\lambda_i J)s_i = 0 \quad i=1,\ldots,q \tag{11}$$

$$t_j^T(A-BKC-\lambda_j J) = 0 \quad j=1,\ldots,q \tag{12}$$

Then[1]:

Theorem 3.2

Suppose $Js_1,\ldots,Js_q$ are linearly dependent. Then (7) is not satisfied. Dually if $t_1^TJ,\ldots t_q^TJ$ are not linearly independent then (7) is not satisfied.

The author has shown[2] in connection with state feedback that the necessary conditions for (7) implied by Theorem 2.2 are also sufficient (see Theorem 4.3 below).

4. THE REACHABLE AND "OBSERVABLE" SUBSPACES

In this section we give simple geometric characterisations of the reachable and "observable" subspaces for a controllable and observable system (1), (2). We shall then relate the possible closed loop eigenvectors (11), (12) to these subspaces and thereby arrive at an upper bound on q in (10) if (7) is to be satisfied. Clearly this upper bound must be smaller than rank J and we re-iterate that it can be strictly smaller.

Definition 4.1

Let $\mathcal{R}$ be a subspace $\mathbb{R}^n$ minimal subject to satisfying

$$J\,\mathcal{R} \subseteq A\,\mathcal{R} \tag{13}$$

$$\mathcal{B} \subseteq A\,\mathcal{R} \tag{14}$$

where $\mathcal{B}$ denotes the space spanned by columns of B. Dually, let σ be a subspace of $\mathbb{R}^n$ minimal subject to satisfying

$$J^T\sigma \subseteq A^T\sigma \tag{15}$$

$$\mathcal{C} \subseteq A^T\sigma \tag{16}$$

where $\mathcal{C}$ denotes the space spanned by the columns of C^T.

It is easy to see that the spaces $\mathcal{R}$, σ are uniquely determined. It is less easy to see that $\mathcal{R}$ is the subspace of states reachable from $x(0) = 0$; indeed the only proof of this fact known to the author uses Theorem 4.1 below and Theorem 2.1 of Pandolfi[5] so is rather roundabout and ungeometrical in character. A more direct geometrical proof would be very valuable.

The subspace σ is introduced, in the first instance, for reasons of mathematical elegance and its system theoretic significance can only be judged at present by assessing the dual of the reachable subspace in relation to observability. On this basis σ would appear to consist of the space of those linear functionals of the state whose values can be inferred from the values of the output function y(t). Again a direct proof of this would be very valuable.

The subspaces $\mathcal{R}$ and σ are closely related to vectors s_i and t_j satisfying (11) and (12). Whatever the matrix K in (11) and (12), these vectors must be in the subspaces.

$$s_i \in \mathcal{S}_J(\lambda_i) = (s\,|\,A-\lambda_i J)s \in \mathcal{B}) \tag{17}$$

$$t_j \in \mathcal{T}_J(\lambda_j) = (t\,|\,(A^T-\lambda_j J^T)t \in \mathcal{C}) \tag{18}$$

Suppose $s \in \mathcal{S}_J(\lambda_i)$ then

$$(A-\lambda_i J)s = Bv$$

so $x(t) = s\,e^{\lambda_i t}$, $u(t) = v\,e^{\lambda_i t}$ satisfy (1). In particular $s = x(0)$ is an allowable initial condition. Since in a controllable system the subspaces of states reachable from 0 and of allowable initial conditions coincide[5,7] we have demonstrated that

$$\mathcal{S}_J(\lambda_i) \subseteq \mathcal{R} \text{ for all } \lambda_i \in \mathbb{C} \tag{19}$$

and dually that

$$\mathcal{J}_J(\lambda_j) \subseteq \sigma \text{ for all } \lambda_j \in \mathbb{C} \tag{20}$$

Inclusions (19) and (20) can also be established by easy linear algebraic arguments; much more subtle linear algebraic arguments are required to prove

Theorem 4.1[2]

$$\mathcal{R} = \sum_{\lambda_i \in \mathbb{C}} \mathcal{S}_J(\lambda_i), \quad \sigma = \sum_{\lambda_i \in \mathbb{C}} \mathcal{J}\, J(\lambda_j) \tag{21}$$

Combining Theorems 3.2 and 4.1 we have

Theorem 4.2

If (4) is satisfied the integer q in (10) must satisfy

$$q \leq \min(\dim J\,\mathcal{R},\ \dim J^T\sigma) \tag{22}$$

Very simple examples show that the expression on the right-hand side of (22) can be strictly smaller than rank J. In the case of state feedback; that is, $C = I$; (22) can be replaced by

$$q \leq r = \dim J\,\mathcal{R} \tag{23}$$

Furthermore equality can be attained in (23) when feedback is applied:

Theorem 4.3[2]

Suppose the system (1) is controllable and $\Lambda = (\lambda_1,\ldots,\lambda_r)$ is a set of $r = \dim J\mathcal{R}$ complex numbers closed under complex conjugation. Then there exists a real m x n matrix K such that the roots of the polynomial

$$\det(A-BK-\lambda J)$$

are $\lambda_1, \lambda_2, \ldots \lambda_r$ and no other elements of $\mathbb{C}$.

The "expected" value of r in (23) is rank J in that examples in which r < rank J need to be specially constructed. Our next result gives a condition on A, B, J which is necessary and sufficient for r to be equal to rank J:

Theorem 4.4

Suppose the system (1) is controllable. Then the condition

$$v^T J = 0, \quad v^T AS_\infty = 0, \quad v^T B = 0 \Longleftrightarrow v = 0 \tag{24}$$

where S_∞ is a matrix whose columns are a basis of ker J, is equivalent to dim $J\mathcal{R}$ = rank J.

Dually, suppose the system (1), (2) is observable. Then the condition

$$Jv = 0, \quad T_\infty Av = 0, \quad Cv = 0 \Longleftrightarrow v = 0 \tag{25}$$

where T_∞ is a matrix whose columns are a basis of ker J^T, is equivalent to dim $J^T\sigma$ = rank J.

5. POLE ASSIGNMENT BY OUTPUT FEEDBACK

Theorem 4.3 gives the solution, at least in terms of an existence theorem, to the state feedback pole assignment. No such complete answer is known for the output feedback pole assignment problem. In this section we describe a partial answer. As always we shall assume that the system (1), (2) is controllable and observable; it greatly simplifies the exposition if we also assume that conditions (24) and (25) are satisfied so that the upper bound in (22) is equal to rank J.

An account of the results of this section without these additional assumptions is available[1].

Theorem 5.1

Suppose the system (1), (2) is controllable and observable, satisfies (24) and (25) in addition

$$m + p > n \tag{26}$$

Then given any set of $\Lambda = (\lambda_1, \lambda_2, \ldots, \lambda_r)$, r = rank J of complex numbers closed under complex conjugation there exists a real m x p matrix K such that the roots of the polynomial

$$\det(A-BKC-\lambda J)$$

are $\lambda_1, \lambda_2, \ldots, \lambda_r$ and no other elements of $\mathbb{C}$.

If (24) and (25) are not satisfied this theorem remains true provided (26) is strengthened somewhat and
$r = \min(\dim J\,\mathcal{R}, \dim J^T \sigma)$.

We give a brief indication of the proof of Theorem 5.1. Define a subspace $\mathcal{C}_o$ of $\mathbb{R}^n$ by

$$\mathcal{C}_o = \{c : A^T c,\ J^T c \ \varepsilon\ \mathcal{C}\}$$

It is easy to see that $\dim \mathcal{C}_o \geq 2p - n$. Because of this and (26) it can be assumed that $\mathcal{C}_o \neq \{0\}$; for either $2p - n > 0$ or $2m - n > 0$ and interchanging B and C^T and transposing A and J does not affect the mathematical problem at hand.

Now suppose there exists a subset $\{\lambda_1, \ldots, \lambda_\ell\}$ of Λ closed under complex conjugation and vectors $s_i \ \varepsilon\ s_J(\lambda_i)$ for $i = 1, \ldots, \ell$ such that

$$Js_1, \ldots Js_\ell \text{ are linearly independent} \tag{27}$$

$$s_i = s_j \text{ if } \lambda_i = \lambda_j \text{ for } 1 \leq i,\ j \leq \ell \tag{28}$$

$$\dim < \mathcal{B}\ ,\ s_1, \ldots s_\ell > = \min(n, m+\ell) \tag{29}$$

$$< s_1, \ldots, s_r > \cap \mathcal{C}_o^{\perp} = \{0\} \tag{30}$$

where "<...>" denotes "the subspace spanned by ...". It is fairly easy to show that such vectors exist, at least in the case in which $\ell = 1$ and $\lambda_1 \ \varepsilon\ \mathbb{R}$.

Now let $S = [s_1, \ldots, s_\ell]$ be the n x ℓ matrix with the columns indicated. For $i = 1, \ldots, \ell$

$$(A-\lambda_i J)s_i = Bv_i$$

for some $v_i \in \mathbb{C}^m$. By (3), CS is of rank ℓ so there exists a real (by (28)) m x p matrix K_1 such that

$$K_1 C s_i = v_i \quad i = 1,\ldots,\ell$$

Then $\lambda_1,\ldots\lambda_\ell$ are J-eigenvalues of $A-BK_1C$ with J-eigenvectors $s_1,\ldots s_\ell$. Moreover K_1 can be chosen so that $\det(A-BK_1C-\lambda J)$ is not identically zero.

Now let U, V, W, P be real matrices of sizes $n \times \ell$, $n\times(n-\ell)$, $n\times(n-\ell)$, $(p-\ell)\times p$, respectively, of full rank and such that

$$U^T JS = I_\ell \;,\; V^T JS = O,\; W^T S = O,\; PCS = O,\; V^T V = W^T W = I_{n-\ell}$$

It is easy to see how these matrices can be constructed and that for any $m\times(p-\ell)$ matrix L

$$\begin{bmatrix} U^T \\ \hline V^T \end{bmatrix} (A-B(K_1+LP)C-\lambda J)\ [S\ W]$$

$$= \begin{bmatrix} \Lambda_\ell - \lambda I_\ell & * \\ O & A^*-B^*LC^*-\lambda J^* \end{bmatrix} \tag{31}$$

where $A^* = V^T(A-BK_1C)W$, $B^* = V^T B$, $C^* = PCW$, $J^* = V^T JW$ and

$$\Lambda\ell = \begin{pmatrix} \lambda_1 & & & O \\ & \lambda_2 & & \\ & & \ddots & \\ O & & & \lambda_\ell \end{pmatrix}$$

Then the crucial fact is

Theorem 5.2

The system

$$J^* x = A^* x^* + B^* u^* \tag{32}$$

$$y^* = C^* x^* \tag{33}$$

has all the properties hypothesised of the system (1), (2) in the statement of Theorem 5.1.

Since the system (32), (33) has smaller dimensional state space than (1), (2) we may argue by induction that a matrix L can be found so that the south-east corner of the matrix (31) is singular when $\lambda = \lambda_{\ell+1}, \lambda_{\ell+2}, \ldots, \lambda_n$ and for no other elements of $\mathbb{C}$. Then $K = K_1 + LP$ is the required output feedback matrix.

6. REFERENCES

1. Farrington, S.N., (1983) MSc Thesis, University of Salford.

2. Fletcher, L.R., Pole assignment in degenerate linear systems, to appear in Linear and Multilinear Algebra.

3. Luenberger, D.G., (1977) Dynamic equations in descriptor form, IEEE Trans. Aut. Control, vol. AC-22, p. 312-321.

4. Manitius, F., (1982) F-controllability and observability of linear retarded systems, Appl. Math. Optim., Vol, 19, p 73-95.

5. Pandolfi, L., (1980) Controllability and stabilisation for linear systems of algebraic and differential equations, J. Optimisation Theory and Applics., vol. 30, p 601-620.

6. Wilkinson, J.H., (1978) Linear differential equations and Kronecker's canonical form, in "Proceedings of the Symposium on Recent Advances in Numerical Analysis at Mathematics Research Centre, Madison, Wisconsin" (C. de Boor and G.H. Golub eds) Academic Press, New York.

7. Yip, E.L. and Sincovec, R.F., (1981) Solvability, controllability and observability of continuous descriptor systems, IEEE Trans Aut. Control, vol. AC-26, p 702-707.

ROBUST POLE ASSIGNMENT BY OUTPUT FEEDBACK

E.K.-W. Chu and N.K. Nichols
(University of Reading)

and

J. Kautsky
(Flinders University of South Australia)

ABSTRACT

The numerical solution of the pole assignment problem by static output feedback for multi-variable time-invariant systems is discussed. A 'robust' solution is determined such that the closed-loop eigenvalues are as insensitive as possible to perturbations in the system data. A numerical algorithm which produces an approximate solution is presented, and theoretical bounds on the errors are derived. A numerical example is given.

1. INTRODUCTION

The problem of pole assignment by output feedback in a multi-input linear control system is essentially underdetermined, with several degrees of freedom in the solution. For a *robust* design it is desirable to choose a feedback such that the assigned poles are as insensitive as possible to perturbations in the coefficient matrices of the closed loop system. A number of formally constructive methods for pole assignment by output feedback are described in the literature [3, 11,13] but these procedures are not in general *stable* for numerical computation and do not necessarily lead to *robust* solutions to the problem.

Recently we have developed reliable numerical methods for determining robust, or well-conditioned solutions to the *state* feedback pole assignment problem [6,7,8]. In this paper we extend these techniques to the problem of robust pole placement by *output* feedback. The algorithm we describe obtains the required feedback by assigning to the closed loop system matrix linearly independent eigenvectors corresponding to the assigned eigenvalues, or poles, such that the modal matrix of eigenvectors is as well-conditioned as possible [15]. This criterion guarantees that the poles are as insensitive to perturbations in the system and gain matrices as is feasible, and also that the resulting feedback gains and corresponding transient response are as reasonably bounded as may be expected, and that a lower bound on the stability margin is maximised.

In the next section the robust pole assignment problem is defined in detail and theoretical considerations are discussed. In Section 3 the numerical algorithm is described. Applications and results are presented in Section 4, and concluding remarks follow in Section 5.

2. ROBUST OUTPUT FEEDBACK

We consider the time-invariant, linear multivariable system with dynamic state equations

$$\dot{\underline{x}} = A\underline{x} + b\underline{u}$$
$$\underline{y} = C\underline{x} \qquad (2.1)$$

where $\underline{x}$, $\underline{u}$, $\underline{y}$ are n, m and p dimensional vectors, respectively, and A,B,C, are real constant matrices of compatible orders. Matrices B and C are assumed, without loss of generality, to be of full rank. To modify the poles of the system, an output feedback $\underline{u} = K\underline{y} + \underline{v}$ may be used, where the feedback matrix K is chosen such that the modified (closed loop) dynamic system

$$\dot{\underline{x}} = (A + BKC)\,\underline{x} + B\underline{v} \qquad (2.2)$$

has the desired poles. The output feedback pole assignment problem for system (2.1) is stated precisely by

<u>Problem 1</u>. Given real matrices (A,B,C) of order ($n \times n$, $n \times m$, $p \times n$) and a set of n self-conjugate complex numbers $L = \{\lambda_1, \lambda_2, \ldots, \lambda_n\}$, find a real $m \times p$ matrix K such that the eigenvalues of $A + BKC$ are λ_j, $j=1,2,\ldots,n$. □

Conditions for the existence of solutions to Problem 1 have been derived elsewhere [4,5,9]. If the system (2.1) is completely controllable and observable and if $m + p > n$, then for any set L of eigenvalues there exists at least one solution K such that the eigenvalues of $A + BKC$ are arbitrarily close to those of L. If the dimensional requirement is not satisfied, then solutions may still exist for certain sets of poles to be assigned, and, in any case, by the use of a dynamic compensator, an enlarged output feedback problem can always be obtained, which does satisfy the requirement [12,14].

Our aim here is to develop methods for finding a feedback K, solving Problem 1, such that the closed loop system is *robust* in the sense that its poles are as insensitive to perturbations as possible. We let $\underline{x}_j$ and $\underline{y}_j$, $j=1,2,\ldots,n$, be the right and left eigenvectors of the closed loop system matrix $M \equiv A + BKC$ corresponding to eigenvalue $\lambda_j \in L$. If M is *non-defective* that is, M has n linearly independent eigenvectors, then the sensitivity of the eigenvalue λ_j to perturbations in the components of A,B,C and K is known[15] to depend upon the magnitude of $c_j \equiv \|\underline{y}_j\|_2 \|\underline{x}_j\|_2 / |\underline{y}_j^T \underline{x}_j| \geq 1$. A general measure and upper bound for the sensitivities is then given by the condition number $\kappa_2(X) \equiv \|X\|_2 \|X^{-1}\|_2$ of the (non-singular) matrix $X = [\underline{x}_1, \underline{x}_2, \ldots, \underline{x}_n]$ of eigenvectors. The sensitivities of the assigned poles can thus be controlled by an appropriate choice of the eigenvectors of the system. We observe that system matrices which are defective are necessarily less robust than those which are non-defective. The robust pole assignment problem is thus formulated as

Problem 2. Given (A,B,C) and L (as in Problem 1), find real matrix K and non-singular matrix X satisfying

$$(A + BKC)X = X\Lambda \quad , \tag{2.3}$$

where $\Lambda = \mathrm{diag}\{\lambda_1, \lambda_2, \ldots, \lambda_n\}$, such that some measure ν of the conditioning of the eigenproblem is optimized. □

We remark that no restriction on the controllability of (A,B) is made. Provided that the uncontrollable modes are included in the set L to be assigned (with their correct multiplicity), then solutions to the feedback problem may exist. Although the uncontrollable modes of the system cannot be affected by the feedback K, the corresponding eigenvectors may be modified, and the conditioning of uncontrollable modes may be improved by an appropriate choice of K.

Conditions under which a specific non-singular matrix X can be assigned to the problem are given by

Theorem 1. Given Λ and X non-singular, then there exists K, a solution satisfying (2.3) if and only if

$$U_1^T(AX - X\Lambda) = 0 \tag{2.4}$$

$$(X^{-1}A - \Lambda X^{-1})V_1 = 0 \tag{2.5}$$

where

$$B = [U_0, U_1] \begin{bmatrix} Z_B \\ 0 \end{bmatrix}, \qquad C = [Z_C, 0] \begin{bmatrix} V_0^T \\ V_1^T \end{bmatrix} \tag{2.6}$$

with $U = [U_0, U_1]$, $V = [V_0, V_1]$ orthogonal, and Z_B, Z_C non-singular. Then K is given explicitly by

$$K = Z_B^{-1} U_0^T (X\Lambda X^{-1} - A) V_0 Z_C^{-1} \quad . \tag{2.7}$$

Proof. The existence of decompositions (2.6) follows from the assumption that B and C are of full rank. From (2.3), K must satisfy

$$BKC = X\Lambda X^{-1} - A.$$

Then pre- and post-multiplication by U^T and V, respectively, gives

$$\begin{bmatrix} Z_B \\ 0 \end{bmatrix} K [Z_C, 0] = \begin{bmatrix} U_0^T \\ U_1^T \end{bmatrix} (X\Lambda X^{-1} - A) [V_0, V_1] \quad ,$$

from which (2.4), (2.5) and (2.7) follow by comparison of components, since X, U and V are all invertible. □

We note that the matrix $Y = [\underline{y}_1, \underline{y}_2, \ldots, \underline{y}_n]$ of left eigenvectors (with an appropriate scaling) is now given by $Y^T = X^{-1}$. An immediate consequence of Theorem 1 is thus

Corollary 1. The right and left eigenvectors, $\underline{x}_j$ and $\underline{y}_j$, of $A + BKC$ corresponding to the assigned eigenvalue $\lambda_j \in L$ must be such that

$$\underline{x}_j \in S_j \equiv N\{U_1^T(A - \lambda_j I)\}, \; \underline{y}_j \in T_j \equiv N\{V_1^T(A^T - \lambda_j I)\}, \tag{2.8}$$

where $N\{\cdot\}$ denotes null space.

A further direct consequence of Theorem 1 is

Corollary 2. The gain matrix K and the transient response of the closed loop system (2.2), where $\underline{x}(0) = \underline{x}_0$ and $\underline{v}(t) \equiv 0$, satisfy the inequalities

$$\|K\|_2 \leq (\|A\|_2 + \max_j\{|\lambda_j|\}\kappa_2(X))/\sigma_m\{B\}\sigma_m\{C\} \quad ,$$

where $\sigma_m\{\cdot\}$ denotes minimal singular value, and

$$\|\underline{x}(t)\|_2 \leq \kappa_2(X) \max_j\{|e^{\lambda_j t}|\} \quad \|\underline{x}_0\|_2 \quad .$$

The second corollary shows that K and $\underline{x}(t)$ can be bounded in terms of the system data and the condition number, $\kappa_2(X)$, of the assigned eigenvectors, and, therefore that minimising the conditioning of the eigenproblem (2.3) also minimizes a bound on the feedback gains and the magnitude of the transient response.

A lower bound on the stability margin of the closed loop system in terms of the conditioning can also be derived. We have

Theorem 2. The return difference $I + G(s) + \hat{\Delta}(s)G(s)$ of the disturbed closed loop system, where $G(s) = -KC(sI - A)^{-1}B$, remains non-singular at $s = i\omega$ for disturbances which satisfy $\|\hat{\Delta}(i\omega)\|_2 < \delta(K)$, where $\delta(K)$ is bounded by

$$\delta(K) \geq \min_j\{\mathrm{Re}(-\lambda_j)\}/\kappa_2(X) \quad \|B\|_2 \|C\|_2 \|K\|_2 .$$

Proof. The proof follows as for Theorems 6-7[8], taking $F = KC$. □

From Corollary 2 and Theorem 2 it can be seen that if the conditioning of the eigenproblem is minimized, then a lower bound on the stability margin is maximised over all feedback matrices which assign the given (stable) eigenvalues.

For given data A, B, C and L, the minimal conditioning that can be achieved is limited, however. If we let the columns of matrix S_j and matrix T_j be orthonormal bases for the spaces S_j and T_j, defined in Corollary 1, and define $S = [S_1, S_2, \ldots, S_n]$, and $T = [T_1, T_2, \ldots, T_n]$, then a lower bound on the achievable value of $\kappa_2(X)$ is given by

$$\kappa_2(X) \geq \max\{\kappa_2(S)/\sqrt{n},\ \kappa_2(T)/\sqrt{n}\} \quad . \tag{2.9}$$

This result follows directly from Theorem 8[8]. Although this lower bound cannot necessarily be realised, it does give an indication of the suitability of the set L of poles for assignment.

3. NUMERICAL ALGORITHM

We now consider the practical implementation of the theory of Section 2 and describe a numerical algorithm for constructing approximate solutions to Problem 2. The procedure consists of three basic steps.

Step A: Compute the decompositions (2.6) of B and C; construct orthonormal bases S_j, $\hat{S}_j$ and T_j, $\hat{T}_j$ for the spaces S_j, T_j, defined by (2.8), and their complements, $j=1,2,\ldots,n$.

Step X: Select $\underline{x}_j \in S_j$ with $\|\underline{x}_j\|_2 = 1$, $j=1,2,\ldots,n$ such that $X = [\underline{x}_1, \underline{x}_2, \ldots, \underline{x}_n]$ is *well-conditioned* and $\underline{y}_j \in T_j$, $\forall_j$

where $Y \equiv [\underline{y}_1, \underline{y}_2, \ldots, \underline{y}_n] = X^{-T}$.

<u>Step K</u>: Find matrix $M \equiv A + BKC$ by solving $MX = X\Lambda$ and compute K from (2.7).

The first and third steps are achieved using standard library software for computing QR or SVD (singular value) decompositions of matrices and for solving systems of linear equations[2]

The key step is Step X. The solution is found by an iteration in which each vector $\underline{x}_j$ is replaced by a new vector of unit length belonging to the subspace S_j such as to minimise the functional

$$F = F_1 + F_2 = \| DX^{-1} \|_F^2 + \sum_j \omega_j^2 \| \hat{T}_j^T \underline{y}_j \|_2^2 \quad , \quad (3.1)$$

where $D = \mathrm{diag}\{d_1, d_2, \ldots, d_n\}$, and $\| \cdot \|_F$ is the Frobenius norm. Since $Y^T = X^{-1}$ and $\| \underline{x}_j \|_2 = 1$, we have $c_j^2 = \| \underline{y}_j \|_2^2$ and $F_1 = \sum_j d_j^2 c_j^2$, and, therefore, minimising F_1 corresponds to minimising a weighted sum of the squares of the condition numbers. Without additional constraints, this can be achieved explicitly by a rank one update of X^{-1}, as for the state feedback problem[8]. In the case of output feedback it is necessary also that the left eigenvectors belong to given subspaces, and since $\| \hat{T}_j^T \underline{y}_j \|_2^2 / \| \underline{y}_j \|_2^2$ measures the minimum distance between $\underline{y}_j$ and a vector in subspace T_j, we require $F_2 = 0$ for a solution to the eigenproblem. The method used is in essence, therefore, a penalty function method for minimising the measure F_1 of the conditioning of X subject to the equality constraint $F_2 = 0$.

The functional F can be minimised *explicitly* at each step of the iteration over all choices of $\underline{x}_j$ belonging to S_j, $j=1,2,\ldots,n$. Details of the procedure are given elsewhere[1]. The computed values of F are non-increasing and the procedure is repeated until the maximum possible reduction in F is smaller than a specified tolerance. The solution obtained does not exactly assign the required eigenvalues (unless F_2 is exactly zero). Provided the measure F is small, however, the computed feedback assigns poles close to those specified and such that the sensitivities of these poles are small. More specifically we have

<u>Theorem 3</u>. Given $\Lambda = \mathrm{diag}\{\lambda_1, \lambda_2, \ldots, \lambda_n\}$ and $X = [\underline{x}_1, \underline{x}_2, \ldots, \underline{x}_n]$

non-singular, such that $\underline{x}_j \in S_j$ and $\|\underline{x}_j\|_2 = 1$, then K defined by (2.7) implies

$$(A + BKC)X - X\Lambda = -EX \quad , \qquad (3.2)$$

where $\|E\|_F^2 \leq \sum_j r_j^2 \|\hat{T}_j^T \underline{y}_j\|_2^2$, with r_j fixed constants.

Proof. From (2,4), (2.6) and (2.7) we obtain

$$BKC = (X\Lambda Y^T - A)V_0V_0^T = (X\Lambda Y^T - A)(I - V_1V_1^T) \quad ,$$

since $\underline{x}_j \in S_j$, and (3.2) follows if we write

$$E = X(\Lambda Y^T - Y^T A)V_1V_1^T \quad . \qquad (3.3)$$

Now the complement of the space T_j is spanned by $\hat{T}_j$ and hence $\|\underline{y}_j^T(\lambda_j I - A)V_1\|_2 = \|\underline{y}_j^T \hat{T}_j^T R_j\|_2$ where R_j is an invertible matrix. Taking norms in (3.3) then gives the required result with $r_j^2 = \|X\|_2^2 \|R_j\|_2^2 \leq n \|R_j\|_2^2$. □

This theorem shows that (Λ, X) gives the eigensystem of a perturbed matrix $A + BKC + E$, and we may deduce[15] that the actual eigenvalues $\tilde{\lambda}_j$ of $M = A + BKC$ satisfy $|\tilde{\lambda}_j - \lambda_j| = n\varepsilon c_j + O(\varepsilon^2)$, where $c_j = \|\underline{y}_j\|_2$ and $\varepsilon = \|E\|_F$. Thus, if F_1 and F_2 are sufficiently small, the computed feedback K assigns eigenvalues close to those specified.

The numerical method described here has been implemented using the system MATLAB [10], and a small package of executive files has been developed to carry out the three basic steps of the algorithm with various options.

4. APPLICATION

The procedure described in Section 3 has been applied to a number of examples. To illustrate the behaviour of the method we give here the results obtained for a simple test problem with dimensions $n = 4$, $m = 2$, $p = 3$:

$$A = \begin{bmatrix} 0 & 1 & 0 & 0 \\ 1 & 1 & 0 & 0 \\ -1 & 0 & 0 & 0 \\ 0 & 0 & 0 & 0 \end{bmatrix} \qquad B = \begin{bmatrix} 0 & 0 \\ 1 & 0 \\ 0 & 0 \\ 0 & 1 \end{bmatrix} \qquad C = \begin{bmatrix} 1 & 0 & 0 & 0 \\ 0 & 0 & 1 & 0 \\ 0 & 0 & 0 & 1 \end{bmatrix} .$$

We assign eigenvalue set $L = \{-1,-2,-3,-4\}$. The weights in functional F are taken to be $d_j = 1$, and $\omega_j = 10^4, \forall_j$.

After five iterations of the numerical procedure we obtain the solution

$$K = \begin{bmatrix} -46.65 & 41.39 & 13.48 \\ 36.32 & -31.69 & -10.92 \end{bmatrix},$$

which assigns the eigenvalues $\tilde{L} = \{-3.800, -2.918, -2.226, -0.9727\}$. The measured distance of the left eigenvectors from the required subspaces is given by $F_2 = 1,841$ and the conditioning $\kappa_2(X) = 778.2$, which is quite large. The maximum error in the eigenvalues is 11%, which reflects the poor conditioning. With weights $d_j = 1$ and $\omega_j = 10^6$, $\forall_j$, we obtain, after five iterations of the process,

$$K = \begin{bmatrix} -47.00 & 41.70 & 13.63 \\ 36.52 & -31.90 & -11.00 \end{bmatrix}$$

which assigns the eigenvalues $\tilde{L} = \{-3.998, -3.000, -2.002, -0.9970\}$, with a maximum error of 0.3%. The conditioning $\kappa_2(X) = 786.7$ is here slightly increased, but the distance measure $F_2 = 18.50$ is improved by $O(10^2)$, which is reflected in the increased accuracy of the pole placements.

Although these solutions are not very well conditioned, the results compare favourably with the solution,

$$K = \begin{bmatrix} -47 & 34 & 10 \\ 49 & -35 & -11 \end{bmatrix},$$

derived elsewhere[5], which has condition number $\kappa_2(X) = 10385.0$. In this case small perturbations in K may be expected to give very much worse errors in the eigenvalues assigned.

For other examples [1] solutions with better conditioning can be achieved, and the computed feedback assigns correspondingly better approximations to the required eigenvalues.

5. CONCLUSIONS

The concept of *robustness* for the problem of pole assignment by feedback in control system design is examined here. Generalisations of results obtained previously for the state feedback problem [8] are developed for the case of output feedback. It is shown that the extra degrees of freedom in the solution may be used to determine a well-conditioned closed loop system such that the sensitivities of the assigned poles to perturbations in the system and gain matrices are minimised. A numerical algorithm for constructing such a feedback is described. It is expected that improvements in the computational techniques can be made, and further investigations are presently being conducted.

6. REFERENCES

1. CHU, E.K-W, and Nichols, N.K. (1984) Robust Pole Assignment by Output Feedback. Dept. of Maths., Univ. of Reading, Num. Anal. Rpt.

2. Dongarra, J.J., Moler, C.B., Bunch, J.R., and Stewart, G.W. (1979) LINPACK Users Guide. SIAM, Philadelphia.

3. Fallside, F. (ed.) (1977) Control System Design by Pole-Zero Assignment, Academic Press.

4. Fletcher, L.R. (1979) "Placement des valeurs propres pour les systemes lineaires multivariables", C.R. Acad. Sci. Paris 289 (A), 499-501.

5. Fletcher, L.R. (1980) "An intermediate algorithm for pole placement by output feedback in linear multivariable control systems", Int. J. Control, 31, 1121-1136.

6. Kautsky, J,. Nichols, N.K., Van Dooren, P., and Fletcher, L.R. (1982) "Numerical methods for robust eigenstructure assignment in control system design", Numerical Treatment of Inverse Problems for Integral and Differential Equations, Birkhauser/Biston, pp. 171-178.

7. Kautsky, J., and Nichols, N.K. (1983) Robust Eigenstructure Assignment in State Feedback Control, Dept. of Maths., Univ. of Reading, Num. Anal. Rpt. NA 2/83.

8. Kautsky, J., Nichols, N.K., and Van Dooren, P. (to appear) "Robust Pole Assignment in Linear State Feedback".

9. Kimura, H. (1975) "Pole assignment by gain output feedback", IEEE Trans. Auto. Contr. AC-20, 509-518.

10. Moler, C.B. (1981) MATLAB Users Guide. Univ. of New Mexico, Dept. of Comp. Sci.

11. Munro, M. (ed.) (1979) Modern Approaches to Control System Design. Peter Peregrines.

12. Munro, N., and Novin-Hirbod, S. (1979) "Pole assignment using full-rank output feedback compensators", Int. J. Syst. Sci., 10, 285-306.

13. Patel, P.V., and Munro, N. (1982) Multivariable System Theory and Design, Pergamon.

14. Seraij, H. (1975) "An approach to dynamic compensator design for pole assignment", Int. J. Control, 21, 955-966.

15. Wilkinson, J.H. (1965) The Algebraic Eigenvalue Problem. Oxford University Press.

A COST FUNCTION DESIGN FOR REDUCED ORDER MODELLING

K. Warwick

(University of Newcastle upon Tyne)

ABSTRACT

A cost function is defined which enables a goodness of fit comparison to be made between the various possible reduced order models of a high order system. The function is based on a polynomial whose parameters determine the error between the system transfer function and its associated model. By including an operator specified weighting factor in the cost function, particular attributes of the system response can be highlighted in the response of its model.

1. INTRODUCTION

The design of controllers for systems of high order and their subsequent analysis can often lead to a considerable amount of expended effort. As an alternative to this, however, a low order model can be found which is, in some previously defined sense, a reasonable approximation to the original system.

Many approaches to the modelling problem have been made[1,8,11], most of these concentrating on a continuous-time system and model description, although discrete time modelling has also been investigated[5,10]. Generally these are based partly or wholly on matching as many power series parameters, between the system and model, as is possible given the order of the system and the desired order of the model. This procedure can be based simply on matching as many time series proportionals as possible, i.e. Padé approximation[7]. However, using this method alone, unstable models can be found despite the stability of the original system. To remove this eventuality the model denominator can be found by means of a Routh approximation technique[2,3,9], the model numerator parameters are then calculated using Padé approximation[4,6]. Some models ensure that a number of Markov parameters are matched between system and model as well as a number of time series proportionals[2,6,11].

There are, it follows, a large number of possible models of reduced order, for any one system transfer function. The only useful method of comparison is to calculate each model, in terms of the parameters it contains, and subsequently to graphically plot the model's time response. Hence, after several model responses have been plotted, these can be visually compared with the system response[1], to determine which model is most appropriate. This can be an extremely time consuming procedure and may lead to an incorrect choice of model due to the response only being considered over a short time period. Further, this lengthy comparison process tends to defeat, to an extent, the original reason for obtaining a lower order model, i.e. an overall reduction in effort.

A new approach to reduced order modelling has recently been obtained[10], which is based on determining the error between system and model responses. This is found in the form of an error polynomial, the coefficients of which tend to zero as the model response tends to that of the system.

In this paper the form of modelling based on the error polynomial is employed such that a cost function may be defined, which is equivalent to a weighted sum of error polynomial coefficients. Two courses of action are then considered. Firstly the cost function can be used to replace graphical methods in making a comparison between models obtained via the various methods available. Secondly, subject to a suitable cost function definition being made, as the cost function magnitude approaches zero, so the response of the model approaches that of the system. By minimizing the specified cost function, the best model, in the sense of the definition made, can be found.

This then presents a different method with which reduced order models can be found and places an emphasis on cost function definition rather than on which system/model parameters, i.e. time series or Markov, are to be matched.

2. LOW ORDER MODELLING

It is assumed that the high order continuous-time transfer function, relating plant input to output, is available and can be described by the expression

$$G(s) = \frac{d_0 + d_1 s + \ldots\ldots + d_{n-1}s^{n-1}}{e_0 + e_1 s + \ldots\ldots + e_{n-1}s^{n-1} + s^n} = \frac{D(s)}{E(s)} \qquad (2.1)$$

where the order of the numerator is stated as being one less than that of the denominator, a generalisation solely for explanation purposes.

The transfer function of (2.1) can also be written in terms of a power series expansion about either s=0, giving

$$G(s) = \sum_{i=0}^{\infty} c_i s^i \qquad (2.2)$$

or about $s=\infty$, in which case

$$G(s) = \sum_{i=1}^{\infty} c_{-i} s^{-i} \qquad (2.3)$$

The c_i parameters in (2.2) are directly proportional to the system time moments[7], whereas the c_{-i} parameters in (2.3) are termed the Markov parameters.

The reduced order model may be described in a similar way to the system. Let the model be given by the transfer function

$$R(s) = \frac{a_0 + a_1 s + \ldots\ldots + a_{k-1}s^{k-1}}{b_0 + b_1 s + \ldots\ldots + b_{k-1}s^{k-1} + s^k} = \frac{A(s)}{B(s)} \qquad (2.4)$$

in which the index $k \leqslant n$. However, where k=n the model is equivalent to the system.

The overall problem of calculating a reduced order model can be summarised as follows. The parameters $\{a_i : i=0,1,\ldots, k-1\}$ and $\{b_i : i=0,1,\ldots,k-1\}$ are chosen such that the model transfer function, R(s), is, in some sense, a good approximation to the system transfer function, G(s).

3. COST FUNCTION DEFINITION

The error between transfer function, G(s), and model, R(s), can be described by the equation

$$\frac{D(s)}{E(s)} = \frac{A(s)}{B(s)} + \frac{W(s)}{E(s)B(s)} \qquad (3.1)$$

where $W(s) = W_0 + W_1 s + \ldots\ldots + W_m s^m$ (3.2)

in which $m = n + k - 1$ (3.3)

Equation (3.1) can be written in a simpler form, as

$$G(s) = R(s) + \lambda(s) \tag{3.4}$$

where $\lambda(s)$ is the rational transfer function denoting any undesired error, and may therefore be used in the definition of the W(s) error polynomial as

$$W(s) = \lambda(s)\ E(s)\ B(s) \tag{3.5}$$

This error polynomial can be viewed in another way if it is considered that by applying an identical input to both system and model, where these are not identical an error will be present if the respective outputs are compared. With the input denoted by U(s), the outputs are denoted by $y_s(s)$ and $y_m(s)$ for system and model respectively.

The error is then

$$v(s) = y_s(s) - y_m(s) \tag{3.6}$$

or

$$v(s) = [G(s) - R(s)]U(s) \tag{3.7}$$

which, by inclusion of (3.4), is the same as

$$v(s) = \lambda(s)U(s) \tag{3.8}$$

Therefore it follows that

$$E(s)\ B(s)\ v(s) = W(s)\ U(s) \tag{3.9}$$

From (3.9) the following points can be made. Firstly the m roots of the W(s) polynomial are also the zeros of the transfer function which relates system and model input to the error v(s). Secondly, if the system is stable, i.e. all the roots of E(s) lie in the left half of the s-plane, then for the overall transfer function, relating input to error, to be stable, the model must also be stable, i.e. all the roots of B(s) must also lie in the left half of the s-plane.

For a kth order model there are 2k degrees of freedom, i.e. 2k unknown model parameters, this can be seen from (2.4). Also, when $n > k$ it follows that the number of parameters in the W(s) polynomial is $m+1 = n+k > 2k$. Hence with the 2k degrees of freedom all (m+1) W(s) parameters cannot be made equal to zero, a situation which would be equivalent to the model and system being identical. However, the nearer each W(s) parameter is to zero then the nearer are the system and model transfer functions to each other.

In so far as the magnitude of each W(s) term is a measure of the difference between system and model, the following cost function, which is proportional to the square of the W_i parameters is defined as

$$\phi'(W) = \sum_{i=0}^{m} \alpha_i W_i^2 \tag{3.10}$$

in which $\alpha_i \geqslant 0$.

The scalar value α_i is included to allow for certain time response necessities. These include placing more of an emphasis on earlier or later parts of the time response.

One amendment must however be made to (3.10), and this is necessary to ensure zero steady state error. In this case W(s)=0 when s=0, which is obtained by setting W_o to equal zero. The cost function therefore becomes

$$\phi(W) = \sum_{i=1}^{m} \alpha_i W_i^2 \tag{3.11}$$

provided that $W_o = 0$.

Once a specific choice of each of the α_i values has been made, therefore, the cost function can be employed as either a test on various possible models or as a design procedure itself. This will be shown in the following sections. For any particular weighting function, the best model is the one which minimizes $\phi(W)$.

4. THE WEIGHTING FUNCTION

As has been shown for discrete time systems[10], sequentially setting the W_i parameters to zero for $i=0,1,\ldots,j$ will match the first $(j+1)$ time series proportionals between system and model. This is equivalent to Padé approximation. Further, sequentially setting the W_i parameters to zero for $i=m,m-1,\ldots,m-\ell$ will match the first $(\ell+1)$ Markov parameters between system and model.

In general, Padé approximation is used to match the time response between system and model as the response tends towards the steady state. Markov parameter matching, however, is used to match the response during the initial time period following a change in the input signal.

This matching of responses for small and long time periods will be reflected in the values of α_i chosen such that for i near either m or unity, the weighting will be much heavier, i.e. α_i will be much larger.

Although there is no real limit to the maximum value of any particular α_i, only a certain number of α_i's can be equal to zero. For 2k parameters in the model, 2k equations are required. Thus only (m+1)-2k, which is equal to n-k, α_i terms can be zero.

For the designs carried out in the next sections, the weighting function is based on a logarithmic scale such that $\alpha_1=\alpha_m=1$; $\alpha_2=\alpha_{m-1}=0.1$; $\alpha_3=\alpha_{m-2}=0.01$;...etc. This function places significant emphasis on initial and steady state response matching.

5. MODEL TESTING

Once several models have been obtained for a particular system it is standard practice to find the inverse Laplace Transform for each model such that a comparison can be made between the time responses of the models to see which one approximates most closely that of the system. This approach can be a problem, especially if the inverse Laplace Transform is difficult to obtain or if time is important.

Using the cost function defined previously, with the log weighting factor, the models can be operated on swiftly using simple algebra in order that a comparison can be made.

As an example, consider the following system[1]:-

$$G(s) = \frac{10 + 3s + 13s^2 + 3s^3}{1 + s + 2s^2 + 1.5s^3 + 0.5s^4} \qquad (5.1)$$

The Padé approximant, of the system, for a second order model is found to be:-

$$R(s) = \frac{10 - 2.9184s}{1 + 0.4082s + 0.2857s^2} \qquad (5.2)$$

leading to an error polynomial:-

$$W(s) = 0.0004s + 0.0005s^3 + 4.3163s^4 + 2.3163s^5 \qquad (5.3)$$

Applying the cost function, as previously described, this is calculated as:-

$$\phi(W) = (0.0004)^2 + 0.01 \times (0.0005)^2 + 0.1 \times (4.3163)^2 + 1.0 \times (2.3163)^2 = \underline{7.2283} \qquad (5.4)$$

This can be compared with the second order model obtained using Routh approximation techniques[3] such that

$$R(s) = \frac{10 + 3s}{1 + s + 0.5s^2} \qquad (5.5)$$

with an error polynomial:-

$$W(s) = -2s^2 + 10s^3 \qquad (5.6)$$

Again, applying the cost function;

$$\phi(W) = 0.1 \times (2)^2 + 0.01 \times (10)^2 = \underline{1.4} \qquad (5.7)$$

By making a comparison of the two cost functions obtained, the Padé approximant model (5.2) is shown to be not as good as the Routh model (5.5). This result agrees with that obtained by plotting the time responses, as shown in Figure 1.

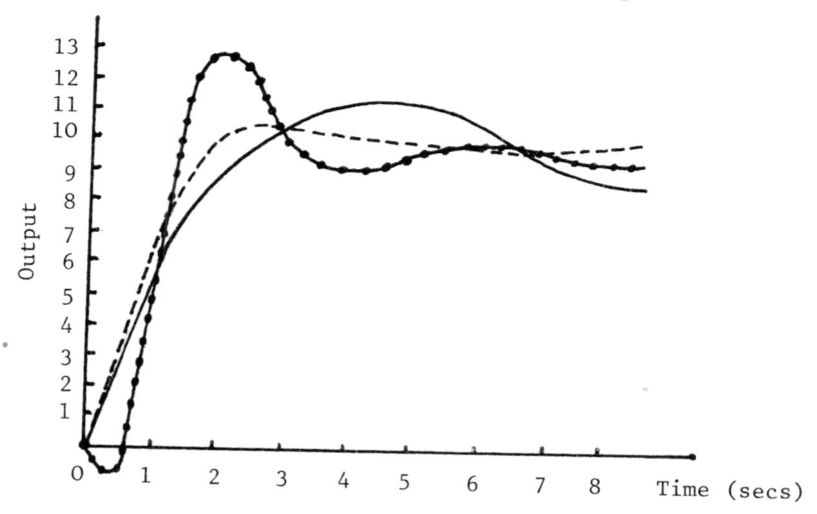

Fig. 1. Step responses for system and models.

—— System (5.1)

-•-•- Model (5.2)

--- Model (5.5)

6 DESIGN PROCEDURE USING THE COST FUNCTION

The reduced (2nd) order model of the system (5.1) is defined in normalised form to be:

$$R(s) = \frac{a_o + a_1 s}{1 + b_1 s + b_2 s^2} \qquad (6.1)$$

The first step is to set the W_o term to be equal to zero.

$$\text{Hence } W_o = d_o - a_o = 0 \qquad (6.2)$$

or, from (5.1): $a_o = 10$

The model therefore becomes:

$$R(s) = \frac{10 + a_1 s}{1 + b_1 s + b_2 s^2} \qquad (6.3)$$

where the parameters a_1, b_1 and b_2 are to be chosen such that the function, $\phi(W)$, is minimized, with the α_i parameters chosen as in the logarithmic function. The cost function is therefore;

$$\phi(W) = W_1^2 + 0.1 \times W_2^2 + 0.01 \times W_3^2 + 0.1 \times W_4^2 + W_5^2 \qquad (6.4)$$

The three equations necessary, such that the unknown parameters may be found are obtained by setting

$$\frac{\partial\phi}{\partial a_1}, \frac{\partial\phi}{\partial b_1}, \frac{\partial\phi}{\partial b_2}:$$

to zero, individually, each partial differential giving the minimum value of (6.4) with respect to the unknown parameter.

In the example chosen, this leads to

$$\begin{bmatrix} 2.4412 & 1 & -0.3581 \\ 1.6164 & 7.98 & -1 \\ 14.1962 & 1 & -1.5103 \end{bmatrix} \begin{bmatrix} b_1 \\ b_2 \\ a_1 \end{bmatrix} = \begin{bmatrix} 1.9268 \\ 3.0732 \\ 10.31 \end{bmatrix} \qquad (6.5)$$

such that

$$R(s) = \frac{10 + 4.1026s}{1 + 1.1153s + 0.6733s^2} \quad (6.6)$$

with $W(s) = 0.0504s - 1.0237s^2 - 3.6864s^3 + 0.9449s^4 - 0.0314s^5$ (6.7)

and $\phi(W) = \underline{0.3335}$

The response of this cost function derived model (6.6) to a step input is compared with the Routh model (5.5) and the original system (5.1) in figure 2.

The design procedure using the cost function method has been introduced by means of a worked example. In practice, however, once W_0 has been equated with zero, the remaining 2k-1 model parameters are found from the 2k-1 equations obtained by calculating the partial differential of the cost function with respect to each of the 2k-1 parameters in turn. Where the model is of low order therefore, the original idea in this paper, only a small number of equations are necessary.

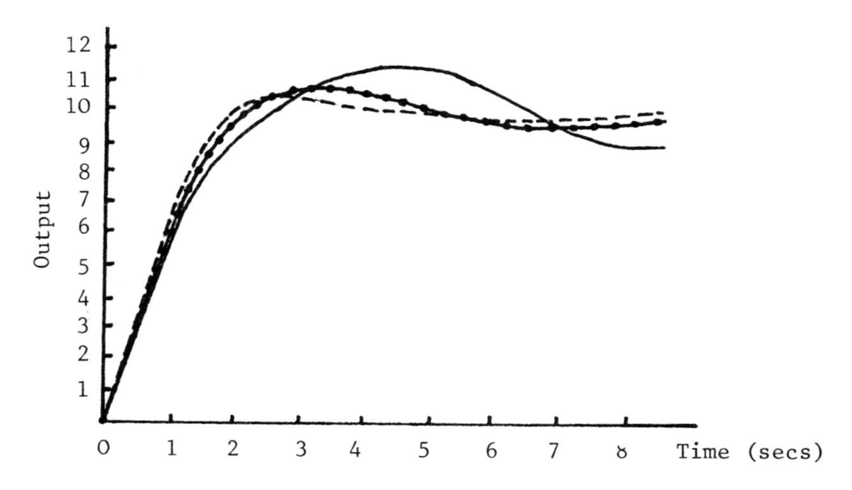

Fig. 2. Comparison of step responses with the new model.

—— System (5.1)

--- Model (5.5)

-•-•- New Model (6.6)

7. CONCLUSIONS

The W(s) polynomial parameters give an indication of the type of modelling carried out[10]. As parameters associated with low order s terms, in W(s), tend to zero so time series proportionals are found to match between model and system. Conversely, the parameters associated with high order s terms give an indication of Markov matching as they tend to zero.

Once the weighting parameters for a particular cost function have been defined, testing of various models obtained against the cost function is far simpler, computationally, than plotting time responses. However, where the particular requirements of the model have been defined by the weighting parameters, the model can be obtained which minimizes this particular cost function, as was shown in the previous section.

Further work in this area is firstly to relate the cost function to the time domain error and secondly to investigate the possibilities of alternative weighting functions to the logarithmic one used in the example.

8. REFERENCES

1. Ashoor, N., and Singh, V. (1982): 'A note on low order modelling', IEEE Trans. on Automatic Control, Vol. AC-27, No. 5, pp. 1124-1126.

2. Chen, T.C., Chang, C.Y., and Han, K.W. (1980): 'Model reduction using the stability equation method and the continued fraction method', Int. J. Control, Vol. 32, No.1, pp. 81-94.

3. Hutton, M.F. and Friedland, B. (1975): 'Routh approximations for reducing the order of linear time-invariant systems', IEEE Trans. on Automatic Control, Vol. AC-20, pp. 329-337.

4. Pal, J. (1979): 'Stable reduced order approximants using the Routh - Hurwitz array', Electronics Letters, Vol. 15, pp. 225-226.

5. Shamash, Y. (1974): 'Continued fraction methods for the reduction of discrete time dynamic systems', Int. J. Control, Vol. 20, No. 2, pp. 267-275.

6. Shamash, Y. (1980): 'Stable biased reduced order models using the Routh method of reduction', Int. J. Systems Sci., Vol.11, pp. 641-654.

7. Shamash, Y. (1975): 'Model reduction using the Routh stability criterion and the Padé approximation technique', Int. J. Control, Vol. 21, No. 3, pp. 475-484.

8. Shamash, Y. (1982): 'Critical review of methods for deriving stable reduced order models', Proc. 6th. IFAC Symposium on Identification and System Parameter Estimation, Washington D.C., pp. 1355-1359.

9. Singh, V. (1981): 'Stable approximants for stable systems: A new approach', Proc. IEEE, Vol. 69, pp. 1155-1156.

10. Warwick, K. (1984): 'A new approach to reduced order modelling', Proc. IEE, Vol. 131, Pt. D, No. 2, pp. 74-78.

11. Westcott, J.H. (1952): 'The frequency response method: its relationship to transient behaviour in control system design', Trans. of Soc. of Instrument Technology, Vol. 4, No. 3, pp. 113-124.

APPROXIMATE INTEGRATION OF A CLASS OF STOCHASTIC DIFFERENTIAL EQUATIONS

C.J. Harris
(Royal Military College of Science, Shrivenham)

and

Y. Maghsoodi
(University of Oxford)

ABSTRACT

In this paper we consider a class of non linear doubly stochastic differential equations and their approximate numerical integration by a form of Taylor series and an application of semi-group ideas.

1. INTRODUCTION

Many dynamical systems are modelled as a precursor to control design by ordinary differential equations; however to incorporate external and internal random disturbances as well as imperfectly known parameters such systems are better modelled by stochastic differential equations [1,2] of the form

$$dx_t = A(t,x_t)dt + B(t,x_t)dW_t + C(t,x_t)dN_t \tag{1.1}$$

where $\{W_t;\ t \geq o\}$ is a vector standard Wiener process, $\{N_t; t \geq o\}$ is a Poisson [15] counting process with intensity rate λ_t and $\{x_t\}$ is the desired state vector solution.

A formal solution to (1.1) is

$$x_t = x_{t_o} + \int_{t_o}^{t} A(s,x_s)ds + \int_{t_o}^{t} B(s,x_s)dW_s + \int_{t_o}^{t} C(s,x_s)dN_s \tag{1.2}$$

The last two integrals cannot be defined as Riemann-Stieltjes integrals with respect to the sample functions of W_s and N_s, since $\{W_t\}$ has unbounded variation and C and N_s may well have sample functions with common discontinuities (ie simultaneous jumps). Nor can these integrals be defined as generalised mean-square Lebesgue-Stieljes integrals since their values would then depend upon the choice of the intermediate points in the finite sums. These integrals have to be interpreted in the Itô or Stratonovich sense.

Consider (1.1) more formally for $t \in [t_o, T]$, $x_{t_o} = x_o$ w.p.1 for A (n x 1), B (n x m), X (n x 1) parameter vectors, and N_t a scalar Poisson process independent of W_t. The equations (1.1) and (1.2) have a unique w.p.1 solution which is a Markov process, if there exist constants α, β such that [1,2]

(i) $||A(x,t)||^2 + ||B(x,t)||^2 + ||C(x,t)||^2 \leq \alpha(1 + ||x||^2)$

(ii) $||A(x,t) - A(y,t)||^2 + ||B(x,t) - B(y,t)||^2 + ||C(x,t) - C(y,t)||^2 \leq \beta(x,y)^2; \forall ||x|| < R, \ ||y|| < R$

(iii) There is a function $Q(h) \geq o$, $\downarrow o$ as $h \to o$ such that

$$||A(x,t) - A(x,t+h)||^2 + ||B(x,t) - B(x,t+h)||^2 + ||C(x,t) - C(x,t+h)||^2 \leq \alpha(1 + ||x||^2)\, Q(h) \tag{1.3}$$

In the Taylor series approximation to (1.2) the moments of the increments of the solution play a fundamental role; in particular the inequality

$$E[(x_{t+h} - x_t)^{2p}] \leq \alpha_p (1 + ||x||^{2p})h \text{ is useful.} \tag{1.4}$$

For many systems we are interested in the moments of the solution to (1.1), in particular the first and second moments. These may be obtained by application of the generalised Itô formula [1] for total derivative of a functional

$\Phi(x) = x_1^{\,n} x_2^{\,s} x_3^{\,q}$ .... and then by taking expectation values. For example the first two moments of (1.2) satisfy the ordinary differential equations

$$dE(x_t) = E\{A(x_t,t)\}\, dt + E\{C(x_t,t)\}\lambda_t dt$$

$$dE(x_i x_j) = E(a_i x_i + a_j x_j) + E([BB^T]_{ij}) + \lambda_t E(x_i c_j + x_j c_i + c_i c_j) \tag{1.5}$$

These equations are only solvable when A,B,C are linear in x, otherwise lower order moments involve solutions to higher order equations; then the only viable solutions to (1.2) are by approximation schemes [3,9,12]

2. APPROXIMATION BY TAYLOR SERIES OF THE INTEGRANDS[3]

Definition 1. a The deterministic function $y(h)$ is said to be of order h^α $(o(h^\alpha))$ iff: $\lim_{h\to o} \frac{|y(h)|}{h^x} = o, \quad \forall\ x < \alpha$

b The random function $x(h)$ is said to be of order of probability $h^\alpha (o_p(h^\alpha))$ iff, given any $\varepsilon > o$:

$$\lim_{h\to o} \text{prob} \left\{ \frac{|x(h)|}{h^x} \geq \varepsilon \right\} = o, \ \forall\ x < \alpha$$

To numerically solve (1.2) by Taylor series, partition the the integration interval $[t_o, T]$ into M subintervals $h = t_{r+1} - t_r$, for $r = 1,\ldots,M-1$

Then the solution to (1.2) over (t_r, t_{r+1}) is

$$x_{r+1} = x_r + \int_{t_r}^{t_{r+1}} A(s,x_s)ds + \int_{t_r}^{t_{r+1}} B(s,x_s)\,dW_s + \int_{t_r}^{t_{r+1}} C(s,x_s)\,dN_s \tag{2.1}$$

$r = o,1,2,\ \ldots\ M-1.,x_r = x(t_r)$.

The integrands in (2.1) can be replaced by their Taylor series about (x_r,t_r); for simplicity consider the scalar version of (2.1) then

$$\begin{aligned} x_{r+1} &= x_r + \bar{a} \int ds + \bar{a}_t \int (s-t)\,ds + \bar{a}_x \int (x_s - x_r)\,ds \\ &+ \int [o(|\Delta s|^2) + o(|\Delta x_s|^2)]ds + \bar{b} \int dW_s + \bar{b}_t \int (s-t_r)\,dW_s \\ &+ \bar{b}_x \int (x_s-x_r)\,dW_s + \int [o(|\Delta s|^2) + o(|\Delta x_s|^2)]dW_s \\ &+ \bar{c} \int dN_s + \bar{c}_t \int (s-t_r)\,dN_s + \bar{c}_x \int (x_s-x_r)\,dN_s + \int [o(|\Delta s|^2) + o(|\Delta x|^2)]dN_s \end{aligned} \tag{2.2}$$

where $\bar{a} = a(x,t_r)$, $\bar{a}_t = \frac{\partial \bar{a}}{\partial t}$, $\bar{a}_x = \frac{\partial \bar{a}}{\partial x}$ etc. Terms which involve (x_s-x_r) can be successively replaced by expressions similar to (2.2), or by the simpler Euler approximation to (x_s-x_r);

$$x_t - x_r = \bar{a}\,(t-t_r) + \bar{b}(W_t-W_r) + \bar{c}(N_t-N_r)$$

$$+ \int_{t_r}^{t} [o(|\Delta s|)+o(|\Delta x_s|)ds] + \int_{t_r}^{t} [o(|\Delta s|)+ o(|\Delta x|)]dW_s +$$

$$\int_{t_r}^{t} [o(|\Delta s|) + o(|\Delta x|)]dN_s \qquad (2.3)$$

To determine the orders of error due to truncation we need[8,15]:

Theorem 1: If the coefficients A,B and C of (1.1) satisfy the existence and uniqueness conditions of (1.3) and are continuous and differentiable up to order 4 in t and x then;

$$\text{(i)} \quad \int[o(|\Delta t|) + o(|\Delta x_t|)]dt = o_p(h^{3/2})$$

$$\int[o(|\Delta t\Delta) + o(|\Delta x_t|)]dW_t = o_p(h) \qquad (2.4)$$

$$\int[o(|\Delta t|) + o(|\Delta x_t|)]dN_t = o_p(h^{3/2})$$

$$\text{(ii)} \quad \int[o(|\Delta t|^2) + o(|\Delta x_t|^2)]dt = o_p(h^2)$$

$$\int[o(|\Delta t|^2) + o(|\Delta x_t|^2)]dW_t = o_p(h) \qquad (2.5)$$

$$\int[o(|\Delta t|^2) + o(|\Delta x_t|^2)]dN_t = o_p(h^2)$$

$$\text{(iii)} \quad \int[o(|\Delta t|^3) + o(|\Delta x_t|^3)]dt = o_p(h^2)$$

$$\int[o(|\Delta t|^3) + o(|\Delta x_t|^3)]dW_t = o_p(h) \qquad (2.6)$$

$$\int[o(|\Delta t|^3) + o(|\Delta x_t|^3)]dN_t = o_p(h^2)$$

where the limits of integration are between t_r and t_{r+1}

We can now apply theorem 1 to the Taylor series (2.2) to obtain a series of approximation integration algorithms. Consider the special case when c = o, the Euler substitution leads to the scalar

Euler-Stochastic Algorithm[16]: $\bar{x}_{r+1} = \bar{x}_r + \bar{a}h + bZ_{1h} + o_p(h)$ (2.7)

where $z_{1h} \sim N(o,h)$.

Alternatively the double Taylor series substitution for $c = o$ leads to the more generalised and accurate

Taylor-Stochastic Algorithm[3,10,11]:

$$x^i_{r+1} = x^i_r + h\bar{a}_i + \sum_j^m \bar{b}_{ij} z_{ij} + \tfrac{1}{2} \sum\sum\sum_{jkp}^{mnm} \bar{b}_{ij,x_k} \bar{b}_{kj} Z_{ip} Z_{ij}$$

$$+ \sum\sum_{jk}^{mn} \bar{b}_{ij,x_k} \bar{b}_{kj} [Z_{ij}^2 - h] + o_p (h^{3/2}) \qquad (2.8)$$

for $i = 1,2,..,n$. These algorithms have found applications in Chemical Reactors[13], Flexible Spacecraft[6] and Biological Systems[4,5].

The Euler stochastic algorithm for the scalar version of (1.2) can be written directly as

$$\bar{x}_{r+1} = \bar{x}_r + \bar{a}\ h + \bar{b}\ Z_{ih} + \bar{c}\ \Delta N_n + o_p(h) \qquad (2.9)$$

The improved accuracy of the Taylor series method (2.8) would suggest an improvement over (2.9) for the scalar Poisson driven system (1.2). However $E\{|\Delta N_t|^P\} = o(\Delta t)$

for all integer p, unlike $E\{|\Delta W_t|^P\} = o(\Delta t^{p/2})$ for the Wiener only driven systems of (2.7) and (2.8), therefore the Taylor series approach is unlikely to give increased accuracy over the Euler algorithm (except by minor correction terms[8]) for the generalised Poisson driven stochastic differential system (1.2).

3. SEMI-GROUP APPROXIMATION ALGORITHMS[7,8,9,10]

The various Taylor series algorithms are lengthy, complex and computationally expensive to implement for generalised multivariable stochastic systems. Simpler algorithms can be generated by 'a priori' selection of the algorithm structure in a series form by terminating it at the appropriate error order level. Milshtein[9,10] used such an approach and the concept of stochastic semi-groups to generate an algorithm identical to the Harris Taylor-series algorithm (2.8) for Wiener driven processes. Here we extend the semi-group approximation approach to the Wiener and Poisson driven stochastic differential equations (1.1).

Definition 2. Let Y be a complex Banach space. For each $t > o$ let S_t: $Y \rightarrow Y$ be a bounded linear operator on Y. If $S_{s+t} = S_s S_t$ then the family of operators $G = \{S_t; t \geq o\}$ construct a one parameter semi-group of bounded linear operators on Y.

Consider the following Banach space:

$Y \triangleq \{f(t,x)$; bounded, real valued, continuous, measurable in t,x and $||f|| = \sup_{t,x} f(x,t)$, and let $p(t_o,x;\ t_o + t,X) =$ prob $\{x_{t_o+t} \in X/x_{t_o} = x\}$ denote the transitional function of the Markov process solution to (1.2). Define S_t: $Y \rightarrow Y$ by the conditional expectation

$$S_t\ f(t_o,x) = E\{f(t_o+t,x_{t_o+t})/x_{t_o} = x\}$$

$$= \int_{R^n} p(t_o,x;\ t_o + t,)\ dy\ f(t_o + t,y) \qquad (3.1)$$

By using the Chapman-Kolmogorov identity it can be shown[2,14] that the Markov transition operators S_t construct a one parameter (contraction) semi-group. To generate an approximation 'power' series for $S_t f$ we require the stochastic infinitesimal operator, I: $Y \rightarrow Y$, of the semi-group S_t [9,10]

$$I\ f(s,x) = \lim_{t \to o+} \left\{ \frac{S_t\ f(s,x) - f(s,x)}{t} \right\}$$

$$= \lim_{t \to o+} \left\{ \frac{E\ f(s+t,x_{s+t})/x_s = x\ - f(s,x)}{t} \right\} \qquad (3.2)$$

The infinitesimal operator for the stochastic differential equation (1.1) completely characterises its solution since we can show[8]

Lemma 1: for the stochastic differential equation (1.1) with $x_{t_o} = x_o$ a constant w.p.1, the infinitesimal operator If(s,x) of a bounded, continuous functional (up to second order in x and t) exists and is given by

$$If(s,x) = \frac{\partial f}{\partial s}(s,x) + \langle \nabla_x f(s,x), A(s,x) \rangle$$

$$+ \tfrac{1}{2}\mathrm{Tr}\,(\nabla^2_{xx} f(s,x)\, B(s,x)\, B^T(s,x) + \lambda_s [f(s,x+C(s,x)) - f(s,x)] \tag{3.3}$$

where <.,.> denotes scalar product.

Lemma 2: If x_t is the unique solution to (1.1) and $f(s,x) \varepsilon Y$ and $I^n f(s,x)$ exist then

$$E\{f(t_o+h, x_{t_o+h})/x_{t_o}=x_o\} = f(t_o,x_o) + h\, I\, f(t_o,x) + \frac{h^2}{2!} I^2 f(t_o,x_o) +$$

$$\frac{1}{(n-1)!} \int^h (h-\tau)^{n-1} E\{I^n f(\tau, x\ )/x_{t_o} = x_o\} d\tau \tag{3.4}$$

We can now use these lemmas to derive mean square error approximation algorithms (1.1). To illustrate the method consider the scalar version of (1.1); we select an algorithm of the form

$$\bar{x}_{t_o+h} = x_o + \alpha_1 h + \alpha_2 \Delta W_h + \alpha_3 (\Delta W_h)^2 + \alpha_4 \Delta N_h + \alpha_5 \Delta N_h \Delta W_h + \ldots\ldots \tag{3.5}$$

(where $\Delta W_h = W_{t_o+h} - W_{t_o}$ etc.) such that

$$E\{(x_{t_o+h} - \bar{x}_{t_o+h})^2/x_o\} = E\{[\Phi(t_o+h, x_{t_o+h}, W_{t_o+h}, N_{t_o+h} \ldots)]^2/x_o =$$

$$o(h^n) \tag{3.6}$$

Since Φ is a function of x, W and N we need to define an augmented state vector X = (x,W,N) before employing Φ in (3.4); in which case (3.4) yields

$$E\{[x - \bar{x}]^2/x_o\} = E\{[\Phi(t_o+h, X_{t_o+h})]^2/x_o\}$$

$$= \sum_{k=o}^{n-1} \frac{h^k}{k!} I^k \Phi(t_o,x_o) + \frac{1}{(n-1)!} \int^h (t_o+h-\tau)^{n-1} E\{I^n \Phi(\tau, X_\tau)\} d\tau \tag{3.7}$$

Subject to the boundedness of $E[I^n \Phi(\tau, X_\tau)]$ the remainder term in (3.7) is $o(h^n)$, thence if the first term on RHS of (3.7) is zero (by selecting the coefficients $\alpha_1, \alpha_2, \ldots$) then (3.6) is satisfied. Then we only need to solve the equations

$$\Phi(t_o,X_o) =, I\Phi(t_o,X_o) = o,\dots, I^{n-1} \quad \Phi(t_o,X_o) = o \tag{3.8}$$

for the coefficients α_1, α_2,... to obtain a mean square error algorithm of order $o(h^n)$ for the scalar version of (1.1). It is trivial to extend (3.5) to the kth step algorithm

$$\bar{x}_{k+1} = \bar{x}_k + \alpha_1 h + \alpha_2 \Delta W_{k+1} + \alpha_3 (\Delta W_{k+1})^2 + \alpha_4 \Delta N_{k+1} + \dots \tag{3.9}$$

A priori knowledge of the Taylor series and Cauchy-Euler approximation series

$$\bar{x}_{k+1} = \bar{x}_k + \alpha_1 h + \alpha_2 \Delta W_{k+1} + \alpha_3 (\Delta W_{k+1})^2 + \alpha_4 \Delta N_{k+1}$$

$$+ \alpha_5 \Delta N_{k+1} \quad \Delta W_{k+1} \quad \alpha_6 (\Delta N_{k+1})^2 + \alpha_7 \Delta Z_{k+1} \tag{3.10}$$

and an augmented state vector $X_t = (x_t, W_t, N_t, Z_t)^T$, where

$$\Delta W_{k+1} = W_{k+1} - W_k \sim N(o,h), \Delta N_{k+1} = N_{k+1} - N_k, \Delta Z_{k+1} = \int^h (W_s - W_k)\, dN_s$$

hence $\Phi = (x - \bar{x} - \alpha_1 h - \alpha_2 \Delta W \dots - \alpha_7 \Delta Z)$ in (3.8) which in turn evaluates the coefficients $\alpha_1,\dots,\alpha_7$ of (3.10) as

$$\bar{x}_{k+1} = \bar{x}_k + (\bar{a} - \tfrac{1}{2}\bar{b}\bar{b}_x)h + \bar{b}\, \Delta W_{k+1} + \tfrac{1}{2}\bar{b}\bar{b}_x \, (\Delta W_{k+1})^2$$

$$+ \tfrac{1}{2}\,(3\bar{c} - \bar{c}_c)\Delta N_{k+1} + (\bar{b}_c - \bar{b})\Delta W_{k+1} \cdot \Delta N_{k+1} + \tfrac{1}{2}(\bar{c}_c - \bar{c})(\Delta N_{k+1})^2$$

$$+ (\bar{b}\bar{c}_x - \bar{b}_c + \bar{b})\; \Delta Z_{k+1} \; ; \; k = 1,2,\dots,M-1 \tag{3.11}$$

where $b_x = \dfrac{\partial b}{\partial x}$, $\bar{b}_c = b(t_k, \bar{x}_k + \bar{c})$ etc

<u>Note</u> (i) When $c = o$ (Wiener driven only process), algorithm (3.11) simplifies to the scalar version of (2.8)

(ii) We can show that the algorithm (3.11) has k^{th} step mean squared error of $o(h^2)$ if the coefficients a,b,c,bb_x, and bc_x are uniformly Lipschitz

Multivariable algorithms may be derived by the same process; consider the all vector coefficient version of (1.1) with independent scalar noise driving process $\{W_t\}$ and $\{N_t\}$. The augmented state vector, X in this case is $X_t = \{x_t, W_t, N_t, Z_t, t\}$ and the resulting algorithm of order $o(h^2)$

$$\bar{x}_{k+1} = \bar{x}_k + (\bar{A} - \tfrac{1}{2}(\bar{B}_x)\bar{B})h + \bar{B}\Delta W_{k+1} + \tfrac{1}{2}(\bar{B}_x)\bar{B}(\Delta W_{k+1})^2 +$$

$$\tfrac{1}{2}(3\bar{C} - \bar{C}_c)\,\Delta N_{k+1} + (\bar{B}_c - \bar{B})\,\Delta N_{k+1}\,\Delta W_{k+1} + \tfrac{1}{2}(\bar{C}_c - \bar{C})(\Delta N_{k+1})^2$$

$$+ ((\bar{C}_x)\bar{B} - \bar{B}_c + \bar{B})\,\Delta Z_{k+1}; \qquad k = 1,2,\ldots, M-1 \qquad (3.12)$$

where B_x is the matrix with ij^{th} element $\partial b_i/\partial x_i$; B_c is a column vector with i^{th} element $b(t_k, \bar{x}_k + \bar{C}_k)$ etc. This algorithm specialises to the Harris-Milshtein algorithm (2.8) when the jump coefficient C vanishes. For convergence to the solution of (1.1) in a finite interval with mean squared error of $o(h^2)$ we only require the uniform Lipschitz continuity of the coefficients used in (3.12).

By a similar, but more complex process, a completely generalised semi-group algorithm of order $o(h^2)$ can be derived[8] for (1.1) of which all the above approximation algorithms are special cases, but space does not allow inclusion here. Additionally the semi-group approach to the approximate integration of generalised stochastic differential equations is not limited to mean squared errors of order $o(h^2)$, algorithms of order $o(h^4)$ have been derived by the authors - the cost being a rapid increase in computational effort. For example[8] the scalar version of (1.1) requires the evaluation of 23 α_i coefficients for an algorithm of order $o(h^4)$!

4. REFERENCES

1. Arnold, L (1974) "Stochastic Differential Equations Theory and Application," J Wiley

2. Gikman, I and Skorokhod, A V (1972) "Stochastic Differential Equations," Springer-Verlag

3. Harris, C J (1979) Int J for Numerical Methods. Vol. 14,37.

4. Harris, C J (1977) Int J Systems Science Vol 8, 393

5. Harris, C J (1978) Proc IEE Vol 125, 441

6. Harris, C J (1978) 7th World IFAC Congress, Helsinki

7. Hille, E and Phillips, R (1957) "Functional Analysis and Semi-Groups," Amer Math Soc. Providence RI

8. Maghsoodi, T (1984) D Phil Thesis Oxford University (Approximate Integration of a class of stochastic differential equations).

9. Milshtein, G N (1974) Theory Prob Applic Vol 19, 3

10. Milshtein, G N (1976) Theory Prob Applic Vol 23, 376

11. Platen, E (1980) Proc IFIP - WG7/1 conf 1978, Springer Verlag. LN in Control and Information Sciences

12. Rao, N J, Borwankar, J D and Ramkrishna, D (1974) SIAM J Control Vol 12, 124

13. Rao, N J, Borwankar, J D and Ramkrishna, D (1974) Chem Engr Science Vol 29, 1193

14. Skorokhod, A V (1965) "Studies in the theory of random processes," Addison Wesley

15. Snyder, D L (1975) "Random Point Processes," J Wiley

16. Wright, D J (1974) Trans IEEE, Vol AC - 19,75

CHAPTER 2

TIME DELAY AND DISTRIBUTED PARAMETER SYSTEMS

EVALUATION OF TIME-DELAY SYSTEM COST FUNCTIONALS

K. Walton and J. E. Marshall

(School of Mathematics, University of Bath)

ABSTRACT

It is useful to have expressions for the cost functional $\int_0^\infty e^2(t)dt$, where $e(t)$ is a function representing error in a control system. The paper shows how the well-known delay-free results may be extended to the time-delay case. A brief survey is given of methods for the calculation of time-delay cost functionals and comparisons are made of approximate methods.

1. INTRODUCTION

The principal aim of this paper is to obtain closed-form expressions for cost functionals arising in systems involving time-delay. Here emphasis will be placed on techniques of evaluation. Applications of the results will be considered in the companion paper[1]. Two typical time-delay systems are shown in Figure 1, where $G(s)$ denotes a delay-free transfer function and τ denotes the time-delay.

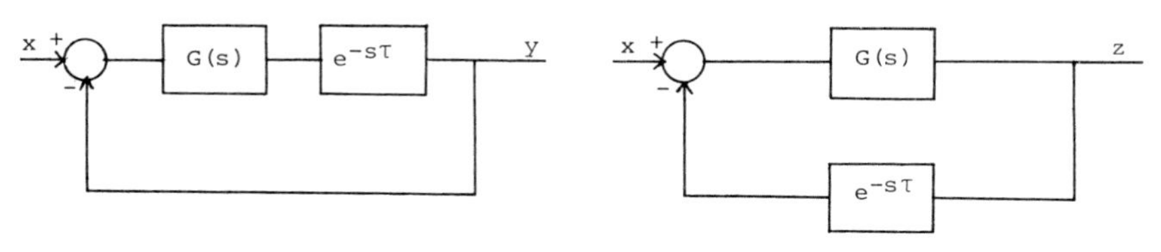

Fig 1 (a) Forward path delay (b) Feedback path delay

In both cases error is defined as input-output and the only cost functional to be considered will be that given by the integral of square error. Thus for the forward path delay case, the cost functional is given by

$$J_F = \int_0^\infty [x(t) - y(t)]^2 dt \tag{1.1}$$

and for the feedback delay case, by

$$J_B = \int_0^\infty [x(t) - z(t)]^2 dt \ . \tag{1.2}$$

For present purposes, attention will be restricted to a step input $x(t) = H(t)$, although the method of evaluation is applicable to more general inputs. It may then be shown[2] that

$$J_F = J_B + \tau \tag{1.3}$$

and consequently in this case only one of the cost functionals need be evaluated and so only J_F will be considered. Assuming that the error $e(t) = x(t) - y(t)$ is reasonably well-behaved and that the cost functional J_F exists, it follows that the steady-state error is zero, that $E(s)$, the Laplace transform of $e(t)$, exists and moreover that all the poles of $E(s)$ lie in the left half of the complex s-plane. It is worth emphasising this fact since it is crucial in what follows. By Parseval's theorem, J_F is then given by

$$J_F = \frac{1}{2\pi i} \int_{-i\infty}^{i\infty} E(s)E(-s)ds \tag{1.4}$$

It is further assumed that the delay-free transfer function $G(s)$ is of the form

$$G(s) = \frac{KP(s)}{Q(s)} \tag{1.5}$$

where $P(s)$ and $Q(s)$ are polynomials and hence, since the Laplace transform of $x(t) = H(t)$ is $X(s) = 1/s$,

$$E(s) = \frac{Q(s)}{s[Q(s) + KP(s)e^{-s\tau}]} \tag{1.6}$$

A further example is that of delayed-state problems involving a single time-delay, an example of which is shown in

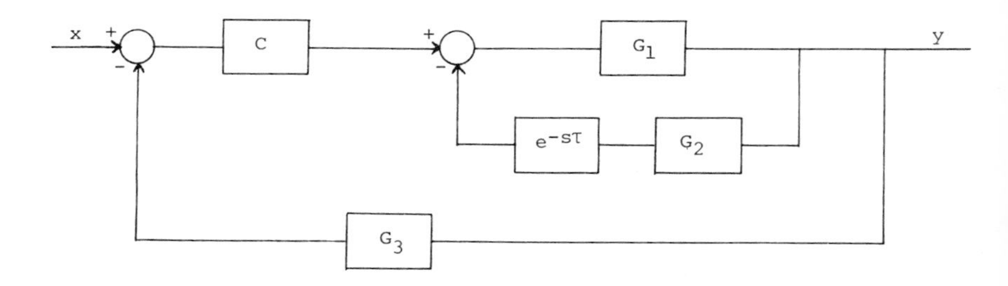

Fig 2 Delayed-state system

Figure 2. For input $x(t)$, output $y(t)$ and error

$e(t) = x(t) - y(t)$, $E(s)$, the Laplace transform of $e(t)$ is given by

$$E(s) = \left\{ \frac{1 + CG_1G_3 - CG_1 + G_1G_2e^{-s\tau}}{1 + CG_1G_3 + G_1G_2e^{-s\tau}} \right\} X(s) \tag{1.7}$$

In all the above $E(s)$ is of the form

$$E(s) = \frac{B(s) + D(s)e^{-s\tau}}{A(s) + C(s)e^{-s\tau}} \tag{1.8}$$

where $A(s)$ to $D(s)$ are polynomials in s and the evaluation techniques to be presented are valid for an $E(s)$ of this general form provided that all the poles of $E(s)$ or equivalently all the zeros of $(A(s) + C(s)e^{-s\tau})$ lie in the left half-plane. To demonstrate the evaluation techniques only one simple example will be considered, namely

$$E(s) = \frac{1}{s + Ke^{-s\tau}} \tag{1.9}$$

This corresponds to $G(s) = K/s$, $P(s) = 1$, $Q(s) = s$ in (1.6).

However, before doing this, consideration is given first to delay-free systems in which $E(s) = B(s)/A(s)$ and the cost functional J is given by

$$J = \int_0^\infty e^2(t)\,dt \tag{1.10}$$

$$= \frac{1}{2\pi i} \int_{-i\infty}^{i\infty} \frac{B(s)B(-s)}{A(s)A(-s)}\,ds \tag{1.11}$$

and a necessary condition for (1.11) to exist is that the degree of the polynomial $B(s)$ be less than that of $A(s)$. Moreover, a necessary condition for the equality of (1.10) and (1.11) is that all the zeros of $A(s)$ lie in the left half-plane or equivalently that the system is stable. On the other hand, if some of the zeros of $A(s)$ lie in the right half-plane (1.11) will still exist however (1.10) will not.

It is well-known[3,4,5,6] that integrals of the type given in (1.11) may be evaluated in terms of the coefficients of $A(s)$ and $B(s)$. The method given in Newton et al[4] is of particular interest since it suggests a technique for the evaluation of integrals arising in systems involving time-delay. The basic idea of this method is to make use of the fact that, by using partial fractions, the integrand may be split into two parts, namely

$$\frac{B(s)B(-s)}{A(s)A(-s)} = \frac{F(-s)}{A(s)} + \frac{F(s)}{A(-s)} \tag{1.12}$$

in which $F(s)$ is a polynomial of degree, at most, one less than that of A(s) and use has been made of symmetry. Thus, if

$$\left.\begin{aligned} A(s) &= a_o s^n + a_1 s^{n-1} + \ldots + a_{n-1}s + a_n \\ B(s) &= \qquad\quad b_1 s^{n-1} + \ldots + b_{n-1}s + b_n \end{aligned}\right\} \tag{1.13}$$

F(s) may be written

$$F(s) = (-)^{n-1}\left\{f_1 s^{n-1} + \ldots + f_{n-1}s + f_n\right\} \tag{1.14}$$

where the factor $(-)^{n-1}$ has been introduced for later convenience and the coefficients f_r may be obtained by equating powers of s in

$$A(s)F(s) + A(-s)F(-s) = B(s)B(-s) . \tag{1.15}$$

The cost functional J is then given by

$$J = \frac{1}{2\pi i}\int_{-i\infty}^{i\infty} \frac{F(-s)}{A(s)}\, ds + \frac{1}{2\pi i}\int_{-i\infty}^{i\infty} \frac{F(s)}{A(-s)}\, ds \tag{1.16}$$

Since, by assumption, all the zeros of A(s) lie in the left half-plane, all those of A(-s) lie in the right half-plane. Consequently, the contour from $-i\infty$ to $i\infty$ for the first integral may be deformed into the infinite semi-circle in the right half-plane, denoted by C_R, since, in doing so, no poles of the integrand are crossed. Similarly, that for the second integral may be deformed into C_L, the infinite semi-circle in the left half-plane. An alternative way of stating this is to say that the first integral is closed on the right and the second on the left. Thus

$$J = \frac{1}{2\pi i}\int_{C_R} \frac{F(-s)}{A(s)}\, ds + \frac{1}{2\pi i}\int_{C_L} \frac{F(s)}{A(-s)}\, ds . \tag{1.17}$$

Evaluating these integrals then yields

$$J = \frac{f_1}{a_o} \tag{1.18}$$

The value of J is thus determined by a_o and f_1, which may be obtained from (1.15) in terms of the coefficients of A(s) and B(s). This will be returned to in §6 when certain implications of this approach will be considered. However at this stage it is the idea of splitting the integrand into two parts, each of which is treated separately by making use of the fact that all the poles of the integrand lie in a half-plane, that is important.

2. THE DIRECT METHOD OF EVALUATION

The problem to be considered is that of the evaluation of the cost functional

$$J \equiv J_F = \frac{1}{2\pi i}\int_{-i\infty}^{i\infty} \frac{ds}{(s+Ke^{-s\tau})(-s+Ke^{s\tau})} \tag{2.1}$$

and the approach adopted is to follow the method for the delay-free case and split the integrand into two parts, each of which may be treated separately. One way to achieve such a split is to consider the zeros of $(s+Ke^{-s\tau})$. At such points

$$-s+Ke^{s\tau} = -\frac{(s^2+K^2)}{s} \tag{2.2}$$

and so (2.1) may be rewritten

$$J = \frac{1}{2\pi i}\int_{-i\infty}^{i\infty}\left\{\frac{1}{s+Ke^{-s\tau}}\left[\frac{1}{-s+Ke^{s\tau}}+\frac{s}{s^2+K^2}\right] - \frac{s}{(s+Ke^{-s\tau})(s^2+K^2)}\right\}ds \tag{2.3a}$$

$$= \frac{1}{2\pi i}\int_{-i\infty}^{i\infty}\left\{\frac{Ke^{s\tau}}{(-s+Ke^{s\tau})(s^2+K^2)} - \frac{s}{(s+Ke^{-s\tau})(s^2+K^2)}\right\}ds \tag{2.3b}$$

Since the term in square brackets in (2.3a) is zero when $(s+Ke^{-s\tau})$ is zero it follows that this part of the integrand has no poles in the left half-plane as can be seen from (2.3b). Also, since all the zeros of $(s+Ke^{-s\tau})$ lie in the left half-plane, it follows that the remainder of the integrand has no poles in the right half-plane.

This suggests that for the first part of the integrand the contour should be closed on the left and for the second part on the right. However, before doing this, consideration must be given to the zeros of (s^2+K^2), namely $s=\pm Ki$. Although these are not poles of the combined integrand, they are poles of the separate parts. Consequently, before separating the integral closing the contours, the contour from $-i\infty$ to $i\infty$ is first deformed in order to avoid passing through these points, and without loss of generality, this is achieved by indenting it to the left of $s=\pm ik$. The contours are then closed and it is noted that the contributions from the semi-circles at infinity are indeed zero. Finally, evaluating the resulting integrals using the theory of residues yields, from (2.3a)

$$J = \sum_{s=\pm iK} \operatorname{res}\left\{\frac{s}{(s+Ke^{-s\tau})(s^2+K^2)}\right\}$$

$$= \frac{1}{2(iK+Ke^{-iK\tau})} + \frac{1}{2(-iK+Ke^{iK\tau})} = \frac{\cos K\tau}{2K(1-\sin K\tau)} \tag{2.4}$$

For the more general case, as in (1.8), when

$$E(s) = \frac{B(s)+D(s)e^{-s\tau}}{A(s)+C(s)e^{-s\tau}} \tag{2.5}$$

an identical method may be used and it is found that

$$J = -\sum_k \underset{s=s_k}{\text{res}} \left(\frac{B(s) + D(s)e^{-s\tau}}{A(s) + C(s)e^{-s\tau}}\right)\left(\frac{B(-s)A(s) - D(-s)C(s)}{A(-s)A(s) - C(-s)C(s)}\right) \tag{2.6}$$

in which $s=s_k$ are the solutions of

$$A(s)A(-s) - C(s)C(-s) = 0 \tag{2.7}$$

Thus the problem is reduced to finding the roots of this polynomial and then evaluating the residues in (2.6)

3. OTHER METHODS OF EVALUATION

Although the above is the most direct method of evaluating time-delay system cost functionals, other methods do exist. For example, Gorecki and Popek[7] were the first to produce closed-form solutions. With minor modifications, their method is briefly as follows.

For the same example considered above, the error e(t) is given by the formula for the inverse Laplace transform, namely

$$e(t) = \frac{1}{2\pi i}\int_{-i\infty}^{i\infty} \frac{e^{st}\, ds}{s + Ke^{-s\tau}} \tag{3.1}$$

Closing the contour in the left half-plane, and noting that the contribution from the infinite semi-circle is indeed zero (as $t \geqslant 0$), the theory of residues yields

$$e(t) = \sum_i \frac{e^{s_i t}}{1 - K\tau e^{-s_i\tau}} \tag{3.2}$$

where the sum is taken over the poles of E(s) or equivalently over all the solutions $s = s_i$ of

$$s + Ke^{-s\tau} = 0 \tag{3.3}$$

and, by assumption, all of these lie in the left half-plane and so have negative real part. Thus

$$J = \int_0^\infty e^2(t)\,dt = \int_0^\infty \sum_i \frac{e^{s_i t}}{1 - K\tau e^{-s_i\tau}} \sum_j \frac{e^{s_j t}}{1 - K\tau e^{-s_j\tau}}\, dt$$

$$= -\sum_i \frac{1}{(1 + \tau s_i)} \sum_j \frac{1}{(1 + \tau s_j)(s_j + s_i)} \tag{3.4}$$

where use has been made of (3.3)

To evaluate this, use is made of the identity

$$\frac{1}{2\pi i} \oint_C \frac{f'(z)\phi(z)\,dz}{f(z)} = \sum_j \phi(a_j) - \sum_k \phi(b_k) + \sum_m \text{res} \frac{f'(c_m)\phi(c_m)}{f(c_m)} \tag{3.5}$$

in which C is a closed curve, $f(z)$ and $\emptyset(z)$ are meromorphic functions within C, a_j and b_k are respectively the zeros and poles of $f(z)$, and c_m are the poles of $\emptyset(z)$ within C. Moreover, it is assumed that the poles of $\emptyset(z)$ and the zeros of $f(z)$ do not overlap. To obtain an expression for the inner sum of (3.4), C is chosen to be the circle at infinity and

$$f(z) = z + Ke^{-z\tau}, \quad \emptyset(z) = -\frac{1}{(z+s_i)(1+\tau z)} \tag{3.6}$$

(3.5) then yields

$$\sum_j \frac{1}{(s_j+s_i)(1+\tau s_j)} = \frac{s_i}{s_i^2+K^2} \tag{3.7}$$

and the cost functional J reduces to

$$J = -\sum_i \frac{s_i}{(1+\tau s_i)(s_i^2+K^2)} \,. \tag{3.8}$$

A similar process to the above then gives

$$J = \sum \text{res} \frac{1-K\tau e^{-z\tau}}{z+Ke^{-z\tau}} \cdot \frac{z}{(1+\tau z)(z^2+K^2)} \tag{3.9}$$

in which the sum is taken over the residues at the three poles $z = \pm iK$ and $z = -1/\tau$. Evaluating these and simplifying then gives

$$J = \frac{\cos K\tau}{2K(1-\sin K\tau)} \tag{3.10}$$

as before.

A third method is obtained by considering the differential equation for $e(t)$. For the above example

$$\frac{de}{dt}(t) = -Ke\ (t-\tau) \tag{3.11}$$

which may be integrated to give

$$e(t) = H(t) - K(t-\tau)H(t-\tau) + K^2 \frac{(t-2\tau)^2 H(t-2\tau)}{2!} - \ldots \tag{3.12}$$

Consequently, the cost functional J is given by

$$J = \int_0^\infty e^2(t)dt = \frac{-1}{K}\int_0^\infty e(t) \frac{d}{dt} e(t+\tau)dt \tag{3.13}$$

making use of (3.11) with t replaced by $(t+\tau)$.

Integration by parts, followed by a change of variable in the integral, then gives

$$J = \frac{e(0)e(\tau)}{K} - \int_0^\infty e(t+2\tau)e(t)dt. \tag{3.14}$$

Repeated applications of this process, making use of (3.11) with a suitable change of variable, and the fact that $e(0) = 1$, results in

$$KJ = e(\tau) - e(3\tau) + e(5\tau) - e(7\tau) + \ldots$$

$$= 1 - \left(1 - 2K\tau + \frac{(K\tau)^2}{2!}\right) + \left(1 - 4K\tau + \frac{(3K\tau)^2}{2!} - \frac{(2K\tau)^3}{3!} + \frac{(K\tau)^4}{4!}\right) - \ldots \quad (3.15)$$

making use of (3.12).

This infinite series may be summed by first noting that the final term in each of the above expressions yields the series for $\cos K\tau$. Secondly, grouping together the penultimate terms yields the series for $\sin 2K\tau$. Although the series is not absolutely convergent, such a re-arrangement may be justified. Hence, continuing in this manner,

$$KJ = \cos K\tau + \sin 2K\tau - \cos 3K\tau - \sin 4K\tau + \cos 5K\tau + \ldots \quad (3.16)$$

which may be easily summed to give

$$J = \frac{\cos K\tau}{2K(1 - \sin K\tau)} \quad (3.17)$$

as before.

4. EXTENSION TO COMMENSURATE DELAYS

For systems involving two time-delays, if one is twice the other then an extension of the direct method may be used to evaluate the cost functional. For example, if

$$E(s) = \frac{1}{s + (1-\alpha)e^{-s\tau} + \alpha e^{-2s\tau}} \quad (4.1)$$

then

$$J = \frac{1}{2\pi i}\int_{-i\infty}^{i\infty} \frac{ds}{(s + (1-\alpha)e^{-s\tau} + \alpha e^{-2s\tau})(-s + (1-\alpha)e^{s\tau} + \alpha e^{2s\tau})} \quad (4.2)$$

As before, the first step is to split the integrand into two parts. This is achieved by finding functions $A(s)$ to $C(s)$ which are independent of τ and such that

$$\frac{1}{(s + (1-\alpha)e^{-s\tau} + \alpha e^{-2s\tau})(-s + (1-\alpha)e^{s\tau} + \alpha e^{2s\tau})}$$

$$\equiv \frac{A(s) + B(s)e^{-s\tau} + C(s)e^{-2s\tau}}{s + (1-\alpha)e^{-s\tau} + \alpha e^{-2s\tau}} + \frac{A(-s) + B(-s)e^{s\tau} + C(-s)e^{2s\tau}}{-s + (1-\alpha)e^{s\tau} + \alpha e^{2s\tau}} \quad (4.3)$$

Moreover it may be shown that these functions are ratios of polynomials in s and that without loss of generality, it is also possible to demand that $C(s) = C(-s)$. Cross-multiplying and equating coefficients of the exponentials in (4.3) gives

$$\left.\begin{aligned} &\alpha A(s) + sC(-s) = 0 \\ &(1-\alpha)A(s) + \alpha B(s) + (1-\alpha)C(-s) + sB(-s) = 0 \\ &-sA(s) + (1-\alpha)B(s) + \alpha C(s) + \alpha C(-s) + (1-\alpha)B(-s) + sA(-s) = 1 \end{aligned}\right\} \quad (4.4)$$

the solution of which is

$$\left.\begin{aligned} &A(s) = -\frac{s(s^2+\alpha^2)}{2\Delta(s)}, \quad B(s) = \frac{(1-\alpha)[s^2+2\alpha s-\alpha^2]}{2\Delta(s)}, \quad C(s) = \frac{\alpha(s^2+\alpha^2)}{2\Delta(s)} \\ &\Delta(s) = s^4 + [2\alpha^2 + (1-\alpha)^2]s^2 + (2\alpha-1)\alpha^2 . \end{aligned}\right\} \quad (4.5)$$

Following the above procedure of indenting the contour to the left of the zeros of $\Delta(s)$ and then closing the contours for the two parts in the appropriate ways then yields

$$\begin{aligned} J &= -\sum \text{res} \left\{ \frac{A(s) + B(s)e^{-s\tau} + C(s)e^{-2s\tau}}{s + (1-\alpha)e^{-s\tau} + \alpha e^{-2s\tau}} \right\} \\ &= \sum \text{res} \frac{(s^2+\alpha^2)}{2\Delta(s)} \left\{ \frac{(1-\alpha)(s-\alpha) + (s^2+\alpha^2)e^{-s\tau}}{(1-\alpha)(s-\alpha) - (s^2+\alpha^2)e^{-s\tau}} \right\} \end{aligned} \quad (4.6)$$

where the sum is taken over the zeros of $\Delta(s)$.

Thus, as before, the problem is reduced to finding the roots of a polynomial and then evaluating residues at these points. Moreover, in principle, this method of evaluation may be used for systems involving an arbitrary number of time-delays provided it is possible to separate the integrand into two parts in the above manner and such that the coefficients involved are ratios of polynomials.

This will be the case provided that all the delays are integer multiples of some fundamental value. For example, if

$$E(s) = \frac{1}{A_o(s) + A_1(s)e^{-s\tau} + \ldots + A_{n-1}(s)e^{-(n-1)s\tau} + A_n(s)e^{-ns\tau}} \quad (4.7)$$

where the $A_r(s)$ are polynomials in s, then, with $z = e^{s\tau}$ it is possible to write

$$\begin{aligned} E(s)E(-s) = &\frac{B_o(s) + B_1(s)z^{-1} + \ldots + B_{n-1}(s)z^{-(n-1)} + B_n(s)z^{-n}}{A_o(s) + A_1(s)z^{-1} + \ldots + A_{n-1}(s)z^{-(n-1)} + A_n(s)z^{-n}} \\ &+ \frac{B_o(-s) + B_1(-s)z + \ldots + B_{n-1}(-s)z^{n-1} + B_n(-s)z^n}{A_o(-s) + A_1(-s)z + \ldots + A_{n-1}(-s)z^{n-1} + A_n(-s)z^n} \end{aligned} \quad (4.8)$$

where the functions $B_r(s)$ are determined by cross-multiplying and equating coefficients of powers of z. As before, it is also possible to demand that $B_n(s) = B_n(-s)$. The functions $B_r(s)$ are found to be ratios of polynomials in s and so by the above procedure, the problem is reduced to finding the roots of a polynomial and evaluating residues.

5. APPROXIMATE METHODS

It is worth noting that the only property of the exponential function that is used in the evaluation technique is that

$$f(-x) = [f(x)]^{-1} \qquad \text{i.e.,} \quad e^{-x} = (e^{x})^{-1} \tag{5.1}$$

Consequently an identical technique will also be applicable for any other function having this property, in particular the diagonal Padé approximations of the exponential. These are of the form

$$f(x) = \frac{1 + a_1 x + a_2 x^2 + \ldots + a_n x^n}{1 - a_1 x + a_2 x^2 + \ldots + a_n (-x)^n} \tag{5.2}$$

and clearly satisfy (5.1). It therefore follows that integrals of the above type may be approximated by replacing the exponentials by diagonal Padé approximations and moreover these approximations will converge to the correct answer as more terms are taken in the Padé approximations. This suggests reasonable approximations for problems involving two arbitrary time-delays, such as occur in the problem of mis-match, may be obtained by using Padé approximations.

An alternative approximate method is to seek a series expression for the cost functional, as in (3.15), which may be truncated to give an approximate expression. Even for systems involving more than one time-delay, a formulation in terms of a delay-differential equation and the use of integration by parts will yield an approximation.

6. DELAY-FREE SYSTEMS

Finally, returning to the delay-free case, it is possible to use the above approach to derive the Routh-Hurwitz stability criteria. In §1 it was shown that if $E(s) = B(s)/A(s)$ and if all the zeros of A(s) lie in the left half-plane then $J = f_1/a_o$ where the coefficient f_1 is obtained from (1.15). To derive these criteria the following theorems are required. The first is

<u>Theorem 1</u> $f_1/a_o > 0$ for all non-zero B(s) if and only if

the zeros of A(s) lie in the left half-plane. The proof of this theorem is in two parts. Firstly, J, as defined in (1.11), is positive for all non-zero B(s) and all A(s) since on the imaginary axis (B(-s)/A(-s)) is the complex conjugate of (B(s)/A(s)). Thus if all the zeros of A(s) lie in the left half-plane $f_1/a_o = J > 0$. To prove the converse particular cases of B(s) are chosen. For example, if $s = X$ is a real root of $A(s) = 0$, choose $B(s) = A(s)/(s - X)$. It then follows from (1.15) that $F(s) = A(-s)/2X(s + X)$ and consequently $f_1 = -a_o/2X$. Thus $f_1/a_o > 0$ implies $X < 0$ and so any real root of $A(s) = 0$ lies in the left half-plane. Similarly it may be shown that any complex root always lies in the left half-plane.

The second theorem required is

<u>Theorem 2</u> $f_1/a_o > 0$ for all non-zero B(s) if and only if $a_1/a_o > 0$ and $f_1^{(1)}/a_o^{(1)} > 0$ for all non-zero b_1 and $B^{(1)}(s)$ where

$$B^{(1)}(s) = B(s) - \frac{b_1}{a_1}\tilde{A}(s) \ , \quad \tilde{A}(s) = a_1 s^{n-1} + a_3 s^{n-3} + \ldots$$
$$A^{(1)}(s) = A(s) - \frac{a_o s}{a_1}\tilde{A}(s) \tag{6.1}$$

and

$$A^{(1)}(s)F^{(1)}(s) + A^{(1)}(-s)F^{(1)}(-s) = B^{(1)}(s)B^{(1)}(-s) \tag{6.2}$$

and $$a_1 f_1 + a_o f_2 = \frac{b_1^2}{2} \ . \tag{6.3}$$

The polynomials $A^{(1)}(s)$ and $B^{(1)}(s)$ are respectively one degree lower than A(s) and B(s) and are precisely the polynomials introduced by Åstrom[5]. Repeated use of this theorem then leads to the Routh-Hurwitz criteria as given in Åstrom[5].

7. CONCLUSIONS

In the above it has been shown how the integral of square error may be evaluated for systems involving a single time-delay, and how it is possible to extend one of the methods to systems involving commensurate delays. For systems involving time-delays other than these, approximate methods must be used. The emphasis throughout has been on techniques of evaluation and not the applications of the results. These will be considered in the companion paper[1].

The only cost functional considered was that of the integral of square error. However, since the method requires only the use of Parseval's theorem and a partial fractions separation, it is possible that the method might extend to other cost functionals, and work on this is currently in progress.

8. REFERENCES

1. Marshall, J.E. and Walton, K. (1984) The analytical design of TDS controllers. This conference
2. Walton, K. and Marshall, J.E. (1984) Closed-form solution for time-delay systems cost functions. Int. J. Control **39**, 1063
3. James, H.M., Nichols, N.B. and Phillips, R.S. (1947) Theory of servo mechanisms. McGraw-Hill.
4. Newton, G.C., Gould, L.A. and Kaiser, J.F. (1957) Analytical design of linear feedback controls. Wiley.
5. Åstrom, K.J. (1970) Introduction to stochastic control theory. Academic Press
6. Jacobs, O.L.R. (1974) Introduction to control theory. Clarendon Press.
7. Gorecki, H. and Popek, L. (1983) Control of systems with time-delay. Third IFAC Symposium on control of distributed parameter systems. Eds: Barbary, J.P. and LeLetty, L. Pergamon.

ANALYTICAL DESIGN OF TIME-DELAY SYSTEM CONTROLLERS

J. E. Marshall and K. Walton

(School of Mathematics, University of Bath)

ABSTRACT

The results of Gorecki and Walton[1,2] on the explicit evaluation of TDS cost functionals are applied to the design of controllers, and to the analysis of mismatch effects in predictor control schemes. Commensurate delay techniques are applied to special cases of mismatch and delayed feed-back control.

1. INTRODUCTION

Parametric optimal control is a well established technique for delay-free systems. It is made possible by the existence of cost functionals evaluated in terms of system and controller parameters. The cost functionals are minimised with respect to certain of these parameters.

Techniques of improvement by mismatch make use of the parameters of sub-plant model and delay model parameters in predictive schemes. These parameters, when regarded as free, give extra design variables and offer the possibility of improvement. These techniques of improvement have been discussed at previous IMA conferences[3,4], and have recently been summarised in the review paper of Hocken, Salehi and Marshall[5]. The optimisation with respect to these extra parameters has hitherto been by way of simulation to obtain graphs of cost function values, or by numerical quadrature using Parseval's theorem in the complex domain[6,7,8,9].

The equivalence of certain time delay systems has the advantage that parameters calculated to optimise one scheme also optimise others. This has enabled some results from predictor control to be applied to delay-free plants[9,10]. This has led to improvement in performance over delay free control schemes. It might be thought counter intuitive, and until recently due to difficulties of implementation, inefficient, to add delay to a delay free scheme. However, it has been known for some while[11] that the augmentation of delay, where it is a practical possibility, can improve dynamic behaviour.

In this paper we explore the use of the recent results[14], where cost functionals for time delay systems have been derived explicitly in terms of system parameters. The results apply directly to the cases for which the error takes the form

$$E(s) = \frac{B(s) + D(S).\exp(-s\tau)}{A(s) + C(s).\exp(-s\tau)} \quad .$$

This expression involves a single delay, τ. However, there are cases of interest involving two delays for which $E(s)$ reduces to this form in non-trivial particular examples.

There are many practical control schemes involving two delays[12,13]. In fact, all predictor schemes involve at least two delays; the plant delay and its model. Such schemes and the related mismatch problems are reviewed later in the paper.

It has been found possible to extend the single delay technique to systems involving two delays which are commensurate, i.e., related by integers. This technique allows analytical solutions for special cases of mismatch in delay. Expressions available for the general case do not reduce to evaluating a finite number of residues, although they are useful for approximations.

It is possible to extend the single delay results, when mismatch in delay is small, to obtain insights into the problems associated with mismatch in two parameters, one of them temporal[13]. These results are an improvement over other approximation techniques as they make use in part of the analytical simplification possible for the single delay case.

We shall introduce application of the single delay method with a simple but surprising example of improvement in performance of a delay free system by the addition of delay in a feedback loop. We then relate this example to the mismatch problem from which it first came to our attention[5]. This demonstrates, in part, the equivalence mentioned earlier. This then leads naturally to a problem of sub-plant mismatch.

The latter part of the paper is concerned with systems with more than one delay. Explicit expressions for cost in the case of commensurate delays are obtained, and illustrated with certain special cases.

2. SYSTEMS WITH A SINGLE TIME DELAY

Consider the system shown in Fig. 1. Denote input and output by $x(t)$, $y(t)$ respectively, and $e(t)$ the error given by $x(t) - y(t)$. We seek to minimise $J(\tau) = \int_0^\infty e^2(t,\tau)dt$, for the step input $x(t) = H(t)$.

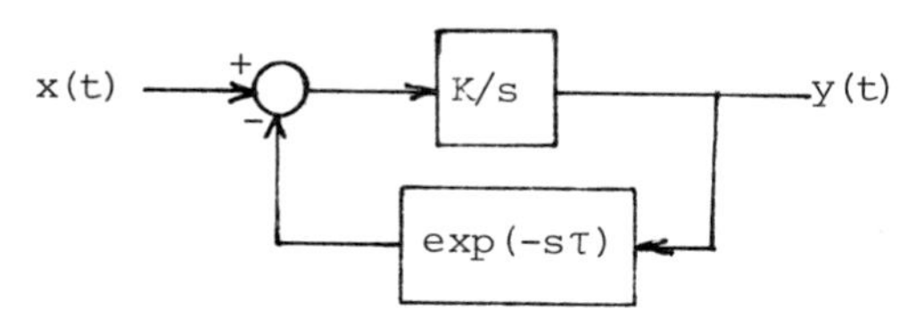

Fig. 1. Delayed feedback

Using results (1.3) and (2.4) of the companion paper[14],

$$J(\tau) = \frac{\cos(K\tau)}{2K(1 - \sin K\tau)} - \tau \quad .$$

Excluding an infinite value of K as impracticable we assume K in the formula is the maximum permitted value. Setting $\frac{\partial J}{\partial \tau} = 0$ it follows that $J(\tau)$ is a minimum at $\tau = \frac{\pi}{6K}$. We note that $J\left(\frac{\pi}{6K}\right) = 0.685J(0)$, an improvement of 31.5% over the delay free value. This example shows that improvement is possible by addition of the delay, whatever the value of K, see Fig. 2.

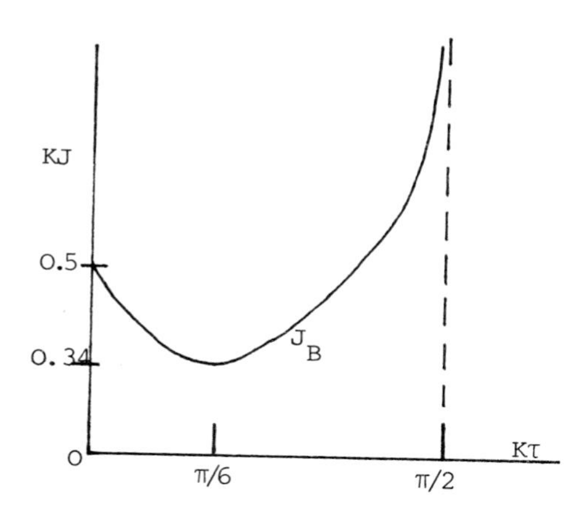

Fig. 2. Costs for 1st order plant with delay τ

The application to plants of higher order but with this structure is discussed in reference 2. For example, when

$G(s) = K(s+1)/s^2$, $J = \int_0^\infty (x(t) - y(t))^2 dt$ is given by

$$J = \frac{1}{2(\alpha^2+\beta^2)} \left(\frac{K+\beta^2 \cos \beta\tau}{K - \beta \sin \beta\tau} - \frac{K - \alpha^2 \cosh \alpha\tau}{K + \alpha \sinh \alpha\tau} \right) - \tau$$

and α^2, $-\beta^2$ are the roots of $y^2 + K^2 y - K^2 = 0$. For $K > 1$ a minimum exists for $\tau > 0$, as $\frac{\partial J}{\partial \tau}$ is negative at $\tau = 0$. The cases $G(s) = \frac{K(s+a)}{s^2}$, and $G(s) = \frac{k}{s(s+a)}$ are to be found there also, together with techniques equivalent to changes of time scale.

Gorecki and Popek[2] have shown that impulsive disturbance, and non-zero (point) initial conditions may be included in the calculations. This is equivalent to making the corresponding modifications to the coefficients A(s), B(s), C(s) and D(s) in the expression for the transformed error. There are no new matters of principle. The corresponding expressions for J, as to be expected, are somewhat complicated. These authors also discuss plants of the form $G(s) = \frac{k}{s+\alpha} e^{-s\tau}$, not including an integration term. In cases such as these where the steady state value of error is non-zero it is essential as they point out to redefine error with respect to the steady-state so that the corresponding cost integrals converge.

3. SYSTEMS WITH TWO DELAYS, BUT REDUCIBLE

Many practical control schemes for time-delay systems are of the predictor type, involving a model (or models) of the controlled plant. Fig. 3 shows such a prediction control scheme and recalls the notation. Mismatch (which may be deliberate or accidental) is said to occur when $\tau \neq \tau'$ or $G \neq G'$, or both.

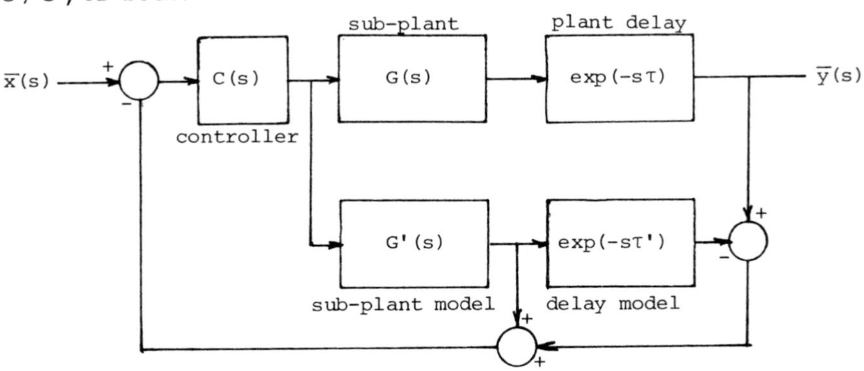

Fig. 3. Predictor control scheme

The error transform

$$E(s) = \overline{x}(s) - \overline{y}(s) = \overline{x}(s) \cdot \frac{1 + CG'(1 - \exp(-s\tau'))}{1 + CG' + C(G\exp(-s\tau) - G'\exp(-s\tau'))}$$

will not, in general, be of the required form due to the existence of two delay terms. Before looking at the general cases it is worthwhile to consider two examples where the reduction is in fact possible. Neither example is in any sense artificial.

Example 1 When $\tau = \tau'$, and $G \neq G'$. The error transform function is given by

$$e(s) = \overline{x}(s) - \overline{y}(s) = \frac{1 + CG'.(1 - \exp(-s\tau))}{1 + CG' + C(G(s) - G'(s))\exp(-s\tau)} \cdot \overline{x}(s)$$

evidently of the required form after a little algebra. When $G = G'$, also, it follows that $J = 1/2K + \tau$, by Smith's principle (when $G = K/s$).

Example 2 $\tau' = 0$. This example arises from deliberate maximal underestimation of plant delay which is appropriate for certain $G(s)$, when τ is small[5].

$$E(s) = \overline{x}\left(1 - \frac{CGT}{1 + CG' + C(GT - G')}\right) \text{ and hence } E(s) = \frac{1}{1 + CGT}\,\overline{x},$$

independent of G', and equivalent to simple unity negative feedback of the plant GT with series controller C, emphasising the need for 'small' τ, and showing for cases where *underestimation* gives improvement that predictor control is not always an obvious choice. It should be stated that it is far more usual for *overestimation* to lead to improvement, particularly for second order sub-plants.

Example 2 illustrates the equivalence of systems. We note that the expression for $E(s)$ is identical to that of the delay free system with delay added in a feedback path, considered earlier.

Example 1 is considered at length in reference 13. The expressions for

$$J = \int_0^\infty (\text{input-output})^2 dt$$

for the case $C(s) = 1$, $G(s) = K/s$, $G'(s) = K'/s$, $\tau = \tau'$, and $x(t) = H(t)$ is

$$J = \frac{K'(1 + K'\tau)}{\lambda^2} - \frac{K^2}{2\lambda^3}\,\frac{(K - K')\sinh\lambda\tau + \lambda}{(K - K')\cosh\lambda\tau + K'} \quad \text{where} \quad \lambda = \sqrt{2KK' - K^2}\,.$$

When $K' < K/2$, λ is purely imaginary and sinh and cosh are replaced by sin, cos respectively, with a change in the sign of the K^2 term, at $\lambda = 0$, J is continuous.

Although the expression for J is not simple its evaluation is obviously much more efficient than the corresponding explicit integration of $e^2(t)$ obtained from simulation, or via Parseval's theorem used numerically. Detailed discussion of the variation of J with K' and τ is given in reference 13, as is an extension via a perturbation method for the case of simultaneous mismatch of subplant.

For completeness we note that J is made smaller by setting $K' < K$, given that τ exceeds a certain value. Further improvement is possible by a simultaneous mismatch in delay. The sign of the mismatch is K' dependent, but in the region of the minimum with respect to K' is equivalent to overestimation, i.e., $\tau' > \tau$.

4. SYSTEMS WITH TWO DELAYS

The application of the technique to systems of more than one delay is obviously desirable. Delay mismatch problems are clearly of this type. They have received much attention via simulation methods[3,5,6,8,10,12].

When $\tau' \neq \tau$ in the system of Fig. 3 the single delay method cannot be used. We distinguish two cases: when the difference $\tau - \tau'$ is small, and when $\tau = mh$, $\tau' = nh$ for unequal positive integers m and n.

When $(\tau - \tau')$ is small it is possible to use perturbation methods, or the Padé approximant for the term $e^{-s(\tau - \tau')}$, although not exact may be made extremely accurate. In general when $\tau - \tau'$ is not small the analytical method is not straightforward to extend. However, and this is obviously useful, there will be cases where τ and τ' are related by a ratio of integers. The case is given for $2\tau' = \tau$ for an explicit evaluation of cost. We choose the example of Fig. 3 again with $G(s) = G'(s) = K/s$.

This example may be considered either as a mismatch problem of a predictor scheme, or as a delay feedback scheme for the delay-free system K/s. Replacing s/K by s, and $K\tau$ by T we have

$$E(s) = \frac{1}{s}\left(1 - \frac{1}{s + 1 - e^{-sT} + e^{-2sT}}\right).$$

Using the technique of Section 4 of the companion paper we obtain

$$J = -\frac{1}{2\pi i}\int_{-i\infty}^{+i\infty}\frac{ds}{s^2}\left\{1 - \frac{1}{s+1-e^{-sT}+e^{-2sT}} - \frac{1}{-s+1-e^{+sT}+e^{+2sT}}\right\}$$

$$+\frac{1}{2s(s^2+1)}\left\{\frac{-s(s+1)-(s+2)e^{-sT}+e^{-2sT}}{s+1-e^{-sT}+e^{-2sT}} - \text{same expression in } -s\right\}$$

from which it follows that

$$J = 1 - T + T^2 - \frac{1}{2}\cdot\frac{\cos T}{1+\sin T},$$

which has a minimum of ~ 0.43, when $k\tau' \sim 0.31$ and $k\tau = 0.62$.

This cost is greater than when $\tau' = 0$, as we would expect due to the constraint $\tau = 2\tau'$. These results are shown qualitatively in Fig. 4.

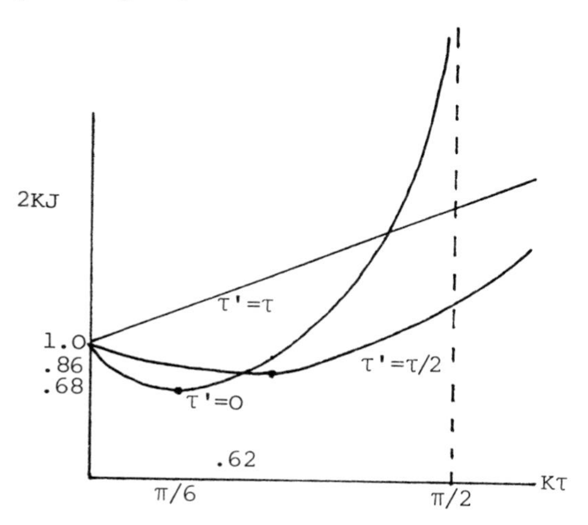

Fig. 4. Costs for various τ'

5. OTHER FEEDBACK STRUCTURES

Consider the feedback arrangement shown in Fig. 5, where

$$\bar{y}(s) = \frac{\bar{x}(s).G(s)}{1 + G(s)\left(1 + \sum_{i=1}^{N}\alpha_i(1-\exp(-sT_i))\right)}$$

We note that the delays are commensurate. Clearly the cases $N=1$, $\alpha_1 = 1$, and $N=2$, $\alpha_1 = -1$, $\alpha_2 = 1$ correspond to problems treated earlier in the paper, and to the feedback structure discussed by Hocken[5].

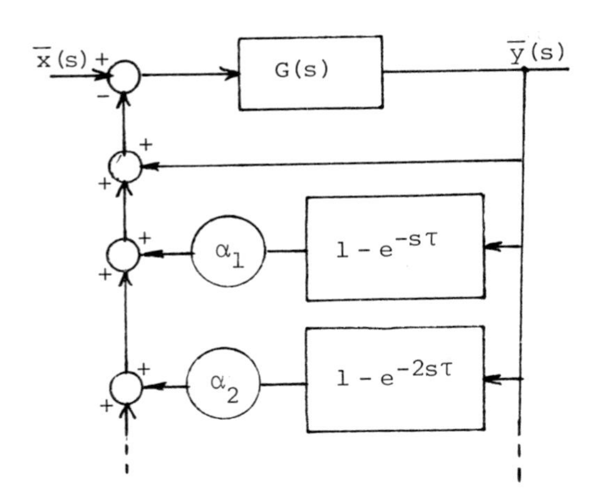

Fig. 5. Commensurate delay feedback scheme

The case $N = 1$, with α_1 and τ free is of interest. The result for $G(s) = K/s$ is that minima are achieved for α taking a large negative value equivalent to a system with positive feedback for $0 < t < \tau$, but which is then stabilised for $t > \tau$ by the feedback due to the delay, provided that $|\tau\alpha|$ does not exceed a certain value.

The commensurate delay analysis extends beyond two delays in an obvious way but the algebraic manipulation increases somewhat. The correction between the structure, and structures related to sampled data control, and linear optimal control justifies further study.

6. CONCLUSIONS

Examples have been given of applications of the newly derived explicit cost functions. These applications are of a preliminary nature and no attempt has been made in the paper to analyse the results of the Ziegler-Nichols[15] and Cohen-Coon[16] results on which much design work is currently based. This and the extension to systems with non-zero initial functions, disturbance and higher order sub-plants has yet to follow. We note that Gorecki has applied some single delay results to first and second order plants with PI and 3-term controllers, and we look forward to the publication of these results.

7. REFERENCES

1. Gorecki, H. and Popek, L. (1983) Control of Systems with

Time-Delay. Third IFAC Symposium on Control of Distributed Parameter Systems. Eds: Barbary, J-P. and LeLetty, L. Pergamon.

2. Walton, K. and Marshall, J.E. (1984) Closed-form Solution for Time-delay Systems Cost Functionals. Int. J. Control **39**, 5, 1063-1071.
3. Garland, B. and Marshall, J.E. (1978) On the Applicability of O.J.M. Smith's Principle. In Recent Theoretical Developments in Control. Ed: Gregson, M.J. Academic Press.
4. Chotai, A. (1981) Parameter Uncertainty and Time-delay System Control. In Third IMA Conference on Control Theory Ed: Marshall et al. Academic Press.
5. Hocken, R.D., Salehi, S.V. and Marshall, J.E. (1983) Time-delay Mismatch and the Performance of Predictor Control Schemes. Int. J. Control, **38**, 2, 433-447.
6. Hocken, R.D. and Marshall, J.E. (1982) Mismatch and the Optimal Control of Linear Systems with Time-delays. Opt. Control. Applns & Methods **3**, 211-219.
7. Hocken, R.D. and Marshall, J.E. (1983) The Effects of Mismatch on an Optimal Control Scheme for Linear Systems with Control Time-delays. Opt. Control Applns & Methods **4**, 47-69.
8. Hocken, R.D., Marshall J.E. and Salehi, S.V. (1983) Time-delay Control: Mismatch Problems. Third IFAC Symposium Control of Distributed Parameter Systems. Eds: Barbary, J-P, LeLetty, L. Pergamon.
9. Marshall, J.E. and Salehi, S.V. (1982) Improvement of System Performance by the Use of Time-delay Elements. IEE Pt D., **127**, 5, 177-181.
10. Chotai, A., Garland, B. and Marshall, J.E. (1981) A Survey of Time-delay System Control Methods. In Control and Its Applications, IEE Conf. Publication. 194, 316-322.
11. The Damping Effect of Time Lag. The Engineer **163**, 1937, 439 (paper by Editorial Staff).
12. Marshall, J.E. Control of Time-delay Systems. Peter Peregrinus. 1979.
13. Walton, K. and Marshall, J.E. (1984) Mismatch in a Predictor Control Scheme. Some Closed Form Solutions. Int. J. Control **40**, 2, 403-419.
14. Walton, K. and Marshall, J.E. (1984) Evaluation of TDS Cost Functionals (This Conference).
15. Ziegler, J.G. and Nichols, N.B. Process Lags in Automatic Control Circuits Trans. ASME **65**, 1943, 433-444.
16. Cohen, G.H. and Coon, G.A. (1973) Theoretical Considerations of Retarded Control. Trans. ASME **75**, 827-834.

EIGENVALUE ESTIMATION IN NON-IDEAL MODAL PROFILE CONTROL OF DISTRIBUTED PARAMETER SYSTEMS

D. Franke

(Hochschule der Bundeswehr Hamburg)

ABSTRACT

The paper is based on infinite-dimensional matrix representations of certain linear distributed parameter control systems. A generalization of Gershgorin's theorems yields Gershgorin disks for all the eigenvalues, thus providing a sufficient stability criterion.

1. INTRODUCTION

The paper recalls the early work by Gould and Murray-Lasso[1] and by Gilles[2] on modal control of distributed parameter systems with distributed feedback. In practical implementation, the use of N actuators and N sensors is based on the assumption that there is negligible eigenfunction content beyond the N-th one.

However, the use of small N will cause the eigenvalues to be shifted from their pre-assigned values of the idealized design. In some situations this may give rise to instability of the actually infinite-dimensional system.

Therefore, a method will be presented which provides eigenvalue estimation for certain infinite-dimensional linear feedback control systems. The idea is based on an extension of Gershgorin's theorems proposed by Franke[3,4].

2. PROBLEM FORMULATION

The paper is concerned with infinite-dimensional sets of first order ordinary differential equations, which arise from modal analysis of certain distributed parameter plants in connection with certain feedback control laws.

Let the linear and time-invariant plant be described by a partial differential equation of the type

$$\partial x(t,\underline{z})/\partial t = Dx(t,\underline{z}) + u(t,\underline{z}), \tag{2.1}$$

where $x(t,\underline{z})$ is the time and space dependent state, defined on a finite and simply connected region $\underline{z}\varepsilon\Omega$. Moreover,

$$u(t,\underline{z}) = \underline{b}^T(\underline{z})\underline{u}(t) \tag{2.2}$$

is the time and space dependent control with specified $\underline{b}^T(\underline{z})$ and lumped N-dimensional control vector $\underline{u}(t)$. D is a self-adjoint spatial differential operator with a discrete spectrum of eigenvalues $\lambda_1, \lambda_2 \ldots$ in Ω, and corresponding orthonormal eigenfunctions $\Phi_1(\underline{z}), \Phi_2(\underline{z}), \ldots$. The boundary conditions of x are assumed to be formally homogeneous by making use of the extended definition of operators[5].

It is assumed that N sensors provide the lumped output vector

$$\underline{y}(t) = \int_\Omega \underline{c}(\underline{z})x(t,\underline{z})d\Omega, \tag{2.3}$$

where $\underline{c}(\underline{z})$ is specified via sensor location.

Modal analysis of such a boundary value problem yields the modal representation

$$\dot{x}_i^*(t) = \lambda_i x_i^*(t) + \underline{b}_i^{*T}\underline{u}(t), \quad i = 1, 2, 3, \ldots, \tag{2.4}$$

$$y_i(t) = \sum_{k=1}^{\infty} c_{ik}^* x_k^*(t), \quad i = 1, \ldots, N, \tag{2.5}$$

where the modes $x_i^*(t)$ are defined by

$$x_i^*(t) = \int_\Omega x(t,\underline{z})\Phi_i(\underline{z})d\Omega, \quad i = 1, 2, 3, \ldots . \tag{2.6}$$

Assume now, for the feedback control of the above plant, a lumped linear feedback law of the type

$$\underline{u}(t) = - \underline{K}\ \underline{y}(t), \tag{2.7}$$

where $\underline{K}$ is a constant gain matrix. Insertion into Eqs. (2.4), (2.5) yields the equation of the over-all control system, which is of the type

$$\dot{\underline{x}}^* = \underline{A}\ \underline{x}^*, \tag{2.8}$$

where $\underline{x}^*$ is the infinite-dimensional vector of modes x_i^*, and matrix $\underline{A}$ has an infinite number of rows.

It should be mentioned that the same type of equation is obtained in case of lumped <u>dynamical</u> feedback instead of Eq. (2.7). Vector $\underline{x}^*$ is then augmented by the vector of the lumped subsystem. The consideration can also be extended to PDE's of second order with respect to time t, e.g. oscillating beams.

Under what conditions will system (2.8) be stable? A sufficient stability criterion has been provided by Franke[3] for <u>single input-single output</u> distributed systems (N=1), based on an extension of Gershgorin's theorems. As a result, Gershgorin disks are obtained for all the eigenvalues of the infinite-dimensional system. The method of order reduction serves as a tool to specify, in a first step, the eigenvalues of the reduced control system.

3. IDEALIZED MODAL PROFILE CONTROL

In case of multiple input-multiple output distributed systems (N>1) the following procedure is proposed which is different from [3]. In the first step, an idealized modal profile controller is designed, following the work by Gould and Murray-Lasso[1] and by Gilles[2]. This method aims at a shifting of the first N dominant eigenvalues.

In the second step, the actual position of the eigenvalues of the infinite-dimensional system will be estimated by means of Gershgorin disks as proposed in [3].

The simplest idealized modal profile controller with proportional gain is

$$u_i^*(t) = \begin{cases} -k_i x_i^*(t), & i = 1, \ldots, N, \\ 0, & i = N+1, \ldots, \end{cases} \tag{3.1}$$

where k_i, $i = 1, \ldots, N$, are the gain factors, and $u_i^*(t)$ are the Fourier coefficients of $u(t,\underline{z})$, where

$$u(t,\underline{z}) = \sum_{i=1}^{N} u_i^*(t)\Phi_i(\underline{z}) \tag{3.2}$$

is used instead of Eq. (2.2).

Because of

$$\dot{x}_i^*(t) = \lambda_i x_i^*(t) + u_i^*(t), \quad i = 1, 2, 3, \ldots, \tag{3.3}$$

the following eigenvalues ρ_i are obtained by idealized modal closed-loop control:

$$\rho_i = \begin{cases} \lambda_i - k_i, & i = 1, \ldots, N. \\ \lambda_i, & i = N+1, \ldots . \end{cases} \tag{3.4}$$

From equation (3.4), the gains k_i can be determined such that

$\rho_1, \ldots, \rho_N$ take specified values. The eigenvalues $\lambda_{N+1}, \ldots$ must of course be presumed to be stable.

4. EIGENVALUE ESTIMATION BY MEANS OF GERSHGORIN DISKS

Bearing in mind Eq. (2.2) for the actual system, the Fourier coefficients of $u(t,\underline{z})$ are

$$u_i^*(t) = \int_\Omega \underline{b}^T(\underline{z})\Phi_i(\underline{z})d\Omega \cdot \underline{u}(t) = \underline{b}_i^{*T}\underline{u}(t), \tag{4.1}$$

hence

$$\underline{u}^*(t) = \underline{B}^*\underline{u}(t), \tag{4.2}$$

where

$$\underline{b}_i^{*T} = \int_\Omega \underline{b}^T(\underline{z})\Phi_i(\underline{z})d\Omega, \quad i = 1, 2, 3, \ldots, \tag{4.3}$$

are the rows of $\underline{B}^*$.

In the same way we obtain, by means of Eq. (2.3),

$$\underline{y}(t) = \underline{C}^*\underline{x}^*(t), \tag{4.4}$$

where

$$\underline{c}_k^* = \int_\Omega \underline{c}(\underline{z})\Phi_k(\underline{z})d\Omega, \quad k = 1, 2, 3, \ldots, \tag{4.5}$$

are the columns of $\underline{C}^*$.

Let the square matrices

$$\hat{\underline{B}} = \begin{bmatrix} \underline{b}_1^{*T} \\ \vdots \\ \underline{b}_N^{*T} \end{bmatrix} \quad \text{and } \hat{\underline{C}} = \left[\underline{c}_1^*, \ldots, \underline{c}_N^* \right] \tag{4.6}$$

be non-singular. Then the idealized modal controller can be implemented approximately by

$$\underline{u}(t) = \hat{\underline{B}}^{-1}\underline{u}^*(t) = -\hat{\underline{B}}^{-1}\mathrm{diag}(k_i)\hat{\underline{C}}^{-1}\underline{y}, \tag{4.7}$$

which determines, with regard to Eq. (2.7), the gain matrix $\underline{K}$:

$$\underline{K} = \hat{\underline{B}}^{-1}\mathrm{diag}(k_i)\hat{\underline{C}}^{-1}. \tag{4.8}$$

What is the effect of $\underline{K}$, when applied to the infinite-dimensional plant?

By means of Eqs. (4.2), (4.4), and (4.7) the following modal representation of the overall control system is obtained:

$$\dot{\underline{x}}^*(t) = [\text{diag}(\lambda_i) - \underline{B}^*\hat{\underline{B}}^{-1}\text{diag}(k_i)\hat{\underline{C}}^{-1}\underline{C}^*]\underline{x}^*(t), \tag{4.9}$$

and therefore, matrix $\underline{A}$ in Eq. (2.8) here takes the form

$$\underline{A} = \left[\begin{array}{c|c} \text{diag}(\rho_i)_N & -\text{diag}(k_i)_N\hat{\underline{C}}^{-1}\overline{\underline{C}} \\ \hline -\overline{\underline{B}}\,\hat{\underline{B}}^{-1}\text{diag}(k_i)_N & \begin{array}{c}\text{diag}(\lambda_i) - \overline{\underline{B}}\,\hat{\underline{B}}^{-1}\text{diag}(k_i)_N\hat{\underline{C}}^{-1}\overline{\underline{C}} \\ \text{from } i=N+1\end{array} \end{array}\right], \tag{4.10}$$

where $\overline{\underline{B}}$ and $\overline{\underline{C}}$ are defined by

$$\underline{B}^* = [\hat{\underline{B}}, \overline{\underline{B}}], \quad \underline{C}^* = [\hat{\underline{C}}, \overline{\underline{C}}]. \tag{4.11}$$

The procedure for estimating the eigenvalues of $\underline{A}$ is the same as in [3,4] and will not be repeated here in detail. The method is applicable whenever the matrix $\underline{A}$ is diagonal dominant in the following sense:

$$\text{(a)} \quad \text{Re}(\lambda_i) \to -c \cdot i^2 \text{ for } i \to \infty \tag{4.12}$$

$$\text{(b)} \quad |a_{ik}| \leq M, \; i \neq k, \tag{4.13}$$

where c and M are positive constants.

5. EXAMPLE

Let the plant be given by

$$\frac{\partial x(t,z)}{\partial t} = \frac{1}{\pi^2} \cdot \frac{\partial^2 x(t,z)}{\partial z^2} + u(t,z), \quad 0 \leq z \leq 1,$$

$$x(t,0) = x(t,1) = 0,$$

$$u(t,z) = b_1(z)u_1(t) + b_2(z)u_2(t),$$

$$b_1(z) = \begin{cases} 1, & 0 \leq z \leq 0.5 \\ 0, & 0.5 < z \leq 1, \end{cases}$$

$$b_2(z) = \begin{cases} 0, & 0 \leq z \leq 0.5 \\ 1, & 0.5 < z \leq 1, \end{cases}$$

$$y_1(t) = \int_0^{0.5} x(t,z)dz, \quad y_2(t) = \int_{0.5}^{1} x(t,z)dz.$$

The plant eigenvalues are

$$\lambda_i = -i^2, \; i = 1,2,3,\ldots$$

The idealized controller is required to shift λ_1 and λ_2 to $\rho_1 = \rho_2 = -6$. Hence we have

$k_1 = \lambda_1 - \rho_1 = 5, \qquad k_2 = \lambda_2 - \rho_2 = 2.$

The following Gershgorin disks are obtained for the eigenvalues of the actual non-ideal control system (N=2):

Dominant disks K_1, K_2:

K_1 centred at $\rho_1 = -6$, with radius $r_1 = \alpha_1 \cdot 5 \cdot \left(\frac{\pi^2}{8} - 1\right)$,

K_2 centred at $\rho_2 = -6$, with radius $r_2 = \alpha_2 \cdot \left(\frac{\pi^2}{8} - 1\right)$.

Non-dominant disks K_i (i = 3,5,7,9,...):

centred at $z_i = \lambda_i - \frac{5}{i^2}$, with radius

$$r_i = 5 \cdot \left[\left(\frac{\pi^2}{8} - 1\right) + \frac{1}{\alpha_1} - \frac{1}{i^2}\right].$$

Non-dominant disks K_i (i = 4,8,12,16,...):

centred at $z_i = \lambda_i$, with radius $r_i = 0$

(since non-controllable and non-observable).

Non-dominant disks K_i (i = 6,10,14,18,...):

centred at $z_i = \lambda_i - \frac{4}{i^2}$, with radius

$$r_i = \frac{4}{\alpha_2} + \frac{\pi^2}{8} - 1 - \frac{4}{i^2}.$$

The free positive parameters α_1, α_2 are determined such that $r_1 = r_2 = 1.5$, hence $\alpha_1 = 1.305$; $\alpha_2 = 6.52$. Fig. 1 illustrates the Gershgorin disks $K_1, \ldots, K_7$. Obviously stability of the closed-loop system is guaranteed.

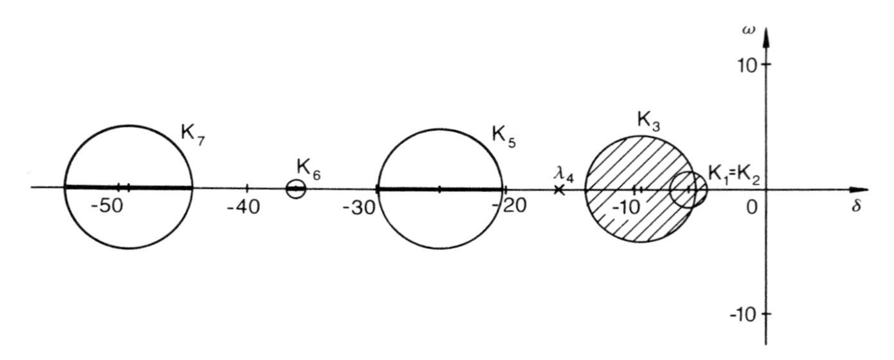

Fig. 1: Gershgorin disks in the example. The hatched domain includes three eigenvalues, each thick section includes one eigenvalue.

6. REFERENCES

1. Gould, L.A. and Murray-Lasso, M.A. (1969), On the Modal Control of Distributed Systems with Distributed Feedback. IEEE Trans. Aut. Control AC-11, pp. 729-737.

2. Gilles, E.-D., (1973), Systeme mit verteilten Parametern. R. Oldenbourg Verlag, München, Wien.

3. Franke, D., (1982), Eigenvalue Estimation by Means of Gershgorin Disks for Infinite-Dimensional Linear Feedback Control Systems. Preprints of the 3rd IFAC-Symposium on Control of Distributed Parameter Systems, Toulouse, pp. X.17-X.20.

4. Franke, D., (1984), On the Influence of Neglected Modes in Distributed Parameter Control Systems. Regelungstechnik 32, pp. 151-156.

5. Brogan, W.L., (1968), Optimal Control Theory applied to Systems described by Partial Differential Equations. In Leondes (Ed.): Advances in Control Systems 6, Academic Press, New York.

ON THE CONTROL OF PARTIAL DIFFERENTIAL EQUATIONS AND ALGEBRAIC GEOMETRY

S.P. Banks

(Department of Control Engineering, University of Sheffield)

ABSTRACT
The classical root locus and Laplace transformation inversion theories are generalized to partial differential equations using algebraic geometry.

1. INTRODUCTION

The control of distributed parameter systems of the form

$$
\begin{aligned}
x &= Ax + Bu \\
y &= Cx
\end{aligned}
\qquad (1.1)
$$

where x belongs to (say) a Hilbert space H, A generates a semigroup T(t) and B, C are (possibly unbounded) operators has been developed recently into a theory which generalizes many of the finite-dimensional results[1,2]. If A generates an analytic semigroup, then the spectral representation of A plays a crucial role in this theory.

In this paper we shall consider systems of the type

$$
\sum_{i+j\leq n} a_{ij}\frac{\partial^{i+1}\phi}{\partial t^{i}\partial x^{j}} = \delta(x)u(t)\ , \qquad x \in [o,\infty). \qquad (1.2)
$$

$$
y(t,x) = \phi(t,x)\ ,
$$

where we have chosen point control at the origin and we have taken C = I for simplicity. Our objective is to study the system (1.2) in the 'frequency domain' and to generalize some recent results on the spectral locus theory of distributed systems[3,4,5,6]. The latter papers have demonstrated the possibility of a distributed root locus theory for many types of systems and even such a theory for certain nonlinear systems.

Assuming zero initial and boundary conditions in (1,2) we may take the joint t and x Laplace transforms to obtain

$$\sum_{i+j\leq n} a_{ij}\,(s_1^i s_2^j)\,\Phi(s_1,s_2) = U(s_2)$$

$$Y(s_1,s_2) = \Phi(s_1,s_2) \; .$$

We define the (generalized) transfer function of this system to be

$$G(s_1,s_2) = \frac{Y(s_1,s_2)}{U(s_2)} = \frac{1}{\sum\limits_{i+j\leq n} a_{ij}(s_1^i s_2^j)}$$

If U is the output of a compensator $kK(s_1,s_2)$ with input V-Y, then the overall feedback system has the transfer function

$$F(s_1,s_2) = \frac{kG(s_1,s_2)K(s_1,s_2)}{1+kG(s_1,s_2)K(s_1,s_2)}$$

$$= \frac{kp(s_1,s_2)}{q(s_1,s_2)+kp(s_1,s_2)} \qquad (1.3)$$

where GK = p/q for $p,q \in \mathbb{C}[s_1,s_2]$ (=set of complex polynomials in two variables).

We shall examine two problems in this paper. Firstly, we shall consider the inversion of the joint Laplace transform of a general rational function p/q and then we shall consider the 'root locus' of the system (1.3). The theory will depend on some elementary results in algebraic geometry which may be found in Kendig[7].

2. POLES AND ZEROS OF RATIONAL FUNCTIONS OF TWO VARIABLES

Consider the rational transfer function

$$G(s) = p(s_1,s_2)/q(s_1,s_2)$$

and write p and q as products of irreducible factors. Then

$$G(s) = \frac{p_1^{\alpha_1}(s_1,s_2)p_2^{\alpha_2}(s_1,s_2) \;\ldots\ldots\; p_m^{\alpha_m}(s_1,s_2)}{q_1^{\beta_1}(s_1,s_2)q_2^{\beta_2}(s_1,s_2) \;\ldots\ldots\; q_n^{\beta_n}(s_1,s_2)} \qquad (2.1)$$

for some natural numbers α_i, β_j. Consider a typical irreducible polynomial $p_i(s_1,s_2)$. The set

$$Z_i = \overline{\{(s_1,s_2)\in\mathbb{C}^2 : p_i(s_1,s_2) = 0\}} \quad (1\leq i\leq m)$$

will be called the zero of the system (2.1) corresponding to p_i, where $\overline{\{ \sim \}}$ denotes closure of a set in projective space $\mathbb{P}^2(\mathbb{C})$. (The reason for defining the zero in $\mathbb{P}^2(\mathbb{C})$ will be seen later). Similarly the sets

$$P_j = \{(s_1,s_2)\varepsilon\mathbb{C}^2 : q_j\ (s_1,s_2) = 0\}\ (1 \leq j \leq n)$$

are called the poles of (2.1). Note that the zero Z_i has multiplicity α_i while the pole P_j has multiplicity β_j. The poles and zeros of a finite dimensional linear system $G(s) = p(s)/q(s)$ are just the roots of the polynomials p and q which have irreducible factors of the form $(s-z_i)$, $(s-p_j)$, respectively. In the present case the poles and zeros are sets in $\mathbb{P}^2(\mathbb{C})$ which are algebraic varieties, i.e. solutions of polynomial equations. The topological structure of these sets has been determined by classical algebraic geometry (Kendig[7]) and sets of this form are compact orientable surfaces of some genus g with a finite number of points identified to other points. (Fig. 1.) (A compact orientable surface of genus g is just a sphere with g handles attached).

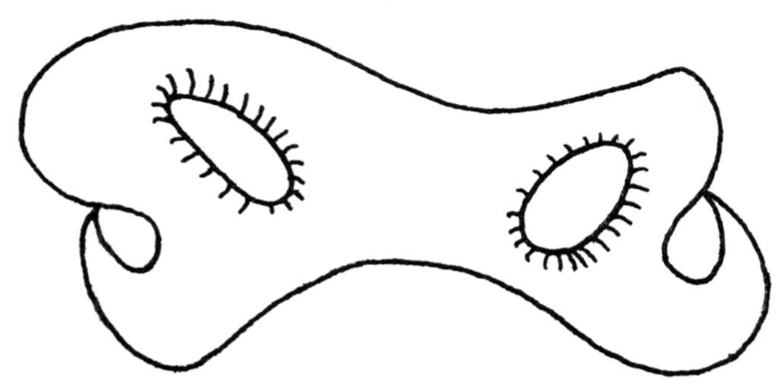

Fig.1. Topological Structure of a Typical Pole or Zero (Genus 2)

We denote by V(p) and affine variety defined by p, i.e.

$$\{ (s_1,s_2)\varepsilon\, \mathbb{C}^2;\ p(s_1,s_2) = 0\}$$

and by $\overline{V(p)}$ its closure in $\mathbb{P}^2(\mathbb{C})$. Then a classical theorem of Bezout states that, for irreducible polynomials $p_1, \ldots, p_m$ we have

$$\deg(\overline{V(p_1)}\cap\ldots\cap\overline{V(p_m)}) = \prod_{i=1}^{m} \deg p_i .$$

3. INVERSION OF THE JOINT LAPLACE TRANSFORM

A basic problem of finite-dimensional linear systems theory is the inversion of the Laplace transform $G(s) = \Pi\,(s-z_i)/\Pi(s-p_j)$ of the system. This can easily be achieved by the expanding G(s) in partial fractions and then using standard inversion formulae. When we consider two parameter systems the inversion problem is much more difficult. In fact, it turns out that we cannot form a 'partial fraction expansion' of a general system $G(s_1,s_2) = p(s_1,s_2)/q(s_1,s_2)$ in the sense of writing

$$G(s_1,s_2) = \sum_{i=1}^{n} \frac{a_i}{q_i(s_1,s_2)}$$

where q_i are the irreducible factors of q (assumed for the moment to be of multiplicity 1). For example, we can never write

$$\frac{1}{s_1^2 - s_2^2} = \frac{a_1}{s_1 - s_2} + \frac{a_2}{s_1 + s_2} \tag{3.1}$$

even if we allow a_1, a_2 to be polynomials.

In order to find conditions under which we can simplify the inversion of the Laplace transformation by partial fraction expansion we must consider the notion of ideal of a ring R with unity. An <u>ideal</u> I in a commutative R with unity is a set such that

$$RI = IR \subseteq I.$$

If $p \in \mathbb{C}[s_1,s_2]$, we define the <u>principal ideal</u> (p) generated by p to be the ideal $p.\mathbb{C}[s_1,s_2] = \{p_1 \in \mathbb{C}[s_1,s_2] : p_1 = p.q$ for some $q \in \mathbb{C}[s_1,s_2]\}$. The <u>sum</u> of two ideals I_1 and I_2 is the set

$$I_1 + I_2 = \{p \in \mathbb{C}[s_1,s_2] : p = p_1 + p_2 \text{ where } p_i \in I_i\}.$$

$I_1 + I_2$ is clearly an ideal.

For any ideal I in $\mathbb{C}[s_1,s_2]$ we consider the <u>variety defined by I</u> $(V(I))$ given by

$$V(I) = \{(s_1,s_2) \in \mathbb{C}^2 : p(s_1,s_2) = 0 \text{ for all } p \in I\}.$$

Finally, for any ideal I in $\mathbb{C}[s_1,s_2]$ we define the <u>radical</u> $\sqrt{I}$ of I by

$$\sqrt{I} = \{p \in \mathbb{C}[s_1,s_2] : p^n \in I \text{ for some natural number } n\}.$$

We then have the following result (Kendig[7]).

<u>Theorem 3.1</u> For any ideals $I_1, I_2 \subseteq [s_1,s_2]$, we have

$$V(I_1 \cap I_2) = V(I_1 . I_2) = V(I_1) \cup V(I_2)$$

$$V(I_1 + I_2) = V(I_1) \cap V(I_2).$$

Moreover, if $V(I_1) \subseteq V(I_2)$ we have

$$\sqrt{I_2} \subseteq \sqrt{I_1} \quad . \quad \square$$

Now suppose that $G(s_1, s_2)$ is a transfer function such that

$$G(s_1, s_2) = \frac{p(s_1, s_2)}{q_1^{i_1}(s_1, s_2) \ldots q_n^{i_n}(s_1, s_n)}$$

i.e. G has poles $q_1, \ldots, q_n$ with respective multiplicities $i_1, \ldots, i_n$.

Then we define

$$q_k' = \prod_{j \neq k} \left\{ q_j^{i_j} \right\} ,$$

and we consider the possibility of writing G in the form

$$G(s_1, s_2) = \sum_{k=1}^{n} \left\{ \frac{p_k(s_1, s_2)}{q_k^{i_k}(s_1, s_2)} \right\} \qquad (3.2)$$

for some polynomials $p_k \in \mathbb{C}[s_1, s_2]$. The next result is straightforward.

Lemma 3.2 G can be written in the form (3.2) if and only if

$$(p) \subseteq \sum_{k=1}^{n} (q_k') \; . \; \square$$

Theorem 3.3 In order that G can be written in the form (3.2), it is necessary that

$$V(q_1) \cap V(q_j) \subseteq V(p)$$

for all pairs of polynomials $q_i, q_j (i \neq j)$ in the denominator of G.

Moreover, it is sufficient that

$$V(q_i) \cap V(q_j) \subseteq V(p_1) \quad (i \neq j)$$

where $p=p_1^{\,m}$ for a suitable polynomial p_1 and an integer m.

Proof From lemma 4.3 we have the necessary condition

$$(p) \subseteq \sum_{k=1}^{n} (q_k') \quad ,$$

and so

$$V(\sum_{k=1}^{n} (q_k')) \subseteq V(p).$$

However,

$$V(\sum_{k=1}^{n} (q_k')) = \bigcap_{k=1}^{n} (V(q_k')) = \bigcap_{j=1}^{n} \left(\bigcup_{i\neq j} V(q_k^{\;i_k}) \right) ,$$

$$= \bigcap_{j=1}^{n} \left(\bigcup_{i\neq j} V(q_k) \right)$$

$$= \bigcup_{i\neq j} (V(q_i) \cap V(q_j)) .$$

For sufficiency note that

$$\bigcup_{i\neq j} (V(q_i) \cap V(q_j)) \subseteq V(p_1)$$

implies that

$$V(\sum_{k=1}^{n} (q_k')) \subseteq V(p_1) \quad ,$$

and so

$$\sqrt{(p_1)} \subseteq \sqrt{\sum_{k=1}^{n} (q_k')} \quad .$$

Hence,

$$p_1^m = p \in \sum_{k=1}^{n} (q_k') \text{ for some m. } \square$$

Corollary 3.4 In order that G can be written in the form (3.2), it is sufficient that

$$V(q_i) \cap V(q_j) \subseteq V(p) \qquad \forall\ i\neq j$$

and

$$\sqrt{\sum_{k=1}^{n} (q_k')} = \sum_{k=1}^{n} (q_k') \quad ,$$

i.e. $\sum_{k=1}^{n} (q_k')$ is self-radical. $\square$

For example, we cannot write $\frac{1}{X_1^2 - X_2^2}$ in the form

$(a_1/(X_1-X_2))+(a_2/(X_1+X_2))$ for any polynomials a_1, a_2, since

$$V(X_1-X_2)\cap V(X_1+X_2) = \{0\} \nsubseteq V(1) = \emptyset .$$

However, we can factorize $\frac{X_1}{X_1^2-X_2^2}$ since the conditions of corollary 3.4 hold;

for $(X_1-X_2)+(X_1+X_2) = (X_1,X_2)=\sqrt{(X_1,X_2)}$.

Consider now the evaluation of the polynomials p_k in (3.2) when G is expressible in this form. Clearly, the p_k's are not unique since if

$$G = \frac{p}{q_1^{i_1} q_2^{i_2}} = \frac{p_1}{q_1^{i_1}} + \frac{p_2}{q_2^{i_2}} ,$$

then

$$G = \frac{p_1 + rq_1^{i_1}}{q_1^{i_1}} + \frac{p_2 - rq_2^{i_2}}{q_2^{i_2}}$$

for any $r\in\mathbb{C}[s_1,s_2]$. However, if

$$G = \sum_{k=1}^{n} \frac{p_k}{q_k^{i_k}} = \sum_{k=1}^{n} \frac{p_k'}{q_k^{i_k}} ,$$

then

$$\sum_{k=1}^{n} (p_k-p_k')\, q_k' = 0$$

Hence (p_k-p_k') is divisible by $q_k^{i_k}$, since q_k' is not divisible by $q_k^{i_k}$ whereas all other q_j' $(j\neq k)$ are divisible by this polynomial.

Therefore,

$$p_k - p_k' \in (q_k^{i_k}) \subseteq \sqrt{(q_k^{i_k})} = (q_k) ,$$

and so p_k is determined modulo (q_k). Hence,

$$p_k \in \mathbb{C}[s_1, s_2] / (q_k)$$

which is the coordinate ring of the affine variety $V(q_k)$, and is denoted by

$\mathbb{C}_{q_k}[s_1, s_2]$, where

$$\bar{s}_1 = s_1 + (q_k) , \quad \bar{s}_2 = s_2 + (q_k) .$$

It follows that

$$p_k = \left[G(s_1, s_2)\, q_k^{i_k}(s_1, s_2)\right]\Big|_{s_1 = \bar{s}_1 , \; s_2 = \bar{s}_2}$$

Example 3.5 Consider the function

$$G(s_1, s_2) = \frac{s_1^2 - 2s_1 s_2^2 - s_1 s_2^3}{(s_1 + s_2)(s_1 - s_2^2)^2}$$

This certainly satisfies the necessary condition for factorisation. If it is factorizable, then put

$$G = \frac{a_1}{(s_1 + s_2)} + \frac{a_2}{(s_1 - s_2^2)^2}$$

Clearly,

$$a_1 = \frac{s_1^2 - 2s_1 s_2^2 - s_1 s_2^3}{(s_1 - s_2^2)^2}\Bigg|_{s_1 = -s_2} = 1$$

$$a_2 = \frac{s_1^2 - 2s_1 s_2^2 - s_1 s_2^3}{(s_1 + s_2)}\Bigg|_{s_1 = s_2^2} = -s_2^3 .$$

The inversion of the Laplace transform can be solved in the case where G is separable into partial fractions by determining the Green's function for the operators

$$q_k\left(\frac{\partial}{\partial t}, \frac{\partial}{\partial x}\right) \qquad 1\leq k\leq n .$$

(see Banks[6] for more details).

4. A ROOT LOCUS THEORY

We shall now consider a feedback system of the form (1.3) and determine the effect of varying k. Recall again the linear finite-dimensional case where we have a system of the form

$$G(s) = \frac{kp(s)}{q(s)+kp(s)} \tag{4.1}$$

in the single complex variable s. Then it is well known (Shinners[8]) that the locus of the zeros of the polynomial q+kp can be regarded as a connected directed graph on $\mathbb{P}^1(\mathbb{C})$ with n branches (almost everywhere) which start on open loop poles and end on open loop zeros (some of which are at '∞').

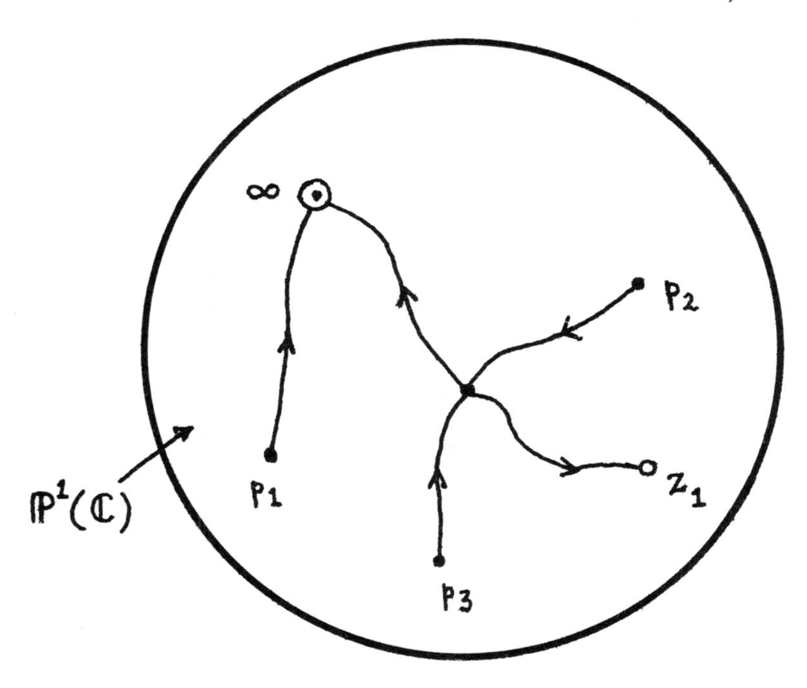

Fig 2. Typical finite-dimensional root locus

Consider now the example

$$G(s) = \frac{s_1^2+s_2}{s_1 s_2} \tag{4.2}$$

in two complex variables. The 'root locus' is specified by the equation

$$s_1 s_2+k(s_1^2+s_2) = 0 \tag{4.3}$$

By continuity of the solutions of polynomial equations it is easy to see that the root locus again begins on open loop poles and ends on open loop zeros. However, system (4.2) has two open loop poles which are topologically spheres in $\mathbb{P}^2(\mathbb{C})$. (For, $s_1 s_2+k(s_1^2+s_2)$ is irreducible and nonsingular and so by the genus formula, Kendig[7]

$$g = (n-1)(n-2)/2 = 0$$

if n=2). Hence for k>o we have a single pole which tends to the single zero $s_1^2+s_2=0$ (in the Hausdorff metric of $\mathbb{C}^2$) as $k \to \infty$. The root locus for distributed systems of the kind (1.2) may therefore have fewer branches than open loop poles, which is quite different from the finite-dimensional case.

The main problem remaining is to count the number of zeros 'at infinity' of an open loop system

$$G(s_1,s_2) = \frac{p(s_1,s_2)}{q(s_1,s_2)} \tag{4.4}$$

In the finite dimensional case where $G(s) = p(s)/q(s)$ this is easily done, since the number of zeros at infinity = n-m where n=deg q, m=deg p (assuming $n \geq m$). However infinity in $\mathbb{P}^2(\mathbb{C})$. is not just a single point as in $\mathbb{P}^1(\mathbb{C})$, but a complete copy of $\mathbb{P}^1(\mathbb{C})$. To obtain $\mathbb{P}^2(\mathbb{C})$ from $\mathbb{C}^2$ we can add a point to each complex 1-space of $\mathbb{C}^2$. Hence the projective 1-space $\mathbb{P}^1(\mathbb{C})$ at infinity in $\mathbb{P}^2(\mathbb{C})$ is parameterised by $\alpha = s_1/s_2$ (apart from $s_2 = 0$). Hence G behaves at ∞ like the system

$$G'(\alpha,s_2) = \frac{p(\alpha s_2,s_2)}{q(\alpha s_2,s_2)} \quad \frac{p'(\alpha,s_2)}{q'(\alpha,s_2)}$$

If $p'_1,\ldots,p'_\delta$ and $q'_1,\ldots,q'_\delta$ are the irreducible components of p' and q', then it is easy to see that p'_i represents a pole at infinity if $p_i'(\alpha,s_2)$ is not independent of s_2 and similarly q'_j represents a zero (again if q'_j is not independent of s_2). The topological structure of the poles and zeros at infinity is again determined by the nature of polynomials p_i',q'_j.

Example 4.1

$$\text{If } G(s_1,s_2) = \frac{s_1^2+s_2}{s_1 s_2}$$

then

$$G(\alpha,s_2) = \frac{\alpha^2 s_2+1}{\alpha s_2}$$

and so the system has one pole and one zero (both spheres) at ∞.

5. REFERENCES

1. Banks, S.P. (1983) State-Space and Frequency-Domain Methods in the Control of Distributed Parameter Systems, Peter Peregrinus.

2. Banks, S.P. (1985) On Nonlinear Systems and Algebraic Geometry, University of Sheffield, To appear in Int. J. Control.

3. Banks, S.P. and Abbasi-Ghelmansarai,F. (1983) Delay Equations, the Left-Shift Operator and the Infinite Dimensional Root Locus, Inst. J.Control, 37, pp235-253.

4. Banks, S.P. and Abbasi-Ghelmansarai, F. (1983) Realisation Theory and the Infinite Dimensional Root Locus, Int.J. Control, 38, pp.589-606.

5. Banks,S.P. and Abbasi-Ghelmansarai, F. (1984) On Systems with Spectral Cluster Points and their Relation to H^p spaces, Int.J.Control, 39, pp. 1295-1308.

6. Curtain, R.F. and Pritchard, A.J. (1978) Infinite Dimensional Linear System Theory, Spring Verlag.

7. Kendig, K. (1977) Elementary Algebraic Geometry, Springer Verlag, GTM No. 44.

8. Shinners, S.M. (1979) Modern Control System Theory and Application, Addison-Wesley.

CHAPTER 3

OPTIMISATION AND OPTIMAL CONTROL

AN EXACT PENALTY FUNCTION ALGORITHM FOR CONSTRAINED CONTROL PROBLEMS WITH DELAY

G. S. Virk

(Sheffield City Polytechnic)

ABSTRACT

An algorithm due to Mayne and Polak[1] is extended to delay systems. The new controls are generated in such a way that differentiability of the functionals w.r.t. the u argument is not required.

1. INTRODUCTION

Mayne and Polak[1] present an exact penalty function algorithm for solving optimal control problems with control and terminal equality constraints. In the presence of feasible points, the existence of solutions is guaranteed by optimising over the space of relaxed controls. We extend these results to delay systems. The extension also includes a number of crucial differences, one of which is that we construct a sequence of relaxed controls rather than of ordinary controls. Both our algorithm and that in Mayne and Polak[1] generate a (possibly) relaxed control in the limit satisfying first order optimality conditions. However, the advantage of our approach appears to be that the construction of the sequence of relaxed controls is much simpler than that of the ordinary controls. The price paid is that we must introduce the sophisticated notion of 'relaxed controls' but Mayne and Polak have to do this anyway to study the convergence properties of their algorithm.

Another distinction is that the new control constructed by our algorithm is a relaxed convex combination[10] of the old control $\underset{\sim}{u}$ and another control $\underset{\sim}{\check{u}}$ found by solving a subproblem,

i.e. $$\underset{\sim}{u}_{new} \triangleq (1-\alpha)\, \underset{\sim}{u} \oplus \alpha\, \underset{\sim}{\check{u}} \qquad (1.1)$$

where $\alpha \in [0,1]$ may be thought of as the step length and $\oplus$ denotes that the relaxed controls $\underset{\sim}{u}$ and $\check{\underset{\sim}{u}}$ are added in a relaxed manner. This method for generating $\underset{\sim}{u}_{new}$ is in contrast to the classical method where the usual search direction, step length is employed. A consequence of (1.1) is that intermediate problems do not need to be linearised about the control variable. Subsequent results necessary in studying the convergence properties of our algorithm are deduced using the 'linear' nature of relaxed controls and differentiability w.r.t. u is not required. Because of this, our procedure presented here gives, in a sense, better approximations (to the nonlinear problem at each iteration) than the method by Mayne and Polak. It seems therefore plausible to expect our algorithm to perform better when the two schemes are implemented. Only theoretical results are presented here because we have not had the opportunity to program our algorithm on test problems. This we hope to do in the near future so that its numerical performance can be compared with that of other schemes.

The paper is set out as follows: In section 2 a brief introduction to relaxed controls is presented together with the control problem P under consideration. The penalised problem P_c is introduced and solved in section 3 and, in section 4, conditions that guarantee the equivalence of the original problem and the penalised problem are presented. In section 5 the algorithm for solving P_c is modified, by insertion of a test function, so that it solves problem P.

2. RELAXED OPTIMAL CONTROL PROBLEMS

Let Ω be a compact and convex subset of R^m and let $\underset{\sim}{V}$ be the space of all probability measures on Ω. Then a relaxed control $\underset{\sim}{u}(t)$ has the representation

$$\underset{\sim}{u}(t) = \int_{\Omega} u(t)\, d(\underset{\sim}{v}(t))u \quad \text{for some probability measures}$$

$$\underset{\sim}{v}(t) \in \underset{\sim}{V} \text{ for every } t$$

Assuming the time interval of interest has been normalised to $T \triangleq [0,1]$, then a relaxed control is any function $\underset{\sim}{u} : T \to \underset{\sim}{V}$

Concerning relaxed controls we observe the following[7,11]:

(a) For any continuous function ϕ: $R^n \times \Omega \times T \to R^p$, the corresponding relaxed function ϕ_r: $R^n \times \underset{\sim}{V} \times T \to R^p$ (called the extension of ϕ to relaxed controls) is defined by

$$\phi_r(x,\underset{\sim}{u}, t) = \int_{\Omega} \phi(x,u,t)\, d\underset{\sim}{u}(u) \tag{2.1}$$

(b) For any function $\nu: R^n \times \underset{\sim}{V} \to R^s$, relaxed controls act in a 'linear' fashion, i.e. if $\underset{\sim}{u}(t) = (1-\alpha)\, \underset{\sim}{u}_1(t) \oplus \alpha\, \underset{\sim}{u}_2(t)$ for all $\alpha \in [0,1]$, then

$$\nu(x, \underset{\sim}{u}) = (1-\alpha)\, \nu(x, \underset{\sim}{u}_1) + \alpha\, \nu(x, \underset{\sim}{u}_2) \text{ for all } x \in R^n \quad (2.2)$$

(c) A relaxed control is said to be measurable if for any polynomial $p(u)$ in u, the function $\zeta : T \to R$ defined by

$$\zeta(t) = p(\underset{\sim}{v}(t)) = \int_{\Omega} p(u)\, d\, \underset{\sim}{v}(t)(u)$$

is measurable.

We let $\underset{\sim}{G}$ be the space of all measurable relaxed control functions.

Definition 1

An infinite sequence $\{\underset{\sim}{u}_i\}_{i=o}^{\infty}$ of relaxed controls in $\underset{\sim}{G}$ converges to $\underset{\sim}{u}^* \in \underset{\sim}{G}$ in the sense of control measures (i.s.c.m.) if, for every continuous function $\phi : \Omega \times T \to R$ and every subinterval Δ of T

$$\int_{\Delta} \phi_r\, (\underset{\sim}{u}_i(t), t)\, dt \to \int_{\Delta} \phi_r (\underset{\sim}{u}^*(t), t)\, dt \quad \text{as } i \to \infty \quad (2.3)$$

We shall be considering the following optimal control problem.

Problem P:

$$\underset{\underset{\sim}{u}}{\text{Min}} \quad h^0(x^{\underset{\sim}{u}}(1)) \quad (2.4)$$

$$\text{s.t. } \dot{x}(t) = f(x(t), x(t-\tau), \underset{\sim}{u}(t), t) \quad \text{for a.a.t} \in T \quad (2.5)$$

$$x(t) = \zeta(t) \text{ for all } t \in [-\tau, 0] \quad (2.6)$$

$$h^j(x^{\underset{\sim}{u}}(1)) = 0 \quad j = 1,2,\ldots,r \quad (2.7)$$

$$\underset{\sim}{u} \in \underset{\sim}{G} \quad (2.8)$$

where a.a. stands for almost all, τ is a positive real number, $T \triangleq [0,1]$ and the function $\zeta : [-\tau, 0] \to R^n$ is assumed to be

---

Note: For all relaxed functions ϕ_r as defined by (2.1), the subscript r will be omitted since it will be apparent from the text whether it is the original function or its relaxed extension that is being considered.

bounded, continuous and to possess a continuous derivative for all $t \in [-\tau, 0]$. We let $x^{\underset{\sim}{u}}$ denote the solution of the delay-differential equation (2.5) due to control $\underset{\sim}{u}$ and initial condition (2.6), etc. We shall assume that Problem P has feasible points,

The following hypothesis is assumed to hold:

Assumption 1

The function $f : R^n \times R^n \times \Omega \times T \to R^n$ and its partial derivatives f_x, f_y, f_{xx}, f_{yy} and the functions $h^j : R^n \to R$, $j = 0, 1, 2, \ldots, r$ and their partial derivatives h^j_x, h^j_{xx} exist and are continuous on their domains (where $y(t)$ represents the delayed argument $x(t-\tau)$).

[Note that the differentiability w.r.t. u hypothesised in Mayne and Polak[1] is not required].

Assumption 2

There exists an $M \in (0,\infty)$ such that

$$\| f(x, y, \underset{\sim}{u}, t)\| \leq M\left\{\|x\| + \|y\| + 1\right\} \tag{2.9}$$

for all $x, y \in R^n$, all $\underset{\sim}{u} \in \underset{\sim}{G}$, all $t \in T$.

where $\|\cdot\|$ denotes the usual Euclidean norm, etc, and

$$\| f(x^1, y^1, \underset{\sim}{u}, t) - f(x^2, y^2, \underset{\sim}{u}, t)\| \leq M\left\{\|x^1 - x^2\| + \|y^1 - y^2\|\right\} \tag{2.10}$$

for all $x^1, x^2, y^1, y^2 \in R^n$, all $\underset{\sim}{u} \in \underset{\sim}{G}$, all $t \in T$.

3. THE PENALISED PROBLEM

A procedure employing an exact penalty function is proposed for solving Problem P. The penalty parameter c will be adjusted automatically to ensure equivalence of the penalised problem and the original problem. To introduce the penalised problem we define $\gamma : \underset{\sim}{G} \to R$ by

$$\gamma(\underset{\sim}{u}) = \max_{j\in 1,2,\ldots,r} \{|h^j(x^{\underset{\sim}{u}}(1))|\} \tag{3.1}$$

$$= \max_{j\in 1,2,\ldots,r} \{h^j(x^{\underset{\sim}{u}}(1))\} \tag{3.2}$$

where we define $h^j(x^{\underset{\sim}{u}}(1))$ for $j\in 1,2,\ldots,r$ by

$$h^{j+r}(x^{\underset{\sim}{u}}(1)) = -h^{j}(x^{\underset{\sim}{u}}(1)) \tag{3.3}$$

Also for $c > o$ we define $\tilde{\gamma}_c : \underset{\sim}{G} \to R$ by

$$\tilde{\gamma}_c(\underset{\sim}{u}) = \frac{h^{o}(x^{\underset{\sim}{u}}(1))}{c} + \gamma(\underset{\sim}{u}) \tag{3.4}$$

Then we define the penalised problem as

Problem P_c:

$$\min_{\underset{\sim}{u}} \quad \tilde{\gamma}_c(\underset{\sim}{u})$$

$$\text{s.t. } \underset{\sim}{u} \in \underset{\sim}{G}$$

To solve Problem P_c we need to know whether γ and/or $\tilde{\gamma}_c$ can be reduced at each $\underset{\sim}{u} \in \underset{\sim}{G}$. For this purpose we define the sequential continuous (i.s.c.m.) functions[10] $\theta : \underset{\sim}{G} \to R$ and $\tilde{\theta}_c : \underset{\sim}{G} \to R$ by:

$$\theta(\underset{\sim}{u}) = \min_{\underset{\sim}{v} \in \underset{\sim}{G}} \max_{j \in 1,2,..,2r} \{h^{j}(x^{\underset{\sim}{u}}(1)) + < h^{j}_{x}(x^{\underset{\sim}{u}}(1)), z^{\underset{\sim}{u},\underset{\sim}{v}}(1)>\} - \gamma(\underset{\sim}{u}) \tag{3.5}$$

$$\tilde{\theta}_c(\underset{\sim}{u}) = \min_{\underset{\sim}{v} \in \underset{\sim}{G}} \max_{j \in 1,2,..,2r} \{h^{j}(x^{\underset{\sim}{u}}(1)) + < \frac{h^{o}_{x}(x^{\underset{\sim}{u}}(1))}{c} + h^{j}_{x}(x^{\underset{\sim}{u}}(1)), z^{\underset{\sim}{u},\underset{\sim}{v}}(1)>\} - \gamma(\underset{\sim}{u}) \tag{3.6}$$

where $<\cdot,\cdot>$ denotes the usual scalar product in R^n and $z^{\underset{\sim}{u},\underset{\sim}{v}} : T \to R^n$ is the solution of

$$\dot{z}(t) = A^{\underset{\sim}{u}}(t)\, z(t) + B^{\underset{\sim}{u}}(t)\, z(t-\tau) + \Delta f(\underset{\sim}{v}, \underset{\sim}{u}) \quad \text{for a.a.} t \in T \tag{3.7}$$

$$z(t) = 0 \qquad \text{for all } t \in [-\tau, 0] \tag{3.8}$$

for all $\underset{\sim}{u}, \underset{\sim}{v} \in \underset{\sim}{G}$, and

$$A^{\underset{\sim}{u}}(t) \triangleq f_x(x^{\underset{\sim}{u}}(t), x^{\underset{\sim}{u}}(t-\tau), \underset{\sim}{u}(t), t) \tag{3.9}$$

$$B^{\underset{\sim}{u}}(t) \triangleq f_y(x^{\underset{\sim}{u}}(t), x^{\underset{\sim}{u}}(t-\tau), \underset{\sim}{u}(t), t) \tag{3.10}$$

$$\Delta f(\underset{\sim}{v}, \underset{\sim}{u}) \triangleq f(x^{\underset{\sim}{u}}(t), x^{\underset{\sim}{u}}(t-\tau), \underset{\sim}{v}(t), t) - f(x^{\underset{\sim}{u}}(t), x^{\underset{\sim}{u}}(t-\tau), \underset{\sim}{u}(t), t) \tag{3.11}$$

Since $\underset{\sim}{G}$ is compact[7], $\theta(\underset{\sim}{u})$ and $\tilde{\theta}_c(\underset{\sim}{u})$ and their corresponding minimising controls in $\underset{\sim}{G}$ exist for all $\underset{\sim}{u} \in \underset{\sim}{G}$.

From the definition of $\tilde{\theta}_c$ we obtain the following result[10].

3.1 Proposition

Suppose $\underset{\sim}{u}^*$ is optimal for problem P_c, $c > o$, then $\tilde{\theta}_c(\underset{\sim}{u}^*) = 0$.

Therefore we define the set Δ_c of desirable points for problem P_c, $c > o$ by

$$\Delta_c \triangleq \{\underset{\sim}{u}^* \in \underset{\sim}{G} : \tilde{\theta}_c(\underset{\sim}{u}^*) = 0\} \tag{3.12}$$

3.2 Algorithm for Solving P_c

Step 0 : $\underset{\sim}{u}_o \in \underset{\sim}{G}$, $c > o$

Step 1 : Set $i = 0$

Step 2 : Compute $x^{\underset{\sim}{u}_i}$ by solving (2.5), (2.6)

Step 3 : Compute $\tilde{\theta}_c(\underset{\sim}{u}_i)$ and find a control $\check{\underset{\sim}{u}}_i \in \underset{\sim}{G}$ which achieves the minimum.

Step 4 : If $\tilde{\theta}_c(\underset{\sim}{u}_i) = 0$ stop

Else define $\underset{\sim}{u}_{\alpha_i} = (1-\alpha_i)\, \underset{\sim}{u}_i \oplus \alpha_i\, \check{\underset{\sim}{u}}_i$

where $\alpha_i \in [0,1]$ is the largest number which

satisfies $\tilde{\gamma}_c(\underset{\sim}{u}_{\alpha_i}) - \tilde{\gamma}_c(\underset{\sim}{u}_i) \leq \alpha_i \tilde{\theta}_c(\underset{\sim}{u}_i)/2$

Step 5 : Set $\underset{\sim}{u}_{i+1} = \underset{\sim}{u}_{\alpha_i}$
Set $i = i+1$
Go to step 2

Algorithm 3.2 has the following convergence properties[10]:

3.3 Theorem

Suppose all the assumptions in section 2 are satisfied. Then Algorithm 3.2 either generates a finite sequence of controls in which case the last element is desirable for P_c, or it generates an infinite sequence and every accumulation point i.s.c.m. (at least one exists) is in Δ_c.

Remark:
The computation of $\tilde{\theta}_c$ and a minimising control $\check{\underset{\sim}{u}}$ at step 3

of Algorithm 3.2 may be performed, using a similar procedure as in Virk[10] or by using a proximity algorithm[2] as in Mayne and Polak[1].

4. EQUIVALENCE OF PROBLEMS P AND P_c

So far we have only considered solving problem P_c and no relation to problem P is made. We now wish to impose conditions to ensure that the two problems are equivalent so that solving P_c automatically solves P. This equivalence can be guaranteed by satisfying two conditions. The first of these is to make sure that solving P_c results in a control that is feasible for P. This is true if for all $\underset{\sim}{u} \in \underset{\sim}{G}$ satisfying $\gamma(\underset{\sim}{u}) > 0$ we can reduce γ i.e. move nearer the feasible region. Obviously this holds if $\theta(\cdot) < 0$ for nonfeasible controls. The second condition is to ensure that a local or global solution to P also solves P_c. This essentially means that the set $W(\underset{\sim}{u})$ defined by

$$W(\underset{\sim}{u}) \triangleq \{(h^o(x^{\underset{\sim}{u}}(1)), h^1(x^{\underset{\sim}{u}}(1)), \ldots, h^r(x^{\underset{\sim}{u}}(1)) : \underset{\sim}{u} \in \underset{\sim}{G}\} \subset R^{r+1}$$

is not tangential to the cost axis at a minimising control and first order necessary conditions for optimality hold[1,10]. To state this in a form compatible to the problem we need to define a few terms. Let $I_1, I_2, \ldots I_{2^r}$ denote the following sets:

$$\begin{aligned} I_1 &= \{1, 2, 3, \ldots, r-2, r-1, r\} \\ I_2 &= \{1, 2, 3, \ldots, r-2, r-1, 2r\} \\ I_3 &= \{1, 2, 3, \ldots, r-2, 2r-1, r\} \\ &\vdots \\ I_{2^r} &= \{r+1, r+2, \ldots, 2r-2, 2r-1, 2r\} \end{aligned} \tag{4.1}$$

These sets have the property that if $j \in I_i$, $j \leq r$ then $j+r \notin I_i$. Define J by

$$J = \{I_i : i = 1, 2, \ldots, 2^r\}$$

For each $I \subset J$ let $\phi^I : \underset{\sim}{G} \to R$ be defined by

$$\phi^I(\underset{\sim}{u}) \triangleq \min_{\underset{\sim}{v} \in \underset{\sim}{G}} \max_{j \in I} \langle h_x^j(x^{\underset{\sim}{u}}(1)), z^{\underset{\sim}{u},\underset{\sim}{v}}(1)\rangle \tag{4.2}$$

These functions are sequentially continuous i.s.c.m.[10]

We now state the above conditions in a more compact form:

Assumption 3

(1) If $\gamma(\underset{\sim}{u}) > 0$, $\underset{\sim}{u} \in \underset{\sim}{G}$, then $\theta(\underset{\sim}{u}) < 0$

(2) If $\gamma(\underset{\sim}{u}) = 0$, $\underset{\sim}{u} \in \underset{\sim}{G}$, then $\phi^I(\underset{\sim}{u}) < 0$ for all $I \subset J$

If these are satisfied it can be shown[10] that problems P and P_c are equivalent for $c \geq c^*$ for some finite c^*. This c^* is given by the magnitude of the multipliers in the necessary conditions of optimality for problem P. We quote these here for convenience, full details are presented in Virk[10].

Proposition 4.1

Let $\underset{\sim}{u}^* \in \underset{\sim}{G}$ be optimal for P. Then there exist multipliers $\psi^1, \psi^2, \ldots \psi^r \in \tilde{R}$ such that

$$\langle h_x^o(x^{\underset{\sim}{u}^*}(1)) + \sum_{j=1}^{r} \psi^j h_x^j (x^{\underset{\sim}{u}^*}(1)), z^{\underset{\sim}{u}^*,\underset{\sim}{v}}(1)\rangle \geq 0 \quad \text{for all } \underset{\sim}{v} \in \underset{\sim}{G} \tag{4.3}$$

We now define the desirable set, Δ, for Problem P by

$$\Delta \triangleq \{\underset{\sim}{u}^* \in \underset{\sim}{G} : \gamma(\underset{\sim}{u}^*) = 0 \quad \text{and (4.3) is satisfied}\} \tag{4.4}$$

5. ALGORITHM FOR SOLVING PROBLEM P

As stated previously the method proposed for solving problem P is to initially solve problem P_c and then to adjust the penalty parameter c to ensure equivalence. An algorithm that solves P_c is presented in Section 3, therefore we only require a procedure for assessing whether c is large enough. This may take the form of a test function $t_c : \underset{\sim}{G} \to R$, so that if $t_c(\underset{\sim}{u}) \leq 0$, size of c is large enough and $t_c(\underset{\sim}{u}) > 0$ requires c to be increased. An algorithm model that uses a test function in this way is presented in Polak[12]. This consists of the two main components stated above, i.e. an algorithm (defined by a point-to-set map $A_c : \underset{\sim}{G} \to 2^{\underset{\sim}{G}}$) for solving P_c, and a test function t_c. We present this model for convenience.

5.1 Algorithm Model

Data : $\underset{\sim}{u}_o \in \underset{\sim}{G}$, $\{c_j\}_{j=0}^{\infty}$ with $c_1 < c_2 < \ldots \lim_j c_j = \infty$

Step 1 : Set $i = 0$, $j = 0$

Step 2 : If $t_{c_j}(\underset{\sim}{u}_i) > 0$ Set $j = j+1$ and repeat Step 2. Else proceed to Step 3.

Step 3 : If $\underset{\sim}{u}_i \in \Delta_{c_j}$ Stop

Else compute a $\underset{\sim}{v}_i \in A_{c_j}(\underset{\sim}{u}_i)$

Step 4 : Set $\underset{\sim}{u}_{i+1} = \underset{\sim}{v}_i$

Set $i = i+1$

Goto Step 2

This algorithm model has the following properties[12]:

5.2 Theorem

If

(i) For each j, algorithm A_{c_j} generates controls $\underset{\sim}{u}^* \in \underset{\sim}{G}$ satisfying $\underset{\sim}{u}^* \in \Delta_{c_j}$

(ii) The test function is sequentially continuous i.s.c.m.

(iii) For each j, if $\underset{\sim}{u}^* \in \Delta_{c_j}$ and $t_{c_j}(\underset{\sim}{u}^*) \le 0$, then $\underset{\sim}{u}^* \in \Delta$.

(iv) For every $\underset{\sim}{u}^* \in \underset{\sim}{G}$, there exists an integer j^* such that if $\{\underset{\sim}{u}_i\}_{j=0}^{\infty}$ is a sequence in $\underset{\sim}{G}$ converging i.s.c.m. to $\underset{\sim}{u}^*$, then there exists an integer i_o such that

$$t_{c_j}(\underset{\sim}{u}_i) \le 0 \qquad \text{for all } i \ge i_o,\ j \ge j^*$$

Then the algorithm either constructs a finite sequence, in which case the last control is desirable for problem P, or it constructs an infinite sequence and every limit point (at least one exists) is desirable for P.

It is obvious that the algorithm presented in section 3 satisfies (i) above. We now consider the test function. We define t_c as follows:

$$t_c(\underset{\sim}{u}) = \tilde{\theta}_c(\underset{\sim}{u}) + \frac{\gamma(\underset{\sim}{u})}{c} \tag{5.1}$$

Since $\tilde{\theta}_c$ and $\gamma(\underset{\sim}{u})$ are sequentially continuous i.s.c.m. so is $t_c(\underset{\sim}{u})$, thus satisfying (ii) above. For this choice of t_c it can be shown that (iii) and (iv) above also hold[10]. Therefore we can now present our algorithm for solving problem P.

5.3 Algorithm

Data : $\underset{\sim}{u}_o \in \underset{\sim}{G}$, $0 < c_o < c_1 < \ldots\ldots \lim\limits_{j} c_j = \infty$

Step 1 : Set $i = 0$, and $j = 0$

Step 2 : Compute $x^{\underset{\sim}{u}_i}$ by solving (2.5) (2.6)

Step 3 : Compute $\tilde{\theta}_{c_j}(\underset{\sim}{u}_i)$ and the minimising control $\check{\underset{\sim}{u}}_i \in \underset{\sim}{G}$

Step 4 : Compute $t_{c_j}(\underset{\sim}{u}_i)$

If $t_{c_j}(\underset{\sim}{u}_i) > 0$ Set $c_j = c_{j+1}$

Set $j = j+1$

and goto Step 3

Else proceed to Step 5

Step 5 : If $\tilde{\theta}_{c_j}(\underset{\sim}{u}_i) = 0$ Stop

Else continue

Step 6 : Define $\underset{\sim}{u}_{\alpha_i} = (1-\alpha_i)\underset{\sim}{u}_i \oplus \alpha_i \check{u}_i$ where $\alpha_i \in [0,1]$ is the largest number which satisfies

$$\tilde{\gamma}_{c_j}(\underset{\sim}{u}_{\alpha_i}) - \tilde{\gamma}_{c_j}(\underset{\sim}{u}_i) \leq \alpha_i \tilde{\theta}_{c_j}(\underset{\sim}{u}_i)/2$$

Step 7 : Set $\underset{\sim}{u}_{i+1} = \underset{\sim}{u}_{\alpha_i}$

Set $i = i+1$

Goto Step 2

Since algorithm 5.3 satisfies all the conditions required in Theorem 5.2, we can deduce the following:

5.4 Theorem

Suppose the assumptions stated in the text are satisfied and that Algorithm 5.3 generates a sequence of controls $\{\underset{\sim}{u}_i\}$. Then

(i) This sequence is finite $\{\underset{\sim}{u}_i\}_{i=0}^{k}$, in which case c_j is increased only a finite number of times as well, and the last control is desirable for P, i.e. $\underset{\sim}{u}_k \in \Delta$.

(ii) The sequence is infinite $\{\underset{\sim}{u}_i\}_{i=0}^{\infty}$ and there exists a $j^* < \infty$ such that $j \leq j^*$ throughout computation (i.e. c_j is still only increased a finite number of times and then remains constant at $c_j{}^*$) and every limit point $\underset{\sim}{u}^*$ i.s.c.m. of this sequence (at least one exists) is desirable for problem P.

6. CONCLUSION

An optimal control problem with terminal equality and

control constraints is considered. An algorithm using an exact penalty function method which solves it together with the convergence results is presented. This work is an extension, together with some modifications and improvements, of the work done by Mayne and Polak[1] to delay systems. The problem is not linearized w.r.t. the control argument[1], and therefore we can expect our approximations to perform better than those of Mayne and Polak.

7. REFERENCES

1. Mayne, D.Q. and Polak, E. (1982), "An Exact Penalty Function Algorithm for Optimal Control Problems with Control and Terminal Equality Constraints", Parts I and II Journal of Optimisation Theory and Applications, Vol 32, pp 211-246 and pp 345-364.

2. Polak, E. (1971), "Computational Methods in Optimisation", Academic Press, New York.

3. Dem'yanov, V.F. and Malozemov, V.N. (1974), "Introduction to Minimax", John Wiley.

4. McShane, E.J. (1967), "Relaxed Controls and Variational Problems", SIAM J. Control, Vol 5, No 3, pp 438-485.

5. Mayne, D.Q. and Maratos, N. (1979), "A First Order Exact Penalty Function Algorithm for Equality Constrained Optimisation Problems", Math. Prog. Vol 16, pp 303-324.

6. Warga, J. (1977), "Steepest Descent with Relaxed Controls", SIAM J. Control, Vol 15, No 4, pp 674-682.

7. Warga, J. (1972), "Optimal Control of Differential and Functional Equations", Academic Press, New York.

8. Williamson, L.J. and Polak, E. (1976), "Relaxed Controls and the Convergence of Optimal Control Algorithms", SIAM J. Control, Vol 14, pp 737-756.

9. Virk, G.S. (1985), "A Strong Variational Algorithm for Delay Systems", Journal of Optimisation Theory and Applications, Vol 45, No 2.

10. Virk, G.S. (1982), "Algorithms for State Constrained Control Problems with Delay", PhD Thesis, Imperial College.

11. Young, L.C. (1969), "Lectures on the Calculus of Variations and Optimal Control Theory", W.B. Saunders Company.

12. Polak, E. (1976), "On the Global Stabilization of Locally Convergent Algorithms", Automatica, Vol 1, pp 337-342.

ACKNOWLEDGEMENTS

The author wishes to thank Dr R B Vinter for his help and encouragement. The financial support of the UK Science and Engineering Research Council is gratefully acknowledged.

FEEDBACK AND MINIMAX OPTIMISATION

T.J. Owens

(University of Strathclyde)

ABSTRACT

It is shown that for the minimisation of weighted sensitivity in the sup norm, when the weighting function has conjugate symmetry, application of a procedure due to Kwakernaak[2] will give the same optimal weighted sensitivity as application of a procedure due to Zames and Francis[1]. Since, the procedure due to Kwakernaak may be applied to more general criteria this equivalence can be utilised to make several straightforward extensions to results of Zames and Francis.

1. INTRODUCTION

Zames and Francis[1] have considered the problem of optimal disturbance attenuation for the feedback system of Fig. 1.

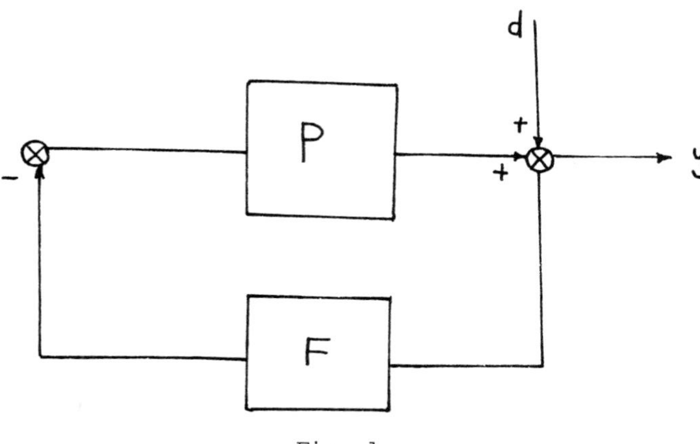

Fig. 1

For the system of Fig. 1 the closed loop response to a disturbance at the plant output is given by $(1+PF)^{-1}$ and is called the sensitivity operator. The initial objective in disturbance attenuation is to make this sensitivity small in some appropriate sense. Since, in practice engineers are interested in disturbance attenuation over specific frequency ranges Zames and Francis[1] consider the problem of making weighted sensitivity small, where the weighted function may be thought of as emphasising a particular frequency range. In the following we denote the sensitivity operator by $S(j\omega)$ and the weighting function by $V(j\omega)$. Zames and Francis[1] seek the controller that minimises (1.1)

$$\sup_{\omega} |V(j\omega)\ S(j\omega)| \tag{1.1}$$

and provide a solution procedure to obtain that controller.

Kwakernaak[2] has considered the problem of design for feedback systems of the type of Fig. 1, with $d \equiv 0$, where the emphasis is eliminating, or at least decreasing, the effect of plant variations. Given a suitable expressed bound on the relative size of the perturbations, the analysis leads to a robustness criterion involving a weighted combination of the sensitivity function and its complement $T(j\omega) = 1 - S(j\omega)$. Kwakernaak[2] seeks the controller which minimises the criterion (1.2)

$$\sup_{\omega} \sum_{k} \left| V_k(j\omega)\ S(j\omega) + W_k(j\omega)\ T(j\omega) \right|^2 \tag{1.2}$$

where the V_k and W_k are suitably chosen weighting functions and provides a solution procedure to obtain that controller.

This paper is motivated by the observation that a controller that minimises the criterion $\sup_{\omega} |V(j\omega)\ S\ (j\omega)|$ (criterion(1.1)) will also minimise $\sup_{\omega} |\ V(j\omega)\ S(j\omega)|^2$ (a special case of criterion (1.2)).

2. PRELIMINARIES

Kwakernaak[2] shows that the solution to the optimisation problem (1.2) is subject to certain restrictions on the weighting functions, contingent on the solution of the following pair of polynomial equations,

$$\left.\begin{aligned} &\omega_1 \phi_+ \theta + \omega_2 \Psi_+ \zeta = \chi \\ &\sum_k \chi_k \chi_k^* = \lambda^2 \chi \chi^* \\ \text{where,}\quad &\chi_k = \alpha_{1k} \omega_{1k} \phi_+ + \alpha_{2k} \omega_{2k} \Psi_+ \zeta \end{aligned}\right\} \quad (2.1)$$

Here, ϕ_+, Ψ_+ are monic polynomials whose roots are the right half plane poles of the plant, respectively, $\deg(\phi_+) = n_+$, $\deg(\Psi_+) = m_+$. The weighting functions are denoted by

$V_k = \dfrac{\alpha_{1k}}{\beta_{1k}}$, $W_k = \dfrac{\alpha_{2k}}{\beta_{2k}}$, where α_{1k}, β_{1k}, α_{2k}, β_{2k} are real polynomials of degree a_{1k}, b_{1k}, a_{2k}, b_{2k}, respectively.

$\omega_1 = \prod_k \beta_{1k}$, $\omega_2 = \prod_k \beta_{2k}$, $\omega_{1k} = \prod_{j \neq k} \beta_{1j}$, $\omega_{2k} = \prod_{j \neq k} \beta_{2j}$ and $\deg(\omega_1) = b_1 = \sum_k b_{1k}$ and $\deg(\omega_2) = b_2 = \sum_k b_{2k}$. χ, θ and ζ are real polynomials to be determined and λ a real constant. For optimality, a solution is sought with χ having all its roots in the left half-plane and $\deg(\chi) \leq x$, $\deg(\theta) \leq t$, $\deg(\zeta) \leq z$, where $x = n_+ + m_+ + b_1 + b_2 + e$, $t = m_+ + b_2 + e$, $z = n_+ + b_1$. e is the smallest non-negative integer such that $a_{2k} \leq b_{2k} + e$, for all k.

By examination of equations (2.1) it can be seen that when Kwakernaak's method is applied imaginary axis poles and zeros do not affect λ, which determines the optimal weighted sensitivity. Therefore, for the Kwakernaak method imaginary axis poles and zeros do not affect the optimal weighted sensitivity. It is shown in Francis and Zames[3] that for the solution procedure of Zames and Francis[1] imaginary axis poles and zeros do not affect optimal weighted sensitivity. Therefore, in the following only plants without poles or zeros on the imaginary axis are considered, without loss of generality.

3. A COMPARISON OF THE KWAKERNAAK PROCEDURE WITH THAT OF ZAMES AND FRANCIS

The following Lemma from Zames and Francis[1] is utilised.

Lemma 1

(a) There is a unique weighted sensitivity function $\tilde{X}$ in H^{∞} which attains the minimum norm, min $\{\|X\| = X = VS\}$, where $\|X\| = \sup_{\omega} |V(j\omega)\ S(j\omega)|$. The necessary and sufficient conditions for $\tilde{X}$ satisfy the interpolation constraints:-

$$\left.\begin{aligned} X(b) &= V(b) \\ \left(\frac{d}{ds}\right)^{p-1} X &= \left(\frac{d}{ds}\right)^{p-1} V \text{ at } s = b \end{aligned}\right\} \tag{3.1}$$

at each distinct zero b, Re(b) > O of multiplicity p; the total of such constraints is m_+

and that it have the (allpass) form

$$\tilde{X} = D \prod_{i=1}^{m_+-1} \left(\frac{c_i - s}{\bar{c}_i + s}\right) \prod_{i=1}^{n_+} \left(\frac{a_j - s}{\bar{a}_j + s}\right) \tag{3.2}$$

in which $\text{Re}(c_i) \geq 0$, D is constant satisfying

$$|D| = \|\tilde{X}\| , \quad \prod_{j=1}^{n_+} (a_j - s) = \phi_+ .$$

(b) If the plant P(s) and weighting V(s) have conjugate symmetry, then so does $\tilde{X}$, and the coefficients in (3.2) are real and occur in conjugate pairs.

The following observations may be made:

(i) In practice $\prod_{j=1}^{n_+} \left(\frac{a_j - s}{\bar{a}_j + s}\right) = \frac{\phi_+}{\phi_+^*}$ since complex poles for a real plant always occur in conjugate pairs.

(ii) If the weighting function has conjugate symmetry, by Lemma 1 part b) $\prod_{i=1}^{m_+-1} \left(\frac{c_i - s}{\bar{c}_i + s}\right)$ may be written $\frac{\theta_+}{\theta_+^*}$, i.e., deg (θ_+)

$= m_+ - 1$, where θ_+ has all its roots in the right half-plane.

(iii) The interpolation constraints (3.1) are equivalent to the constraint that $\tilde{X} - V$ has a factor of Ψ_+.

It will be shown below that given the characterisation of the solution in Lemma 1, the solution can be obtained by solving (2.1), when the weighting function has conjugate symmetry.

$P = \frac{\Psi}{\phi}$, $F = \frac{\sigma}{\rho}$, $VS = \frac{\alpha_{11}\rho\phi}{\beta_{11}(\rho\phi+\sigma\Psi)}$. By (3.2) and utilising observation (ii) $\frac{\alpha_{11}\tilde{\rho}\phi}{\beta_{11}(\tilde{\rho}\phi+\tilde{\sigma}\Psi)} = \frac{D\phi_+\theta_+}{\phi_+^*\theta_+^*}$. Hence

$$\alpha_{11}\tilde{\rho}\phi\theta^*_+\phi^*_+ = D\theta_+\phi_+\beta_{11}\ (\tilde{\rho}\phi+\tilde{\sigma}\Psi).$$

$$\left(\frac{1}{D}\ \alpha_{11}\theta_+^*\phi_+^*\right)(\tilde{\rho}\phi) = \beta_{11}\theta_+\phi_+\ (\tilde{\rho}\phi) + \Psi_+\ (\beta_{11}\theta_+\phi_+\tilde{\sigma}\Psi_-\Psi_o) \quad (3.3)$$

The interpolation constraints require

$$X - V = \frac{D\theta_+\phi_+}{\theta^*_+\phi^*_+} - \frac{\alpha_{11}}{\beta_{11}} = \frac{D\beta_{11}\theta_+\phi_+ - \alpha_{11}\theta_+^*\phi_+^*}{\beta_{11}\theta_+^*\phi_+^*} = \Psi_+f \quad (3.4)$$

Hence, $\frac{1}{D}\alpha_{11}\phi_+^*\theta_+^* = \beta_{11}\phi_+\theta_+ + \Psi_+\zeta$ (3.5)

Multiplying by $(\tilde{\rho}\phi)$ gives

$$\left(\frac{1}{D}\ \alpha_{11}\phi_+^*\theta_+^*\right)(\tilde{\rho}\phi) = \beta_{11}\phi_+\theta_+\ (\tilde{\rho}\phi) + \Psi_+\zeta\ (\tilde{\rho}\phi) \quad (3.6)$$

Comparing (3.3) and (3.6) it can be seen that the interpolation constraints require $\beta_{11}\theta_+\phi_+\tilde{\sigma}\Psi_-\Psi_o = \zeta(\tilde{\rho}\phi)$. Cancelling $(\tilde{\rho}\phi)$ in (3.3) it can be seen that the characterisation of the solution given by Lemma 1 implies (3.5) be solved for the optimal controller; in this case equations (2.1) are

$$\left.\begin{aligned} &\beta_{11}\phi_+\theta + \Psi_+\zeta = \chi \\ &\chi_1 \chi_1^* = \lambda^2 \chi \chi^*, \text{ where } \chi_1 = \alpha_{11}\phi_+\theta \end{aligned}\right\} \tag{3.7}$$

Utilising the observation by Zames[4] that α_{11} may be assumed without loss of generality to have all of its roots in the left half-plane, equations (3.7) reduce to

$$\beta_{11}\phi_+\theta + \Psi_+\zeta = \frac{1}{\lambda}\alpha_{11}\phi_+^*\theta^* \tag{3.8}$$

Thus, (3.5) can be seen to be equivalent to (3.8) subject to the restriction that utilising D, θ_+ and the corresponding ζ, obtained by solving (3.8), application of Kwakernaak's[2] method leads to a $\tilde{\chi}$ of the form (3.2) when the plant has no poles or zeros on the imaginary axis. Kwakernaak[2] gives the optimum controller as

$$\tilde{F} = \frac{\phi_-\zeta}{\Psi_-\beta_{11}\theta_+} \tag{3.9}$$

Hence, $\tilde{S} = \dfrac{1}{(1+\tilde{F}P)} = \dfrac{\beta_{11}\theta_+\phi_+}{(\beta_{11}\phi_+\theta_+ + \Psi_+\zeta)}$ and

$$|\tilde{x}| = |V\tilde{S}| = \frac{\alpha_{11}\phi_+\theta_+}{(\beta_{11}\phi_+\theta_+ + \Psi_+\zeta)} = \frac{D\phi_+\theta_+}{\phi_+^*\theta_+^*} \tag{3.10}$$

Thus, it has been demonstrated that the two methods are equivalent for the minimisation of optimal weighted sensitivity when the weighting function has conjugate symmetry.

4. NUMERICAL EXAMPLE

For $P(s) = (s-1)^2/(s+1)^2(-s+2)$, $V(s) = (s+1)/(10s+1)$. Using an eigenvalue eigenvector approach Kwakernaak[5] solves equations (2.1) to give $\lambda = -1.24688$, $\theta = -0.391346+s$, and $\zeta = 1.41041+10.8020s$. Notice that the optimal θ has its root in the right half-plane as the result of section 3 tells us it must have.

Hence, $\tilde{X} = \frac{\lambda\phi_+\theta_+}{\phi_+^*\theta_+^*} = -1.24688 \; \frac{(-s+2)\;(-0.391346+s)}{(s+2)\;\;(-0.391346-s)}$

Applying the theory of Sarason[6], Francis and Zames[3] obtain

$$\tilde{X} = -1.247 \; \frac{(s-2)\;(s-0.391)}{(s+2)\;(s+0.391)}$$

5. EXTENSIONS TO THE ZAMES AND FRANCIS PROCEDURE

The comparison of section 3 can be used to make straightforward extensions to Zames and Francis[1] procedure. Consider the optimisation criterion

$$\sup_{\omega} \sum_{k} \left| V_k(j\omega)S(j\omega) \right| \tag{5.1}$$

This is equivalent to the criterion $\sup_{\omega} \left(\sum_{k} |V_k(j\omega)S(j\omega)|\right)^2$ (5.2)

Now, $\sup_{\omega} \left(\sum_{k} |V_k S|\right)^2 = \left(|V_1| + \;.\;.\; + |V_k|\right)^2 |S|^2$, if $V_i = \frac{\alpha_{1i}}{\beta_{1i}}$ then,

for Kwakernaak's method optimising (5.1) is equivalent to optimising $\sup_{\omega} |VS|^2$ if V has conjugate symmetry, where

$V = \frac{\alpha_{11}}{\beta_{11}} + \;.\;.\; + \frac{\alpha_{1k}}{\beta_{1k}} = \frac{\alpha_{11}\omega_{11} + \;.\;.\; + \alpha_{1k}\omega_{1k}}{\omega_1}$. Note: V will certainly have conjugate symmetry if each of the individual V_k's had conjugate symmetry. Therefore (5.1) can be solved by substituting $(\alpha_{11}\omega_{11} + \;.\;.\; + \alpha_{1k}\omega_{1k})$ for α_{11} and ω_1 for β_{11}

in (3.5) i.e. $\frac{1}{D}(\alpha_{11}\omega_{11} + \;.\;.\; + \alpha_{1k}\omega_{1k})\;\phi_+^*\theta_+^* - \omega_1\phi_+\theta_+ + \Psi_+\zeta$ (5.3)

and solving (5.3) for D, θ_+ and ζ. By Lemma 1 there always exists such a solution and it is unique. Examination of (2.1) shows that for this case Kwakernaak's[2] method is equivalent to solving (5.3) subject to $\theta = \theta_+$.

Consider now the optimisation criterion

$$\sup_{\omega} |W(j\omega)T(j\omega)| \tag{5.4}$$

observing that

$$VS = \frac{\alpha_{11}\rho\phi}{\beta_{11}(\rho\phi+\sigma\Psi)} \quad \text{and} \quad WT = \frac{\alpha_{21}\sigma\Psi}{\beta_{21}(\rho\phi+\sigma\Psi)}$$

it can be seen that the two optimisation problems are symmetric. Hence, the following Lemma may be stated.

Lemma 2

(a) There is a unique weighed complementary sensitivity function $\tilde{Y}$ in H^{∞} which attains the minimum norm, $\min\{\|Y\| : Y=WT\}$, where $\|Y\| = \sup_{\omega} |W(j\omega)T(j\omega)|$. The necessary and sufficient conditions for $\tilde{Y}$ to be that function are that $\tilde{Y}$ satisfy the interpolation constraints:-

$$\left.\begin{aligned} Y(a) &= W(a) \\ \left(\frac{d}{ds}\right)^{q-1} Y &= \left(\frac{d}{ds}\right)^{q-1} W \text{ at } s = a \end{aligned}\right\} \tag{5.5}$$

at each distinct pole a, $\mathrm{Re}(a) > 0$, of multiplicity q, the total of such constraints is n_+:

and have the (allpass) form

$$\tilde{Y} = D \prod_{i=1}^{n_+-1} \left(\frac{c_i - s}{\bar{c}_i + s}\right) \prod_{j=1}^{m_+} \left(\frac{b_j - s}{\bar{b}_j + s}\right) \tag{5.6}$$

in which $\mathrm{Re}(c_i) \geq 0$, D is a constant satisfying $|D| = \|\tilde{Y}\|$,

$$\prod_{j=1}^{m_+} (b_j - s) = \Psi_+ .$$

(b) If the plant P(s) and weighting function W(s) have conjugate symmetry then so does $\tilde{Y}$, and the coefficients (5.6) are real or occur in conjugate pairs.

Utilising Lemma 2 results completely analogous to those for (5.1) can be obtained for the criterion $\sup_{\omega} (\sum_{k} |W_k(j\omega)\ T\ (j\omega)|)^2$. The comparison of Section 2 can also be used to provide an alternative way of establishing the existence of unique optimal solutions to Kwakernaak's[2] procedure for certain classes of problem.

6. CONCLUSIONS

It has been shown, that for the optimisation of weighted sensitivity when the weighting function has conjugate symmetry application of Kwakernaak's[2] procedure gives the same optimal weighted sensitivity as application of Zames and Francis[1] procedure. The immediate benefit of this result is that Kwakernaak's[2] eigenvalue eigenvector algorithm developed to solve the above problem may be used to apply Zames and Francis[1] design methodology. The procedure used by Zames and Francis[1] to obtain the optimal controller results in a different expression for every extra right half-plane zero the plant has. This result, together with the generality of Kwakernaak's[2] expression (1.2) has enabled certain extensions of Zames and Francis[1] work to be made in a straightforward manner.

7. ACKNOWLEDGEMENTS

The author thanks Professor H. Kwakernaak of the Technical University Twente, The Netherlands, for many helpful discussions during his visits to Twente; Professor M. Grimble and Dr. M. Johnson of the Industrial Control Unit, The University of Strathclyde, Scotland, for their many helpful comments; The United Kingdom Science and Engineering Research Council for their support on the robustness research project.

8. REFERENCES

1. Zames, G., and Francis, B. (1983) "Feedback, minimax sensitivity and optimal robustness". IEEE Trans. Aut. Control, Vol. Ac-28, No. 5, pp. 585-600.

2. Kwakernaak, H. (1983) "Robustness optimisation of linear feedback systems". Proc. 22nd CDC. San Antonio, Texas.

3. Francis, B., and Zames, G. (1984) "On optimal sensitivity theory for SISO feedback systems". IEEE Trans. Aut. Control, Vol. AC-2, No. 1, pp. 9-16.

4. Zames, G., (1981) "Feedback and optimal sensitivity; Model reference transformations, multiplicative seminorms, and approximate inverses". IEEE Trans. Aut. Control, Vol. AC-26 No. 2, pp. 301-320.

5. Kwakernaak, H. (1984) "Minimax frequency domain performance and robustness optimisation of linear feedback systems" To appear: IEEE Trans. Auto. Control, Vol. AC-29, No. 11.

6. Sarason, D., (1967) "Generalised interpolation H^{∞}". Trans. AMS 127, pp. 179-203.

LIE ALGEBRAS AND THE SINGULAR CONTROL PROBLEM

J.L. Kotsiopoulos
(Department of Electrical Engineering, Imperial College, London)

and

D.J. Bell
(Control Systems Centre, UMIST, Manchester)

ABSTRACT

It has been widely recognized that for real-analytic nonlinear systems, Lie brackets determine behaviour[3]. In optimal control theory, necessary conditions of any order can be expressed as conditions on the subspaces of Lie algebras of vector fields[6,9] and of Hamiltonian functions, and this fact can be exploited in deriving necessary conditions on singular-nonsingular junctions. This last topic is examined in section 4 of this paper, while sections 1,2,3 serve as a comprehensive background by surveying important results in the area, under a unified framework.

1. INTRODUCTION AND MOTIVATION

Let M be an (n+1)-dimensional, real-analytic manifold with an analytic vector field X_u defined on it, such that any of its local representations with respect to any chart is a jointly smooth function of u and x, with $u: \mathbb{R} \to \mathbb{R}: t \to u(t)$ Lebesgue measurable on a time-interval $[t_o, t_f]$, $x \in M$ and $x_o = t$ (i.e. time is one of the dimensions of M). We take U as the set of all u's such as above and we define the corresponding flow γ_u as a dynamical system on M. Although more general definitions exist[3], this will suffice for our purposes. Furthermore, the set of maps (or the multi-map):

$$\Gamma: M \times \mathbb{R} \times U \to M \text{ s.t. } \Gamma = \{\gamma_u | u \in U\} \qquad (1.1)$$

will be called a dynamical polysystem on M with U as the control-set. We shall restrict U to $\Omega = \{u | |u(t)| \leq K \in \mathbb{R}^+\}$ and

X_u to control-linear vector fields such that $X_u(x) = f(x) + u(t)g(x)$, $f,g \in V(M)$, the set of all analytic vector fields on M, $u \in \Omega$. Thus Γ becomes a control-linear dynamical polysystem with reference control u_o, written as:

$$\dot{x} = X_{u_o}(x) + u(t)Y(x),\ u \in \Omega,\ t \in [t_o,t_f],\ x(t_o) = p \quad (1.2)$$

where $Y = g$ and $u(t)$ is the difference between the control and the reference control $u_o(t)$, with Ω changed accordingly to Ω'. For simplicity of notation we shall now drop the subscript of X_{u_o} and consider that the reference solution is given for $u(t) \equiv 0$.

The space V(M) is a Lie algebra denoted by $\mathcal{L}(X,Y)$, $X,Y \in V(M)$ under the Lie bracket operation $[X,Y] \in \mathcal{L}(X,Y)$, and the reference solution of (1.2) is the diffeomorphism $\exp(s-t_o)X\colon M \to M$: $x(s) = \exp(s-t_o)X(p)$, at $t=s$, belonging to the Lie group of $\mathcal{L}(X.Y)$. Without loss of generality, we assume from now on that $t_o = 0$.

If $u(t) \neq 0$ in $[0,t_f]$, then by a theorem of Chen[4] the solution can be decomposed uniquely as:

$$x(t) = (\exp tX)\circ y(t;u)(p)$$

where $$\dot{y}(t;u) = u(t) \sum_{i=0}^{\infty} \frac{(-t)^i}{i!} ad^i X(Y)(y),\ y(0) = p, \quad (1.3)$$

and $$ad^i X(Y) = [X, ad^{i-1}X(Y)],\ adX(Y) = [X,Y].$$

Since $y(t;u)$ is the solution curve of a symmetric analytic vector field $Z_u = ue^{-tadX(Y)}$, (ref.[2]), it follows[3] that Z_u, $u \in \Omega$ have integral manifolds $I(Z_u) \subset I(L(\mathcal{L}^1))$ where $\mathcal{L}^1 = \{\pm ad^i X(Y)\ i = 0,1,\ldots\}$ and $L(\mathcal{L}^1)$ is the Lie subalgebra of $\mathcal{L}(X,Y)$ spanned by $\mathcal{L}^1$. Therefore:[1,4]

$$\Gamma(p,t,\Omega) \subset (\exp tX)\circ I(L(\mathcal{L}^1))(p),\ t \in [0,t_f] \quad (1.4)$$

Equation (1.4) shows that the "attainable" or "reachable" set of Γ is a composition of diffeomorphisms belonging to the Lie group of $\mathcal{L}(X,Y)$ and moreover that we can define groups of diffeomorphisms (under the composition ($\circ$) operation) starting at any point of the reference solution and corresponding to admissible control variations with tangent vectors belonging to $\mathcal{L}(X,Y)$ along the reference trajectory. Hermes[5] defines such a group as:

$$(\exp\sigma_{2k}(s)X)\circ(\exp r_{2k}(s)(X+Y))\circ(\exp\sigma_{2k-1}(s)X)\circ(\exp r_{2k-1}(s)(X-Y))\circ$$

$$\circ\ldots\circ\ (\exp\sigma_1(s)X)\circ(\exp r_1(s)(X-Y))\circ\exp\left(-\sum_{i=1}^{2k}(\sigma_i(s)+r_i(s))X\right)$$

$$=\exp(sW+O(s)) \tag{1.5}$$

where $W\in E_X^+\subset\mathcal{L}(X,Y)$ and is such that k exists with r_i, σ_i $(i=1,\ldots,2k)$ positive rational polynomials $[0,\ \varepsilon]\to\mathbb{R}$, $\varepsilon>0$, where $O(s)$ denotes higher powers of s. E_X^+ is convex[5] and each W is tangent to $\Gamma(p,t,\Omega)$ at $x(t) = \exp tX(p)$, for $s = 0$. If W is zero at $x(t)$, then we denote by W the first nonzero tangent vector at $x(t)$, obtained either after h differentiations of (1.5)[6] or after h reparameterisations of r_i, σ_i followed by a 1st differentiation each time[5]. In the later case, composition of exp's is done easily by the use of the Baker-Campbell-Hausdorff formulae[5]. W's constructed in this way belong to important finite-dimensional subspaces of $\mathcal{L}(X,Y)$, which give necessary conditions of optimality through the Maximal Principle[6].

2. THE OPTIMAL CONTROL PROBLEM AND THE MAXIMAL PRINCIPLE

Consider a dynamical polysystem Γ on M, s.t.(1.2). If the reference control $u(t)\equiv 0$ minimizes the functional $\phi_0(x(t_f))$ s.t. a linearly independent set of C^∞- boundary conditions $\phi_i(x(t_f)) = 0$, $i = 1,\ldots,m$, then the reference trajectory $x(t) = \exp tX(p)$ lies on the boundary of $\Gamma(p,t,\Omega')$, $t\in[0,t_f]$ and moreover:

The Maximal Principle (MP)[6]: There exists a non-trivial adjoint vector $\lambda: t\to(\lambda_0(t),\ \lambda_1(t),\ldots,\lambda_n(t))^T\in T^*M$ (the co-tangent bundle of M), which defines a Hamiltonian system on M with Hamiltonian function:

$$H_X:\ (x,\lambda)\to\lambda^T(t)X(x(t)),\ x(t)=\exp tX(p),\ t\in[0,t_f] \tag{2.1}$$

and with λ and x as canonical local coordinates[7,8]. In addition to the well-known Hamilton's equations, the following are true:

$$\dot{\lambda}^T(t) = -\frac{\partial}{\partial x} H_X(x(t),\lambda(t)), \quad \lambda(t_f) = \sum_{i=0}^{m} v_i \frac{\partial}{\partial x}\phi_i(x(t_f)), v_0 \geq 0 \tag{2.2}$$

$$H_{X+uY}(x(t), \lambda(t)) \geq H_X(x(t), \lambda(t)) = 0 \text{ a.e. in } [0,t_f], \ \forall u \in \Omega \tag{2.3}$$

$$H_W(x(t), \lambda(t)) \geq 0, \text{ a.e. in a subinterval } \Delta \text{ of } [0,t_f], \ \forall W \in E_X^+ \tag{2.4}$$

where u is continuous[6] on Δ and where the notation H_Z is taken as:

$$H_Z: (x,\lambda) \to \lambda(t)^T Z(x(t)), \ Z \in V(M)$$

with $(\lambda,x) = (\lambda(.), \exp(.)X(p))$ the solution curves of the Hamiltonian vector field X_H corresponding to X. With this structure, every H_Z is defined as a Hamiltonian function along the solution curves of X_H, on the symplectic manifold T*M, and through a symplectic form on it, an isomorphism I can be constructed[7] between the Lie algebra of all Hamiltonian vector fields on T*M and the set $H \triangleq \{H_Z | Z \in V(M)\}$. Thus a Lie algebra can be defined (denoted as $\mathcal{L}(H_{Z_1}, H_{Z_2}), Z_1, Z_2 \in V(M)$) through the homomorphism I, under the Poisson bracket $\{.,.\}$ operation:

$$\{H_X, H_Y\} \triangleq I([I^{-1}(H_X), I^{-1}(H_Y)]), \text{ which becomes}^{7,8}$$

$$\{H_X, H_Y\}(x,\lambda(t)) = \lambda^T(t)[X(x), Y(x)] \tag{2.5}$$

in canonical local coordinates. It is also readily verified that:

$$\frac{d}{dt} H_Z(x(t),\lambda(t)) = \{H_X, H_Z\}(x(t), \lambda(t)), \ x(t) = \exp tX(p), Z \in V(M) \tag{2.6}$$

We call H_X the <u>Hamiltonian function of the extremal system</u> $S \triangleq (X, H_X, \lambda)$ corresponding to a reference control u_0 (or $u(t) \equiv 0$) and satisfying the first order necessary conditions of the MP(i.e. excluding (2.4)).

3. CONSEQUENCES OF THE MAXIMAL PRINCIPLE

For a control-linear Γ, equation (2.3) yields the bang-bang optimal control $u(t) = -K\mathrm{sgn}\phi(t)$ (with $\phi(t) \equiv \lambda^T(t)g(x(t))$ as the switching function). However, it is easily shown through repeated use of (2.5) and (2.6)[6,9], that for $u \in \mathrm{int}.\Omega'$, an S exists, if and only if $\dim.\mathcal{L}^1 < \eta$, and is thus called a singular extremal system (S is unique when $\dim.\mathcal{L}^1 = \eta-1$).

Proposition 3.1. Let S be an extremal system on Γ s.t. $\dim.\mathcal{L}^1 < \eta$. Then the smallest integer q (where $h = 2q-1$ in Section 1) for which[6]:

$$[Y, \mathrm{ad}^{2q-1}X(Y)] \notin \mathcal{L}^1 \text{ a.e.at } x(t) = \exp tX(p),\ t \in \Delta \tag{3.1}$$

is a property of S and Δ called the local order of S, which satisfies[11]:

$$\frac{\partial}{\partial u_o}\frac{d^{2q}}{dt^{2q}}\frac{\partial H}{\partial u_o}(x(t), \lambda(t)) = \lambda^T(t)[Y, \mathrm{ad}^{2q-1}X(Y)](x(t)),\ \text{a.e. in } \Delta \tag{3.2}$$

Moreover, if there exists an integer $p \leq q$, called the polysystem order, s.t.

$$[g, \mathrm{ad}^i f(g)](x) \equiv 0,\ \forall x \in M,\ i=1,..,2p-1 \tag{3.3}$$

then $\frac{d^{2q}}{dt^{2q}}\frac{\partial H}{\partial u_o}(x(t), \lambda(t))$ is a polynomial in u of degree $2(q-p) + 1$ and coefficients in $\mathcal{L}(H_f, H_g)(x(t), \lambda(t))$.

The control u_o of Proposition 3.1, such that $u_o \in \mathrm{int}.\Omega$ is called a singular control on S. The distinctions on the notion of order, arise from work by Lewis[12].

If $\mathcal{L}^{2,2q-1} \triangleq \{W | W = [\mathrm{ad}^j X(Y), \mathrm{ad}^i X(Y)],\ i+j = 2q-1\}$, then[5]

$$W(x(t)) = (-1)^i[\mathrm{ad}^{2q-1}X(Y), Y](x(t)) + L(x(t))$$

where $L(x(t)) \in \mathcal{L}^1(x(t))$, and a control variation can be constructed whose first nonzero tangent vector is $W \in \mathcal{L}^{2,2q-1}$. Indeed, Krener[6] proved that this is possible with $W = (-1)^q [Y, \mathrm{ad}^{2q-1}X(Y)]$, which through (2.4) gives the well-known

Generalised Legendre-Clebsch (GLC) condition as:

$$H_W(x(t),\ \lambda(t)) \equiv \lambda^T(t)(-1)^q[Y,\ ad^{2q-1}X(Y)](x(t)) \geq 0 \text{ a.e. in } \Delta \tag{3.4}$$

for an optimal extremal system S. Using (3.2) another expression can also be obtained for $H_W(x(t),\lambda(t)) \geq 0$.

4. SINGULAR-NONSINGULAR JUNCTIONS

The results and the tools developed in the previous sections will be used in deriving necessary conditions for optimality at singular-nonsingular junctions. We shall restrict ourselves to piecewise analytic controls around the junction, thus allowing the use of Taylor expansions in the neighbourhood of the junction point. Also, following the existing literature[10,13], we shall simplify calculations by assuming p=q. The removal of this assumption is the subject of future work by the authors[16].

The following theorem is a generalisation of previous results[10,13,14,15] and is along the lines of the McDanell and Powers conjecture (Theorem 2 in [13]).

<u>Junction Theorem</u>: Let t_c be a point at which singular ($t_c - \varepsilon$, $\varepsilon > 0$) and nonsingular ($t_c + \varepsilon$, $\varepsilon > 0$) subarcs of an optimal extremal system S in Γ are joined, and let p be both the local order and the polysystem order. Assume the control u (where $u_0 \triangleq u$ in this section) is piecewise analytic in a neighbourhood of t_c, and let $u^{(r)}(t_c)$, $r \geq 0$ be the lowest order derivative of u which is discontinuous at t_c, and $\beta^{(m)}(t_c)$, m=0,1,2 be the lowest order time-derivative of the GLC expression

$$\beta(t) \equiv \frac{\partial}{\partial u}\frac{d^{2p}}{dt^{2p}}\frac{\partial H}{\partial u}(x(t),\ \lambda(t)) \tag{4.1}$$

which is nonzero at t_c. Then

$$(-1)^{p+m}\beta^{(m)}(t_c) > 0 \tag{4.2}$$

and p+r+m is an odd integer if and only if

$$(-1)^{p+m}L_p^m(t_c) > 0 \tag{4.3}$$

$$\text{where } L_p^m(t_c) \triangleq \frac{\partial}{\partial u} \frac{d^m}{dt^m}\left(\frac{\partial H}{\partial u}\right)^{(2p)}(t_c) + R_p^m(r)\beta^{(m)}(t_c) \tag{4.4}$$

$$R_p^0 = 0,\ R_p^1 = r,\ R_p^2 = rp + \binom{r+2}{2} - 1 \tag{4.5}$$

and $\beta^{(m)}(t_c)$ is interpreted as $\frac{1}{2}\left\{\beta^{(m)}(t_c^+) + \beta^{(m)}(t_c^-)\right\}$ if it is discontinuous. Equations (4.4) and (4.5) hold for all integer triads (m,p,r), with the exception of $(2,1,0)$ which gives for (4.4):

$$\begin{aligned} L_1^2(t_c) &= \ddot{\beta}(t_c^-) + \ddot{\beta}(t_c^+) + \lambda^T(t_c).\ [g,\ ad^3 f(g)] \\ &+ (u(t_c^+) + u(t_c^-))[g,[g,\ ad^2 f(g)]] \\ &+ u(t_c^+)u(t_c^-)[g,[g,[g\ adf(g)]]]\ (t_c) \end{aligned} \tag{4.6}$$

Proof Outline: Since the complete proof is in [11], lengthy details will be omitted. Following [13] we substitute continuous terms in the Taylor series expansion of the switching function ϕ around t_c. By defining

$$\frac{d^{2p}}{dt^{2p}}\phi(t) \triangleq \phi(t)^{(2p)} \triangleq \alpha(t) + u(t)\beta(t),\ \ddot{\beta}(t_c) \neq 0,\ \beta(t_c) = \dot{\beta}(t_c) = 0$$

we have on the singular side:

$$\phi^{(2p)}(t) \equiv 0 \Rightarrow \alpha(t) \equiv -\beta(t)u_s(t) \Rightarrow \alpha^{(r+j)}(t)$$

$$\equiv -\sum_{i=0}^{r+j}\binom{r+j}{i}\beta^{(r+j-i)}(t)u_s^{(i)}(t) \tag{4.7}$$

and by substituting the continuous terms in the expansion of $\phi(t_c + \varepsilon)$ we get[11]:

$$\phi(t_c+\varepsilon) = \frac{\varepsilon^{2p+r+m}}{(2p+r+m)!}\ (u^{(r)}(t_c^+) - u^{(r)}(t_c^-))L_p^m(t_c) + O(\varepsilon^{2p+r+m}) \tag{4.8}$$

First order necessary conditions can be applied to (4.8)[13] and (4.3) is readily obtained. Condition (4.2) follows by expanding β around t_c and applying the GLC condition.

All that is left now is the derivation of (4.8) through (4.7) and through the techniques of the previous sections. Indeed, substitution of (4.7) into the expansion of $\phi(t_c+\varepsilon)$, results in terms of the form:

$$\alpha^{(r+i)}(t_c^+) - \alpha^{(r+i)}(t_c^-) \text{ and } \beta^{(r+i)}(t_c^+) - \beta^{(r+i)}(t_c^-),\ i = 1,2$$

which are polynomials in $u(t_c)$, $u^{(1)}(t_c),\ldots,u^{(r+i-1)}(t_c)$. Using the Leibniz rule for repeated differentiation of products and equation (2.6) we get[11]:

(i) <u>p=1</u>: $\alpha^{(r+i)}(t_c^+) - \alpha^{(r+i)}(t_c^-) =$

$$\left\{\frac{\partial}{\partial u}\ddot{\alpha}(t_c) + (r+i-2)\frac{d}{dt}\left\{\lambda^T(t_c)[g,[f,[f,g]]](t_c)\right\}\right\}\cdot$$
$$\cdot\left\{u^{(r+i-2)}(t_c^+) - u^{(r+i-2)}(t_c^-)\right\} + \lambda^T(t_c)[g,[f,[f,g]]](t_c)\cdot$$
$$\cdot\left\{u^{(r+i-1)}(t_c^+) - u^{(r+i-1)}(t_c^-)\right\}$$

$$\beta^{(r+i)}(t_c^+) - \beta^{(r+i)}(t_c^-) =$$

$$\left\{\frac{\partial}{\partial u}\ddot{\beta}(t_c) + (r+i-2)\frac{d}{dt}\left\{\lambda(t_c)^T[g,[g,[f,g]]](t_c)\right\}\right\}\cdot$$
$$\cdot\left\{u^{(r+i-2)}(t_c^+) - u^{(r+i-2)}(t_c^-)\right\}+$$
$$\lambda^T(t_c)[g,[g,[f,g]]](t_c)\left\{u^{(r+i-1)}(t_c^+) - u^{(r+i-1)}(t_c^-)\right\}$$

(ii) <u>p>1</u>: $\alpha^{(r+i)}(t_c^+) - \alpha^{(r+i)}(t_c^-) =$

$$\left\{\lambda^T(t)[\mathrm{ad}^{2p}f(g),\ \mathrm{ad}f(g)](t_c) + (r+2)p\ddot{\beta}(t_c)\right\}\cdot$$
$$\cdot\left\{u^{(r+i-2)}(t_c^+) - u^{(r+i-2)}(t_c^-)\right\}$$

$$\beta^{(r+i)}(t_c^+) - \beta^{(r+i)}(t_c^-) =$$

$$\lambda^T(t)[g,[g,\ \mathrm{ad}^{2p}f(g)]](t_c)\left\{u^{(r+i-2)}(t_c^+) - u^{(r+i-2)}(t_c^-)\right\}$$

In case (ii) expression are simpler because of the use of two Lie bracket identities due to Bortins[15]. Equation (4.7) is obtained by substituting to the series for $\phi(t_c+\varepsilon)$ and assembling the resulting terms[11].

Special cases arise when r=0,1. Equation (2.6) suggests that $\alpha^{(m)}(t_c)$, $\beta^{(m)}(t_c)$ might be discontinuous, and this is according to the following table:

	p=1	p>1
r=1	$\ddot{\alpha}(t_c^{\pm})$, $\ddot{\beta}(t_c^{\pm})$	$\ddot{\alpha}(t_c^{\pm})$
r=0	$\dot{\alpha}(t_c^{\pm})$, $\dot{\beta}(t_c^{\pm})$	$\dot{\alpha}(t_c^{\pm})$, $\dot{\beta}(t_c^{\pm})$
	$\ddot{\alpha}(t_c^{\pm})$, $\ddot{\beta}(t_c^{\pm})$	$\ddot{\beta}(t_c^{\pm})$

(4.9)

More lengthy derivation[11] shows that except for (m,p,r) = (2,1,0) where (4.6) applies, the results are unchanged if we interpret $\beta^{(m)}(t_c)$ as $\frac{1}{2}(\beta^{(m)}(t_c^+) + \beta^{(m)}(t_c^-))$, and this completes the proof.

Results obtained by previous authors are easily deduced from the above through the Lie bracket identities of Bortins[15]. Indeed[11]:

$$m=0:\ L_p^0(t_c) = \beta(t_c)$$

$$m=1:\ L_p^1(t_c) = (p+r+1)\dot{\beta}(t_c)$$

which yield the following:

Corollary:[10,13,14,15] Under the conditions of the junction theorem for m = 0,1, p+r+m is an odd integer, except for (m,p,r) = (1,1,0) where (4.2) and (4.3) reduce to:

$$-\mathrm{sgn}\left\{\dot{\beta}(t_c^+)\dot{\beta}(t_c^-)\right\} = (-1)^{p+r+m}$$

Finally, to illustrate the form of $L_m^2(t_c)$ for $r \neq 0$, we give it explicitly as:

$$L_p^2(t_c) = \left\{(r+2)p + \binom{r+2}{2} + 1\right\}\ddot{\beta}(t_c) + \lambda^T(t_c)\Big\{[g,ad^{2p+1}f(g)]$$

$$-\frac{p+1}{p}[f,[g,ad^{2p}f(g)]] - u\,[f,[g,[g,ad^{2p-1}f(g)]]]\Big\}(t_c).$$

where the convention $ad^0 f(g) \triangleq g$ is used throughout.

5. REFERENCES

1. Hermes, H. (1976) "Local Controllability and Sufficient Conditions in Singular Problems. II". SIAM J. Contr. Opt., 14, 1049-1062.

2. Varadarajan, V.S. (1974) "Lie Groups, Lie Algebras and their Representations". Prentice-Hall.

3. Sussmann, H.J. (1983) "Lie Brackets, Real Analyticity and Geometric Control" in 'Differential Geometric Control Theory', R. Brockett, R.S. Millman, H.J. Sussmann (Eds), Birkhauser, 1-117.

4. Sussmann, H.J. and Jurdjevic, V. (1972) "Controllability of Nonlinear Systems". J. Diff.Eqns., 12, 95-116.

5. Hermes, H. (1978) "Lie Algebras of Vector Fields and Local Approximation of attainable sets". SIAM J. Contr. Opt., 16, 715-727.

6. Krener, A. (1977) "The High Order Maximal Principle and its Application to Singular Extremals". SIAM J. Contr., 15, 256-293.

7. Arnold, V.I. (1980) "Mathematical Methods of Classical Mechanics", Springer.

8. Abraham, R. and Marsden, J.E. (1978) "Foundations of Mechanics", Benjamin/Cummings.

9. Hermes, H. (1976) "Local Controllability and Sufficient Conditions in Singular Problems. I" J. Diff. Eqns., 20, 213-232.

10. Gabasov, R. and Kirillova, F.M. (1972) "High Order Necessary Conditions for Optimality". SIAM J. Contr,. 10, 127-168.

11. Kotsiopoulos, J.L. (1983) "Lie Groups and Lie Algebras in Optimal Control Theory", MSc Dissertation, UMIST.

12. Lewis, R.M. (1980) "Definitions of Order and Junction Conditions in Singular Optimal Control Problems". SIAM J. Contr. Opt., 18, 21-32.

13. McDanell, J.P. and Powers, W.F. (1971) "Necessary Conditions for Joining Optimal Singular and Nonsingular Subarcs". SIAM J. Contr., 9, 161-173.

14. Bell, D.J. and Boissard, M. (1979) "Necessary Conditions at the Junction of Singular and Nonsingular Control Subarcs". Int. J. Control, 29, 981-990.

15. Bortins, R. (1983) "On Joining Nonsingular and Singular Optimal Control Subarcs". Int. J. Contr., 37, 867-872.

16. Kotsiopoulos, J.L., and Bell, D.J. "Necessary conditions for Optimality of Singular-Nonsingular Junctions". Work in progress.

EXPLICIT OPTIMAL PATHS FOR REGULATING A STABILISED SYSTEM

N.M. Christodoulakis
(Engineering Department, University of Cambridge)*

ABSTRACT

Typically the control problems in the Optimal Control framework is solved by specifying desired paths for output and input variables and then by optimising a quadratic objective function for the weighted deviations between actual and desired levels. The paper suggests a decomposition of the design into two stages: In the first, a dynamic controller can be designed using 'classical' control methods as to guarantee some essential dynamic properties for the closed-loop system. With this controller fixed, the second stage the typical quadratic cost function is minimised.

Adopting a sensible parametrisation for the output desired levels, optimal solutions are obtained explicitly, without involving the recursive investigation of the way that specifications of the objective function shape the final solution. The problem is analysed in the discrete-time case.

1. INTRODUCTION

In Figure 1, a deterministic, discrete-time, linear system described by a set Σ_1 of time-invariant state-space matrices, $A_1B_1C_1D_1$, relates m instruments to m targets, through a m×m transfer function matrix G(z). An r-vector of secondary targets, or states, $\underline{w}$ may be also defined to be determined by $\underline{x}$ through a set Σ_2 of $A_2B_2C_2D_2$ matrices that correspond to an rxm transfer function S(z). The aim is to derive a dynamic

---

*Present address: Department of Applied Economics, University of Cambridge.

controller K(z), in order to drive the system towards a desired path for the main and secondary targets.

When controller K(z) has been derived, the values of $\underline{x}$, $\underline{y}$ and $\underline{w}$ are uniquely determined by the reference level $\underline{u}_{ref}(t)$. The following relations hold in the frequency- and time-domain respectively:

$$\begin{aligned} \underline{x}(z) &= H_x(z)\underline{u}(z) & & \underline{x}(t) = \Phi_x(t,\underline{u})\underline{u}(t) \quad (1.1a)\\ \underline{y}(z) &= H_y(z)\underline{u}(z) & \Rightarrow\ & \underline{y}(t) = \Phi_y(t,\underline{u})\underline{u}(t) \quad (1.1b)\\ \underline{w}(z) &= H_w(z)\underline{u}(z) & & \underline{w}(t) = \Phi_w(t,\underline{u})\underline{u}(t) \quad (1.1c) \end{aligned}$$

where t is a positive integer denoting the time period. The closed-loop operators are obtained through the well-known matrix expressions:

$$\begin{aligned} H_x(z) &= [I + KG]^{-1}K & \longleftrightarrow\ & \Sigma_3(A_3B_3C_3D_3) \quad (1.2a)\\ H_y(z) &= [I + GK]^{-1}GK & \longleftrightarrow\ & \Sigma_4(A_4B_4C_4D_4) \quad (1.2b)\\ H_w(z) &= [I + GK]^{-1}S & \longleftrightarrow\ & \Sigma_5(A_5B_5C_5D_5) \quad (1.2c) \end{aligned}$$

In the control literature there exist both analytic expressions and computer software to obtain the above operators in state space form in terms of the given system Σ_1, Σ_2. Defining a single vector $\underline{\xi}(t)$, of dimension (m+m+r), to represent all the variables

$$\underline{\xi}'(t) = [\underline{x}'(t)\ \underline{y}'(t)\ \underline{w}'(t)] \quad (1.3)$$

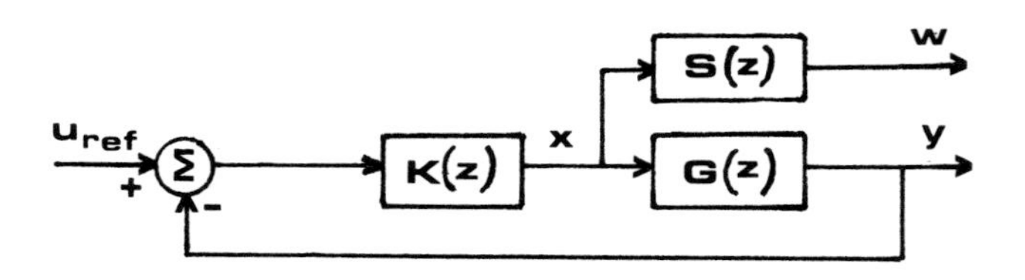

Fig. 1. System configuration

and considering that the desired path for them is given by $\xi_{des}(t)$ over a time-horizon of N periods, the quadratic objective function sought to be minimised is generally written as[3]

$$J_N = \sum_{t=1}^{N} [\underline{\xi}(t) - \underline{\xi}_{des}(t)]' P(t) [\underline{\xi}(t) - \underline{\xi}_{des}(t)] \quad (1.4)$$

where P(t) is a symmetric, positive definite penalty matrix, and (′) denotes transposition.

The solution offered by an Optimal Control Approach depends obviously on the weighting matrix and the specified desired paths[1]. If these specifications are not 'realistic', it is very likely that a satisfactory controller action $\underline{x}(t)$ cannot be found, and then a new design must be attempted in which a less 'ambitious' desired path for $\underline{\xi}(t)$ is traded off with a 'better' dynamic performance of the controlled system.

Alternately, the designer may opt for a Frequency Domain Approach to derive a controller K(z) for the system G(z), such that a number of dynamic properties are ensured. These include stability, good tracking, robustness and effective disturbance rejection, while the incorporation of integral control action in K(z) can always guarantee zero steady-state errors[4]. The problem however remains to find a reference level $\underline{u}_{ref}(t)$ such that the same cost function (1.4) is optimally satisfied. This optimisation problem cannot be dispensed with, because if it is simply set

$$\underline{u}_{ref}(t) = \underline{Y}_{des}(t) \tag{1.5}$$

the main target will be asymptotically reached, but the value of J_N may go very high as the desired paths for $\underline{x}$ and $\underline{w}$ may confront the desired path for $\underline{y}$. The paper presents a method of evaluating an optimal $\underline{u}_{ref}(t)$ that minimises the objective function.

2. PARAMETRISING THE REFERENCE LEVEL

The 'pure-delay' operator is defined by the function:

$$\pi(t,\lambda) = \begin{cases} 1, & \text{for } t \geqslant \lambda \\ 0, & \text{for } t < \lambda \end{cases} \tag{2.1}$$

for any positive integer λ, and then the set of time-functionals is constructed as:

$$F(t,\lambda) = \{f_i = \pi(t,\lambda).(t-\lambda)^{i-1},\ i=1,2\ldots\} \cup \{f_0 = \delta(t-\lambda)\} \tag{2.2}$$

that obviously represents the time-signals of impulse, step, ramp etc, delayed by various λ. When the desired path $\underline{\xi}_{des}(t)$, for all the variables involved in the quadratic objective function (1.4) has been specified, we try to approximate each j-entry (j=1,...,m) of the main-target desired path, $\underline{y}_{des}(t)$, by a linear combination of time-functionals from $F(t,\lambda)$.

Next, a suitable parametrisation for the sought reference vector, $\underline{u}_{ref}(t)$, is considered. Since the ideal case would be to find an optimal demand vector precisely equal to the specified desired target path, $\underline{y}_{des}(t)$, it becomes intuitively clear that we should include the same time-functionals used for $\underline{y}_{des}(t)$, in the parametrisation of $\underline{u}_{ref}(t)$. In such a case only the coefficients of the functionals remain to be determined through the optimisation procedure, while the pattern of the reference vector will be the same as that of the desired path for the main target.

Let us introduce the notation:

- Inputs-Outputs: $j = 1,2,\ldots, m$
- To each j there correspond ν_j delays $\lambda_{j\rho}$, $\rho = 1,2,\ldots, \nu_j$
- To each delay $\lambda_{j\rho}$ there correspond $k_{j\rho}$ functionals ranked in ascending order: $s_1, s_2, \ldots, s_{k_{j\rho}}$.

For each entry j, create the vector

$$\psi_j(t,\lambda_j) = [f_{s_1}(t,\lambda_{j1})\ldots f_{s_{k_1}}(t,\lambda_{j1})\ldots f_{s_{p_\nu}}(t,\lambda_{j\nu})]' \tag{2.3}$$

$$\longleftarrow\text{------ delay } \lambda_{j1} \text{ -----}\longrightarrow$$

and, by simply repeating the delay operator for every delay $\lambda_{j\rho}$ as many times as the number of functionals used for this particular delay, create the vector

$$\gamma_j(t,\lambda_j) = [\pi(t,\lambda_{j1})\ldots\pi(t,\lambda_{j1})\ \ldots\ \pi(t,\lambda_{j\nu})]' \tag{2.4}$$

$$\longleftarrow\text{--- delay } \lambda_{j1} \longrightarrow$$

The following matrix is then created

$$\psi(t,\lambda) = \text{diag}\,\{\psi_j'(t,\lambda_j)\} \tag{2.5}$$

where λ is a vector meant to include all the delays and is considered as an argument merely to underline the fact that the matrix ψ depends on these delays.

The reference vector is then written as

$$\underline{u}_{ref}(t) = \psi(t,\lambda)\underline{\mu} \tag{2.6}$$

where $\underline{\mu}$ is a constant vector of the levels of the corresponding functionals in $\psi(t,\lambda)$, and has a dimension

$$n = \sum_{j=1}^{m} \sum_{\rho=1}^{\nu_j} k_{j\rho} \tag{2.7}$$

If the vector $\underline{\mu}$ is considered as row-partitioned:

$$\underline{\mu} = [a_{11} \ldots a_{1k_{\nu 1}} \vdots \ldots \vdots a_{m1} \ldots a_{mk_{\nu m}}]' = [\underline{\alpha}'_1 \ \ldots \ \underline{\alpha}'_m]'$$

the j-th column-block will include the coefficients of functionals that correspond the the j-entry. Clearly, the optimisation problem has now to determine a value for $\underline{\mu}$.

3. EXPLICIT OPTIMAL SOLUTIONS

The nice property of the above formulation is that when the inputs belong to the family $F(t,\lambda)$, the time-domain operators $\Phi(t,u)$ can be obtained as explicit functions of time, which means that we can find an operator $\Omega(t,\lambda)$ such that

$$\underline{\xi}(t) = \Omega(t,\lambda)\underline{\mu} \tag{3.1}$$

where the matrix Ω is formed by stacking:

$$\Omega(t,\lambda) = [\Omega_x(t,\lambda) \ \Omega_y(t,\lambda) \ \Omega_w(t,\lambda)]' \tag{3.2}$$

In what follows the evaluation of Ω_x is given in terms of the state-space matrices of system Σ_3, and the specified delays λ. The same formula will obviously be applicable for the other two operators Ω_y and Ω_w, so the description below makes use of a generic notation instead of one with particular subscripts.

Denoting by $h_x(t)$ the impulse response of the system transfer function matrix $H_x(z)$, the response $x(t)$ to any input $u(t)$ is given by the convolution:

$$\underline{x}(t) = h_x(t)*\underline{u}(t) = \sum_{\ell=0}^{t} h_x(\ell)\underline{u}(t-\ell) \tag{3.3}$$

Bearing in mind that $h_x(t)$ is given by the Markov Parameters of the system as [2]

$$h_x(0) = D \ , \text{ for } t = 0 \tag{3.4.a}$$

and $$h_x(t) = CA^{t-1}B, \text{ for } t = 1,2,\ldots,\ell,\ldots \tag{3.4.b}$$

we can write

$$\underline{x}(t) = D\underline{u}(t) + C\sum_{\ell=1}^{t} A^{\ell-1}B\underline{u}(t-\ell) \tag{3.5}$$

For simplicity we consider x(t) as a superposition of responses $\underline{x}^j(t)$ coming from the corresponding scalar inputs $u_j(t)$, i.e.

$$\underline{x}(t) = \sum_{j=1}^{m} \underline{x}^j(t) = \sum_{j=1}^{m} \{D_j u_j(t) + C \sum_{\ell=1}^{t} A^{\ell-1} B_j u_j(t-\ell)\} \qquad (3.6.a)$$

where D_j and B_j denote the j-th column vectors of matrices D and B.

For $u_j(t)$ given by the linear combination (2.6), we have that

$$\underline{x}^j(t) = D_j \underline{\psi}_j^{/}(t,\lambda_j)\underline{\alpha}_j + C \sum_{\ell=1}^{t} A^{\ell-1} B_j \underline{\psi}_j^{/}(t-\ell)\underline{\alpha}_j$$

$$= D_j \underline{\psi}_j^{/} \underline{\alpha}_j + C \sum_{\rho=1}^{\nu_j} \sum_{i=1}^{k_{j\rho}} \sum_{\ell=1}^{t} A^{\ell-1} B_j \pi(t-\ell,\lambda_{j\rho}) \cdot (t-\ell-\lambda_{j\rho})^{s_i-1} a_{j\rho_i}$$

(3.6.b)

Taking into account that $\pi(t-\ell,\lambda_{j\rho}) = 0$ for $t-\ell < \lambda_{j\rho}$, i.e. for $\ell > t-\lambda_{j\rho}$

we define the following matrix

$$R_{s_i}(\tau) = \sum_{\ell=1}^{\tau} (\tau-\ell)^{s_i-1} A^{\ell-1} \qquad (3.7)$$

with

$$\tau = \tau_{j\rho} = t - \lambda_{j\rho}$$

For $s_i = 0$ (corresponding to the impulse functional), and for $s_i = 1$ (corresponding to the step one) we have the trivial results for any delay $\lambda_{j\rho}$:

$$R_0(\tau) = A^{\tau-1} \qquad (3.8.a)$$

and

$$R_1(\tau) = (A - I)^{-1}(A^{\tau} - I) \qquad (3.8.b)$$

and any subsequent $R_k(\tau)$ can be obtained through the sequential rule

$$R_{k+1} = \tau R_k - \frac{d}{dA}(AR_k) \qquad (3.9)$$

for any $k = 1,2,\ldots,s_i,\ldots$ The operator $\frac{d}{dA}$ denotes matrix differentiation with respect to a matrix. A proof of this rule is given in the Appendix.

As an example, $R_2(\tau)$ is found to be

$$R_2(\tau) = (A - I)^{-2}(A^{\tau} - I) - \tau(A - I)^{-1} \quad (3.10)$$

and so forth. In the most practical cases the evaluation of a few R_k would suffice.

Obviously, to each functional in expression (2.6) there corresponds a matrix $R_k(\tau)$. Furthermore it is noted that every matrix $R_k(\tau_{j\rho})$ is non-zero only in the region where the corresponding delay operator $\pi(t,\lambda_{j\rho})$ allows. Thus the following matrix, partitioned in square blocks, is finally created for each $j = 1,2,\ldots,m$:

$$Q_j(t,\lambda_j) = \underline{\gamma}'(t,\lambda_j) \circledast [\; R_{sj1}(\tau_{j1}) \ldots R_{sk_{j1}}(\tau_{j1}) \ldots R_{sk_{\nu j}}(\tau_{\nu j})\,] \cdot B_j$$

$$\longleftarrow \text{1st delay} \longrightarrow \quad (3.11)$$

where $\circledast$ denotes the block-wise Hadamard product of a vector by a matrix, in which the (i^{th}) element of the first is multiplied with the (i^{th}) partition of the second. Also the vector B_j is supposed to post-multiply every partition of the above block matrix. If the dimension of matrix A is assumed to be $d \times d$, matrix Q_j is of dimension $d \times n_j$.

Finally stacking the above operators side by side, we define:

$$Q(t,\lambda) = [\, Q_1 \;\; Q_2 \;\ldots\; Q_m \,] \quad (3.12)$$

of dimensions $d \times n$.

After the above matrix has been created, all the summations referred to in expressions (3.6.a) and (3.6.b) can be summarised in the following simple way:

$$\underline{x}(t) = [\, D\Psi(t,\lambda) + CQ(t,\lambda)\,]\underline{\mu}$$

which leads to the concise expression for $\Omega_x(t)$:

$$\Omega_x(t) = D\Psi(t,\lambda) + CQ(t,\lambda) \quad (3.13)$$

Repeating the above procedure for all sub-systems Σ_3, Σ_4, Σ_5, we create from (3.2) the matrix operator Ω of dimension $(m+m+r) \times n$. The optimisation problem (1.4) is now written as

$$\min_{\mu} J_N = \sum_{t=1}^{N} [\,\Omega(t,\lambda)\underline{\mu} - \underline{\xi}_{des}(t)]' P(t) [\,\Omega(t,\lambda)\underline{\mu} - \underline{\xi}_{des}(t)] \quad (3.14)$$

and yields an optimal $\underline{\mu}^*$ given by

$$\underline{\mu}^* = [\, \sum_{t=1}^{N} \Omega'(t,\lambda) P(t) \Omega(t,\lambda)]^{-1} [\, \sum_{t=1}^{N} \Omega'(t,\lambda) P(t) \underline{\xi}_{des}(t)] \quad (3.15)$$

while the optimal cost can be evaluated from

$$J_N^* = -[\sum_{t=1}^{N} \underline{\xi}_{des}' P\Omega][\sum_{t=1}^{N} \Omega' P\Omega]^{-1}[\sum_{t=1}^{N} \Omega' P\underline{\xi}_{des}] \qquad (3.16)$$

with the arguments omitted for simplicity.

The explicit form of the above solution allows any kind of investigation regarding how the penalty matrix or the desired trajectory affect the optimal value of $\underline{\mu}^*$. When they include a set of adjustable parameters $\{\beta\}$, $\underline{\mu}^*$ can be further optimised.

The optimal reference vector is found from

$$\underline{u}^*_{ref}(t) = \Psi(t,\lambda)\underline{\mu}^* \qquad (3.17)$$

and the optimal levels of $\underline{x}$, (and similarly for $\underline{y}$ and $\underline{w}$), are

$$\underline{x}^* = \Omega_x(t,\lambda)\underline{\mu}^* \qquad (3.18)$$

The algorithm must be repeated when the designer wishes to change the pattern of the desired main-target path. Changes on the penalty matrix, P(t), or on the desired levels for the instruments and the secondary targets require only the final solution formula to be evaluated. Observe that keeping the controller K(z) fixed preserves the dynamic properties of the controlled system under any changes of that kind.

4. REMARK

It must be noted here that the above result is not applicable for continuous systems, because an expression for the time domain operator $\Omega(t)$ that corresponds to the transfer function matrix H(z) cannot be found explicitly in terms of the state space matrices.

5. REFERENCES

1. Doyle, J.C. and Stein, G. (1981) Multivariable Feedback Design: Concepts for a Classical/modern Synthesis, IEEE Trans. on Automatic Control, AC-26, pp 4-16.

2. Kailath, T. (1980) Linear Systems, Prentice-Hall, New Jersey.

3. Kwakernaak, H. and Sivan, R. (1972) Linear Optimal Control Systems, Wiley, Interscience, New York.

4. MacFarlane, A.G.J. (1979) Complex Variable Methods for Linear Multivariable Systems, Taylor & Francis, London.

APPENDIX

A sequential rule for R_k

For any constant matrix A and a positive integer τ, we define the operator:

$$R_k(\tau) = \sum_{\ell=1}^{\tau} (\tau-\ell)^{k-1} A^{\ell-1}$$

Then we have:

$$R_{k+1}(\tau) = \sum_{\ell=1}^{\tau} (\tau-\ell)(\tau-\ell)^{k-1} A^{\ell-1} = \tau R_k - \sum_{\ell=1}^{\tau} \ell(\tau-\ell)^{k-1} A^{\ell-1}$$

and by writing

$$\sum_{\ell=1}^{\tau} \ell(\tau-\ell)^{k-1} A^{\ell-1} = \frac{d}{dA} \sum_{\ell=1}^{\tau} (\tau-\ell)^{k-1} A^{\ell} = \frac{d}{dA} (AR_k)$$

the rule follows immediately:

$$R_{k+1} = \tau R_k - \frac{d}{dA}(AR_k)$$

for any $k = 1,2,3,...$

The expression for R_1 is easily calculated independently as

$$R_1 = (A - I)^{-1}(A^{\tau} - I)$$

and so any other term can follow.

REGULARISATION OF A MINIMUM ENERGY PROBLEM

C.M. Dorling

(University of Sheffield)

ABSTRACT

A modified form of the linear quadratic regulator problem with saturating control inputs is considered. The proposed modification involves the inclusion of an additional non-quadratic state penalty term in the cost functional to be minimised. This extra term is designed to regularise the problem in the sense that the optimal control may be expressed as a closed form in the state. This feedback control is characterised and shown to be optimal. An example is included, and possible application to the sub-optimal control of the original linear quadratic regulator is discussed.

1. INTRODUCTION

The problem of minimising the cost functional

$$J(\underline{u}) = \frac{1}{2}\int_{0}^{\infty} [\langle\underline{x}(t),Q\underline{x}(t)\rangle + \langle\underline{u}(t),R\underline{u}(t)\rangle]\, dt \tag{1.1}$$

for an autonomous linear regulator,

$$\dot{\underline{x}} = A\underline{x} + B\underline{u} \qquad (\underline{x}\varepsilon\mathbb{R}^{n}, \underline{u}\varepsilon\mathbb{R}^{m}); \tag{1.2a}$$

$$\underline{x}(0) = \underline{x}^{o}; \quad \underline{x}(t)\to\underline{0} \text{ as } t\to\infty, \tag{1.2b}$$

has been widely studied in the case of the controls u_i being unbounded. It is well known that the optimal control for this linear-quadratic problem (LQP) has the linear state feedback form

$$\underline{u} = -R^{-1}B^{T}P\underline{x}, \tag{1.3}$$

where P is the unique solution of the algebraic (or steady-state) Riccati-type matrix equation

$$PA + A^{T}P - PBR^{-1}B^{T}P + Q = 0 . \tag{1.4}$$

(This assumes that the pairs (A,B) and ($Q^{\frac{1}{2}}$,A) are stabilizable and detectable, respectively[1]; and that R is positive definite.) However, if the control variables are restricted to a compact restraint set: for example, the unit m-cube

$$U = \{\underline{u} \in \mathbb{R}^m : |u_j| \leqslant 1 \quad (j=1,2,\ldots,m)\}; \tag{1.5}$$

this simple linear feedback form is lost, as is readily seen in the scalar-controlled case[2,3].

It is the failure to establish a closed form solution, even for the scalar problem, which motivates the approach taken in this paper: namely, the inclusion of a state penalty term in the cost functional which regularises the problem in the sense of permitting an explicit state feedback valid at all points of the state space to be found.

2. PROBLEM STATEMENT

Consider the problem of minimising the cost functional

$$J(\underline{u}) = \frac{1}{2}\int_0^\infty [\langle \underline{x}(t), Q\underline{x}(t)\rangle + \langle \underline{u}(t), R\underline{u}(t)\rangle + 2\sum_{j=1}^m F_j(\underline{x}(t))]dt, \tag{2.1}$$

subject to (1.2) and (1.5). It is assumed that Q is symmetric and positive semi-definite ($Q>0$) and that R is diagonal:

$$R = \mathrm{diag}\{r_j : \; j=1,\ldots,m\} \quad (r_j \geqslant 0), \tag{2.2}$$

that A is an asymptotically stable matrix, and that ($Q^{\frac{1}{2}}$,A) is observable. The penalty functions $F_j : \mathbb{R}^n \to [0,\infty)$ which will be selected below are required to be continuously differentiable.

The approach to the solution of this problem is via the Maximum Principle of Pontryagin *et al.*[4] The Hamiltonian is given by

$$H(\underline{x},\underline{u},\underline{p}) = \langle \underline{p}, A\underline{x}\rangle - \frac{1}{2}\langle \underline{x}, Q\underline{x}\rangle - \sum_{j=1}^m F_j(\underline{x}) + \langle \underline{p}, B\underline{u}\rangle - \frac{1}{2}\langle \underline{u}, R\underline{u}\rangle, \tag{2.3}$$

where $\underline{p}(.)$ denotes the adjoint vector. Let $\underline{u}^*(t)$ $(0\leqslant t<\infty)$ be a control generating a trajectory $\underline{x}^*(t)$ of the system (2.1),(1.5) from $\underline{x}(0) = \underline{x}^o$ which approaches the state space origin (at least) asymptotically as $t\to\infty$. Then for $(\underline{x}^*(.),\underline{u}^*(.))$ to be an optimal pair, it is necessary that there should exist a corresponding absolutely continuous function satisfying the adjoint equation

$$\dot{\underline{p}}(t) = -A^T\underline{p}(t) + Q\underline{x}(t) + \sum_{j=1}^m \nabla F_j(\underline{x}(t)), \tag{2.4}$$

almost everywhere (a.e.) on $[0,\infty)$; and, moreover, that $\underline{u}^*(t)$ should maximise the Hamiltonian (2.3) for almost all (a.a) $t\varepsilon[0,\infty)$. Letting $\underline{b}^j$ denote the jth column of B, (2.3) will be maximised with respect to the control vector $\underline{u}$ when

$$h(\underline{u},\underline{p}) = \sum_{j=1}^{m} h_j(\underline{u},\underline{p}) = \sum_{j=1}^{m} \langle \underline{p},\underline{b}^j\rangle u_j - \frac{r_j}{2}u_j^2 , \quad (2.5)$$

is maximised with respect to (w.r.t.) each $u_j \varepsilon [-1,1]$ $(j=1,\ldots,m)$. Clearly, any extremal control (in open loop form) therefore satisfies

$$u_j(t) = \text{sat}\,\{\langle \underline{p}(t),\underline{b}^j\rangle / r_j\} , \quad (2.6)$$

for a.a. $t\varepsilon[0,\infty)$, where sat(.) denotes the saturation function defined by

$$\text{sat}(y) = \begin{cases} y & |y| \leqslant 1 \\ y/|y| & |y| \geqslant 1 \end{cases} . \quad (2.7)$$

A linear feedback form for this extremal control is now established. Initially it will be derived formally, but subsequently this formal approach will be justified by verifying the optimality of the determined control.

3. DETERMINATION OF AN EXTREMAL FEEDBACK CONTROL

In order to establish the desired feedback form, a constant symmetric matrix K is now sought such that the adjoint and state vectors are related by

$$\underline{p}(t) = -K\underline{x}(t) \quad \forall t\varepsilon[0,\infty) . \quad (3.1)$$

Differentiating (3.1), and substituting the state and adjoint equations (1.2a) and (2.4), it follows that

$$(KA + A^T K + Q)\underline{x} = -\sum_{j=1}^{m} (\nabla F_j(\underline{x}) + K\underline{b}^j u_j) . \quad (3.2)$$

For each $j = 1,\ldots,m$, let F_j be defined by

$$F_j(\underline{x}) = s_j.[\langle K\underline{b}^j,\underline{x}\rangle - \tfrac{1}{2}r_j.s_j] = h_j(-s_j,-K\underline{x}) \quad (3.3a)$$

where

$$s_j = \text{sat}\{\langle K\underline{b}^j,\underline{x}\rangle / r_j\} . \quad (3.3b)$$

Then, from (2.7) and (3.1)

$$F_j(\underline{x}) = K\underline{b}^j.\text{sat}[\langle K\underline{b}^j,\underline{x}\rangle / r_j] = -K\underline{b}^j u_j , \quad (3.4)$$

and the right hand side (RHS) of (3.2) is zero, leading to the Lyapunov equation

$$KA + A^T K + Q = 0 \quad (3.5)$$

with the unique positive definite solution

$$K = \int_0^\infty \exp(A^T t).Q.\exp(At)dt . \quad (3.6)$$

Thus, the (formally established) extremal feedback control is

$$u_j(\underline{x}) = -\mathrm{sat}\{\langle K\underline{b}^j, \underline{x}\rangle / r_j\}. \tag{3.7}$$

It is now necessary to show that this control will in fact drive the state to the state space origin as $t \to \infty$, and that it does so optimally.

4. STABILITY AND OPTIMALITY OF THE EXTREMAL CONTROL

Consider the positive definite, continuously differentiable function $V: \mathbb{R}^n \to [0, \infty)$ defined by

$$V(\underline{x}) = \frac{1}{2}\langle \underline{x}, K\underline{x}\rangle \tag{4.1}$$

as a candidate Lyapunov function for the system (1.2a) under control (3.7). On taking the total time derivative of (4.1), utilising (3.5) and (3.7), and dropping explicit dependence on t,

$$\dot{V}(\underline{x}) = -\frac{1}{2}\langle \underline{x}, Q\underline{x}\rangle - \sum_{j=1}^{m} \langle K\underline{b}^j, \underline{x}\rangle . \mathrm{sat}\{\langle K\underline{b}^j, \underline{x}\rangle / r_j\} . \tag{4.2}$$

But $Q>0$, so that $\langle \underline{x}, Q\underline{x}\rangle$ is non-negative for $\underline{x} \neq \underline{0}$; while each term of the sum in (4.2) is non-negative. Hence $\dot{V} \leqslant 0$ on all trajec-jectories of the controlled system. Moreover, the observability condition imposed on the problem ensures that $\langle \underline{x}(t), Q\underline{x}(t)\rangle$ cannot remain at zero for an infinite period unless $\underline{x} \equiv \underline{0}$, so that the origin is rendered globally asymptotically stable by the extremal control. (Conditions ensuring asymptotic stability when $(Q^{\frac{1}{2}}, A)$ is unobservable may also be determined[5].)

Turning now to the optimality of the control (3.7), the Lyapunov function (4.1) is also a solution of the Bellman-Hamilton-Jacobi (BHJ) equation for this problem:

$$\min_{u \varepsilon U} \{\langle \nabla V, A\underline{x} + B\underline{u}\rangle + \frac{1}{2}\langle \underline{x}, Q\underline{x}\rangle + \sum_{j=1}^{m} F_j(\underline{x}) + \frac{1}{2}\langle \underline{u}, R\underline{u}\rangle\} = 0. \tag{4.3}$$

Substituting for V and utilising the Lyapunov equation (3.5), the BHJ equation (4.3) becomes

$$0 = \sum_{j=1}^{m} [F_j(\underline{x}) + \min_{u \varepsilon U} \{\langle K\underline{b}^j, \underline{x}\rangle u_j + \frac{r_j}{2} u_j^2\}]$$

$$= \sum_{j=1}^{m} [F_j(\underline{x}) - \max_{|u_j| \leqslant 1} h(u_j, -K\underline{x})] . \tag{4.4}$$

However, in view of (3.3) each maximum in (4.4) is precisely $F_j(\underline{x})$, as required; and moreover, the control (3.7) agrees exactly with the control required to obtain this maximum. Thus, by the standard BHJ results, the extremal control (3.7) is optimal for the problem (1.2), (1.5), (2.1); and the minimum cost of a trajectory from $\underline{x}(0) = \underline{x}^o$ which approaches the origin as $t \to \infty$ is given by

$$V(\underline{x}^o) = \frac{1}{2}\langle \underline{x}^o, K\underline{x}^o \rangle \ . \tag{4.5}$$

5. EXAMPLE: DAMPED OSCILLATOR

As an example of the above results, consider the cost functional (1.1) and system (1.2) with $n = 2$, $m = 1$ (scalar control), $R = r_1 = 1$, and

$$A = \begin{bmatrix} 0 & 1 \\ -1 & -0.2 \end{bmatrix}, \quad b = \begin{bmatrix} 0 \\ 1 \end{bmatrix}, \quad Q = I_2 \ . \tag{5.1}$$

In this case, the Riccati equation (1.4) has the solution

$$P = \begin{bmatrix} a\sqrt{2} - 0.2 & \sqrt{2} - 1 \\ \sqrt{2} - 1 & a - 0.2 \end{bmatrix}, \quad a = \sqrt{2\sqrt{2} - 0.96} \tag{5.2}$$

Figure 1 shows the (numerically-determined) partition of the state space into the regions of saturated (S) and linear (L) and non-linear (N) unsaturated control.

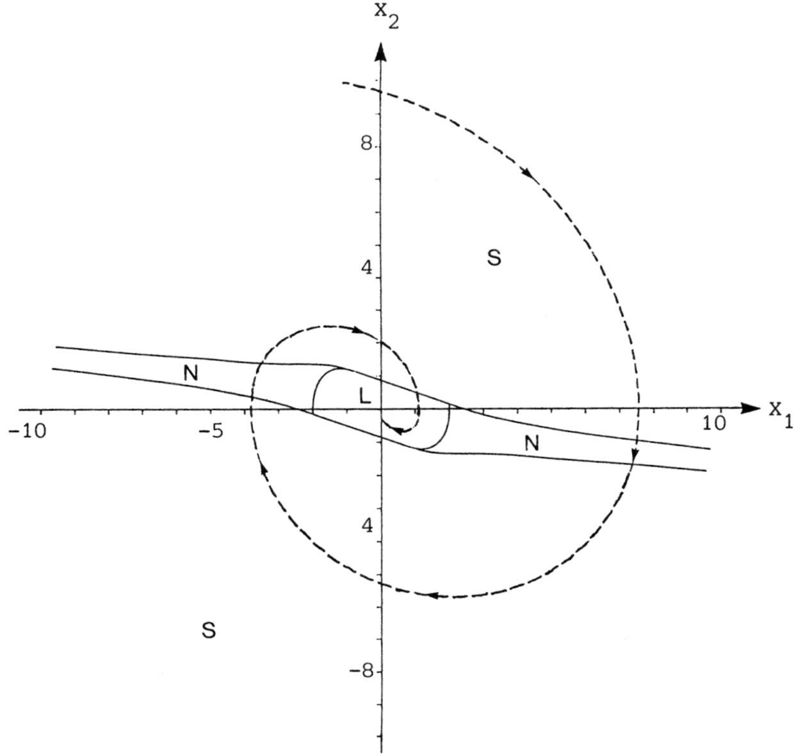

Fig.1 Regions of optimal control for LQP.

For this example, consider now replacing the cost functional (1.1) with (2.1). The Lyapunov equation (3.5) has solution

$$K = \frac{1}{2}\begin{bmatrix} 10.2 & 1 \\ 1 & 10 \end{bmatrix}, \tag{5.3}$$

giving the optimal feedback control

$$\underline{u}(\underline{x}) = -\mathrm{sat}\{\tfrac{1}{2}g(\underline{x})\}\ , \quad g(\underline{x}) = x_1 + 10x_2 \tag{5.4}$$

and the state penalty function $F = F_1$

$$F(\underline{x}) = \begin{cases} \tfrac{1}{2}(|g(\underline{x})| - 1 & |g(\underline{x})| \geq 2 \\ 1/8\ g^2(\underline{x}) & |g(\underline{x})| \leq 2 \end{cases} \tag{5.5}$$

When the control (5.4) is saturated, the trajectories are simply the paths of the damped oscillator under ±1 control. On the other hand, on the unsaturated strip the optimal trajectories are the unforced responses of the system

$$\dot{x}_1 = x_2\ ;\quad \dot{x}_2 = -1.5x_1 - 5.2x_2\ . \tag{5.6}$$

Figure 2 shows the form of the optimal state portrait for the regularised problem. It can be seen from the portrait that all paths approach the origin asymptotically to the line $x_2 = -(2.6 - \sqrt{5.26})x_1$, representing the eigenvector corresponding to the slow mode of system (5.6).

The original motivation of this study was to try to simplify the control structure of the LQP with bounded controls. A possible application of the preceding work might be its use as a sub-optimal control for the LQP. Although analytical work has not produced any results to date, simulation results for the above example seem interesting.

Two sub-optimal control schemes were implemented for this system: $u_p(\underline{x}) = -\mathrm{sat}\langle P\underline{b},\underline{x}\rangle$ and $u_k(\underline{x}) = -\mathrm{sat}\langle K\underline{b},\underline{x}\rangle$. The control law u_p agrees exactly with the optimal control in the linear region about the origin, while u_k produces cost increases of more than 100% of the optimal. However, for larger initial states, the second control produces a smaller increase in the cost of a trajectory, as is shown in Figure 3. These results suggest that an improved suboptimal control might be obtained by applying u_k at points away from the origin, and then switching over to u_p as the origin is approached. This area of sub-optimal control of the LQP is receiving further attention.

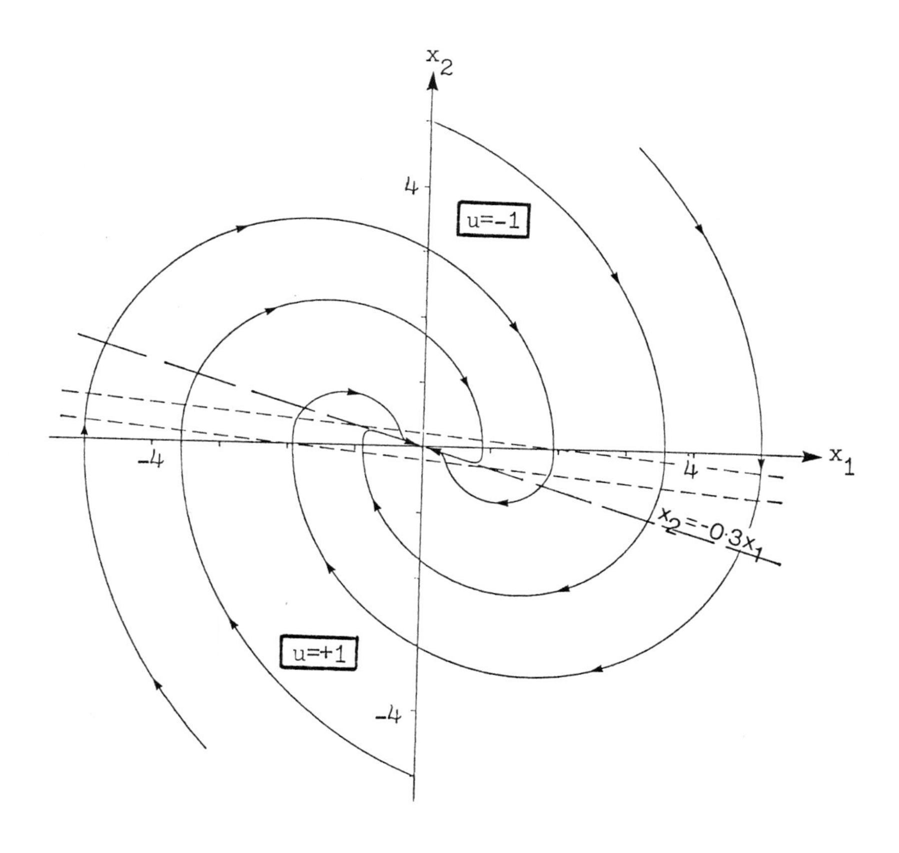

Fig. 2 Optimal state portrait for regularised problem.

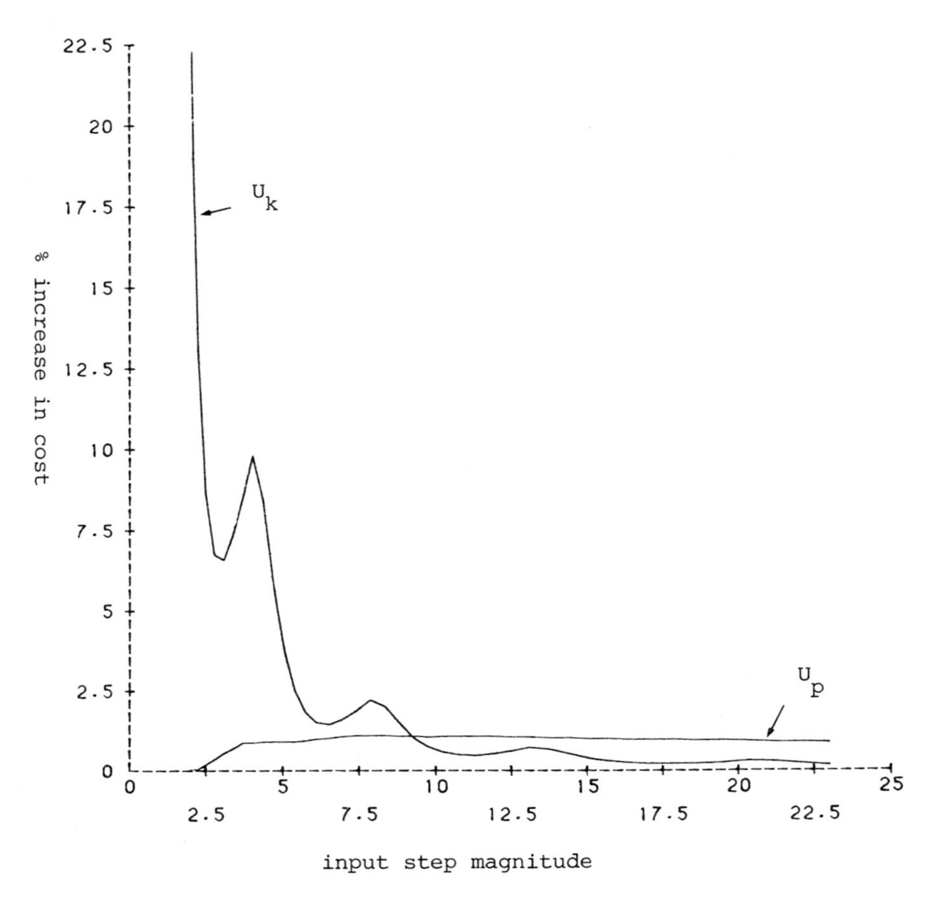

Fig 3 Increase in cost by sub-optimal controls.

6. CONCLUSION

The motivation for the work in this paper has been provided by the failure to find an analytic solution to the minimisation of a quadratic cost functional when the control variables are bounded. The approach employed here is to include a non-quadratic state penalty term in the cost integrand in such a way that the resulting control may be expressed in a simple feedback form. This simple form has been shown to be asymptotically stable, and optimal w.r.t. the new cost functional.

The generalisation of this autonomous problem to the time-varying finite time-interval case does not appear to present any great difficulties, with the ideas presented in this paper being readily extended to include time-dependence. Similar regularisation approaches in the case of $R \equiv 0$ have been made by Frankena and Sivan[6] and by Ryan[7]. If the above results are (formally) taken to the limit as $R \to 0$, (with the limit of the saturation function taken to be the signum function) the results of Ryan[7] are obtained. A similar study of the minimisation of a cost functional with a fuel-type penalty term has also been made[5,8].

7. REFERENCES

1. Wonham, W.M. (1974) "Linear Multivariable Control: A Geometric Approach", Springer-Verlag, Berlin.
2. Johnson, C.D. and Wonham, W.M. (1965) On a Problem of Letov in Optimal Control, J. bas Engng,Trans. ASME, Ser. D 87 81-89.
3. Tcharan, A. (1966) On a Class of Constrained Control, Linear Regulator Problems J. bas Engng, Trans. ASME, Ser. D 88 385-391.
4. Pontryagin, L.S., Boltyanskii, V.G., Gamkrelidze, R.V. and Mischenko, E.F. (1962) "The Mathematical Theory of Optimal Processes", Wiley, New York.
5. Dorling, C.M. (1983) "Aspects of Non-linear Optimal Control Theory", Ph.D Thesis, University of Bath, 1983.
6. Frankena, J.F. and Sivan, R. (1979) A Non-linear Optimal Control Law for Linear Systems, Int. J. Control 30 159-178.
7. Ryan, E.P. (1982) Optimal Feedback Control of Saturating Systems, Int. J. Control 35, 521-534.
8. Dorling, C.M. and Ryan, E.P., On the Synthesis of Optimal Discontinuous Saturating Controls for Linear Systems, to appear.

OPTIMAL CONTROL PROBLEMS IN TIDAL POWER GENERATION

N.R.C. Birkett and N.K. Nichols

(University of Reading)

ABSTRACT

Mathematical control theory is applied to develop a global technique for maximizing energy generation from a tidal power scheme. Optimal time-dependent strategies for controlling flow through a tidal barrier are determined, using dynamic system models. Estuaries with variable geometric parameters are treated, and non-linear relationships between discharge and head difference at the barrier are examined. A numerical algorithm is presented and comparative results for various schemes are given.

1. INTRODUCTION

The problem of harnessing tidal energy has attracted considerable interest over the past years. In Great Britain investigation has concentrated on the generation of power from tidal flow in the Severn Estuary, and a recent report [9] has concluded that a barrage across the River Severn could be both technically and economically feasible. The evaluation of a tidal power scheme requires the accurate calculation of both component costs and total energy production, and poses an extremely complicated optimization problem. In this paper a global technique using the mathematical theory of control is described for determining the maximum average energy generated from a tidal power scheme. This approach simultaneously takes into account both the dynamic nature of the estuarine flow and fixed items of plant, such as turbines, sluices, barrier sites, etc., while optimizing the engineering control parameters.

In earlier studies of tidal power schemes [9,16] plant items and control parameters are optimized using a simplified model in which the surface elevation of the water in the enclosed tidal basin is assumed everywhere constant, and dynamic effects of flow are ignored. The optimized parameters are then used with dynamic models of one and two dimensional flow to obtain energy absorption estimates. Simplified dynamic models are used by other investigators [1,6,10] to determine strategies for control of flow through the barrier. Optimal constant controllers are determined in the linear case [6], and non-linear, switching controls are determined [1,10] under the assumption that the flow velocities and head difference at the barrier vary harmonically with the tidal period. The latter technique is limited in that non-linear head flow properties and higher harmonic effects introduced by the controls are ignored.

A global technique for optimizing the control parameters, taking into account both the estuarine dynamics and plant characteristics simultaneously is described by Birkett & Nichols [3] and Birkett, Count & Nichols [4]. Optimal control theory is applied to the full tidal power problem and the maximum average energy output is calculated for the simple flat surface model and for a linear dynamic model of flow in a rectangular estuary of constant depth and constant cross sectional area. It is assumed that the flow through the generating turbines is directly proportional to the head difference at the barrier.

The extension of this optimal control technique to more realistic models is described here. Estuaries with variable geometric parameters are treated, and the relationship between discharge across the barrier and head difference are represented by non-linear functions. Various power generation schemes are simulated, including both two-way and ebb generation schemes, and schemes with dual controls for turbines and sluices. In the next section the mathematical model of the system is defined and the corresponding optimal control problem is formulated In §3 a numerical method for determining the optimal control strategy is developed, and a computational algorithm given. Generalisations to dual control schemes are described in §4, and results are given in §5 for various models.

2. THE MATHEMATICAL MODEL

2.1 The Optimal Control Problem

We assume that the tidal basin is long relative to its depth, and model the fluid dynamics in the tidal basin by the one-dimensional linearized shallow water equations [15]:

$$\left.\begin{aligned} b(x)\eta_t &= -(A(x)u)_x \\ u_t &= -g\eta_x - pu/h(x) \end{aligned}\right\} x \in [0,\ell] \ , \qquad (2.1)$$

where $b(x) > 0$, $A(x) > 0$, $h(x) > 0$ are the mean breadth, mean vertical cross-sectional area and mean height of the channel, respectively, $\ell >> h(x)$ is the length of the channel, $g > 0$, $p > 0$ are gravitational acceleration and linear friction constants, respectively, $\eta(x,t)$ is the water elevation above mean height and $u(x,t)$ is the horizontal component of fluid velocity.

To simplify the discussion, we assume first that the flow across the barrier (at $x=0$) is controlled through turbines only, and that the influx velocity is proportional to the head difference between the surface elevation in the basin, $\eta(0,t)$, and the tidal elevation, $f(t)$, which is imposed on the seaward side of the barrier and is taken to be periodic with period T. (Generalizations to non-linear head flow conditions and dual controls are described in Section 4). At the upstream end of the basin (at $x=\ell$) zero flow is assumed, and the boundary conditions are thus given as

$$u(0,t) = \alpha(t)\,[f(t) - \eta(0,t)], \quad u(\ell,t) = 0, \qquad (2.2)$$

where the control function $\alpha(t)$ is bounded such that

$$0 \leqq \alpha(t) \leqq a_0, \quad \forall\, t \in [0,T]. \qquad (2.3)$$

We also require that the functions η, u are periodic in time with period T, such that

$$\eta(x,0) = \eta(x,T), \quad u(x,0) = u(x,T) \quad . \qquad (2.4)$$

The instantaneous power output of the turbines is assumed proportional to flow times head difference and the average power generated by the flow is thus given by $\rho g A(0) E/T$ where

$$E = \int_0^T \alpha(t)\,[f(t) - \eta(0,t)]^2\,dt \quad . \qquad (2.5)$$

The optimal control problem is to find $\alpha(t)$ to maximize the functional E, subject to the system equations (2.1), boundary boundary conditions (2.2) and (2.4) and constraint (2.3). The admissible controls $\alpha(t) \in U_{ad}$ are assumed to satisfy (2.3) and to be "sufficiently smooth" for the initial value problem (2.1)(2.2) with smooth *initial* data $[\eta(x,0), u(x,0)]$ to be well-posed.

2.2 Analysis of the System Equations

For the control problem to be well defined the system equations (2.1) together with boundary conditions (2.2) and (2.4) must have a unique solution, for any given (non-trivial) admissible control $\alpha(t)$. It is necessary, therefore, that the boundary conditions are consistent [11] with the system equations (2.1).

Here the time periodic conditions (2.4) effectively replace the usual initial conditions associated with a hyperbolic system. To show that these conditions are natural to impose, in the sense that, in the limit as $t \to \infty$, the solution of the system equation (2.1), with any given initial state, converges to a unique "steady-state" periodic solution satisfying conditions (2.4), we use an iterative process. The iteration is given by

$$\underline{z}^{m+1}(0) = \underline{z}^m(T) \equiv G\underline{z}^m(0) \quad , \tag{2.6}$$

where $\underline{z}(t) = [\eta(x,t), \ u(x,t)]^T$ is the solution of the problem (2.1)-(2.2) with initial data $\underline{z} = \underline{z}(0)$ at $t=0$. If the operator G is a contraction, then iteration (2.6) converges to a unique fixed point $\underline{z}^*$ such that the solution of problem (2.1)-(2.2) with initial data $\underline{z}(0) = \underline{z}^*$ satisfies $\underline{z}(T) = G\underline{z}^* = \underline{z}^*$ and is the required periodic solution satisfying (2.4).

To show that G is a contraction, for any (non-trivial) admissible control $\alpha(t)$, we denote the difference between solutions to the initial-boundary value problem by $\underline{z}(t)$ $\delta\underline{z}(t) = \underline{z}(t) - \hat{\underline{z}}(t)$ and define the norm $\|\cdot\|_s$ on $L_2^2[0,\ell]$[12] by

$$\|\underline{z}(t)\|_s^2 = \int_0^\ell (b(x)\eta^2(x,t) + \mathbf{A}(x)u^2(x,t)/g + k_\alpha(x)\eta(x,t)u(x,t))\,dx, \tag{2.7}$$

where $k_\alpha(x) > 0$ has certain properties. From (2.1)-(2.2) we can then obtain the inequality

$$\frac{d}{dt}\|\delta\underline{z}(t)\|_s^2 \leq -Q_0 \|\delta\underline{z}(t)\|_s^2 + Q_1(t) \quad , \tag{2.8}$$

where $Q_0 > 0$ and $\int_0^T Q_1(t)\,dt \leq 0$. It follows that

$$\|\delta\underline{z}(T)\|_s^2 \equiv \|G\delta\underline{z}(0)\|_s^2 \leq e^{-Q_0 T}\|\delta\underline{z}(0)\|_s^2 \quad , \tag{2.9}$$

which implies the result, since G is linear. We conclude that for all sufficiently smooth data the model system equations (2.1), (2.2), (2.4) are well-posed.

2.3 Necessary Conditions for the Optimal Solution

Necessary conditions for the solution of the optimal control problem (2.1)-(2.5) are derived by the Lagrange technique. The Lagrange functional associated with the problem is defined by

$$L(\alpha) = \int_0^T \alpha(t)[f(t)-\eta(0,t)]^2 + \gamma(t)(u(0,t)-\alpha(t)[f(t)-\eta(0,t)])\,dt$$
$$+ \int_0^T\int_0^\ell \lambda(x,t)[-b\eta_t-(Au)_x] + \mu(x,t)[-u_t-g\eta_x-ph/h]\,dx\,dt. \tag{2.10}$$

Taking variations and using integration by parts, we can show that the first variation of the Lagrangian is given by

$$\delta L(\alpha,\delta\alpha) \equiv \langle \nabla E(\alpha),\delta\alpha\rangle = \int_0^T ([f(t)-\eta(0,t)]^2 + A(0)\lambda(0,t)[f(t)-\eta(0,t)])\,\delta\alpha(t)\,dt, \tag{2.11}$$

provided the state variables $\eta(x,t)$, $u(x,t)$ satisfy the system equations (2.1), (2.2), (2.4) and the adjoint variables $\lambda(x,t)$, $\mu(x,t)$ satisfy the equations:

$$\left.\begin{aligned} b(x)\lambda_t &= -g\mu_x \quad, \\ \mu_t &= -A\lambda_x + p\mu/h \;, \end{aligned}\right\} \quad x \in [0,\ell],\; t \in [0,T], \tag{2.12}$$

with boundary conditions

$$g\mu(0,t)-\alpha(t)A(0)\lambda(0,t) = 2\alpha(t)[f(t)-\eta(0,t)], \quad \mu(\ell,t) = 0, \tag{2.13}$$

and

$$\lambda(x,0) = \lambda(x,T), \quad \mu(x,0) = \mu(x,T) \;. \tag{2.14}$$

(We note that the adjoint equations are well-posed and time-periodic solutions can be generated by an iterative process analogous to (2.6)).

For the control $\alpha(t)$ to be optimal it is necessary that the first variation of the Lagrangian be non-positive for all admissible variations $\delta\alpha$.[7] The optimal control $\alpha(t)$ must, therefore, satisfy

$$\alpha(t) = \begin{cases} a_0 & \text{if } \nabla E(\alpha) \equiv [f(t) - \eta(0,t)]^2 + A(0)\lambda(0,t)[f(t)-\eta(0,t)] > 0, \\ 0 & \text{otherwise,} \end{cases} \tag{2.15}$$

and the control is expected to contain jump discontinuities at zeros of the function gradient $\nabla E(\alpha)$. We cannot guarantee in this case that the optimal solution does not contain singular arcs, and in practice we find that there exist subintervals in time over which $\nabla E(\alpha) \equiv 0$ and the optimal control $\alpha(t)$ takes interior values in the constraint set $[0,a_0]$.

Existence of an optimal control cannot be easily proved, but an iterative process for determining a control which satisfies the necessary conditions can be constructed.

3. THE NUMERICAL METHOD

3.1 The Optimization Technique

To determine an admissible control α with corresponding responses and adjoints satisfying the necessary conditions (2.1)-(2.4) and (2.12)-(2.15), we consider first an up-gradient iteration procedure used previously for the flat basin model[3]. Given an approximation α^k to the optimal, this method produces a new approximation $\alpha^{k+1} = \alpha^k + \theta^k(\tilde{\alpha}^k - \alpha^k)$, where $\theta^k \varepsilon (0,1]$ and

$$\tilde{\alpha}^k = \{a_0 \ \underline{\text{if}} \ \nabla E(\alpha^k) \geqq 0, \ \ 0 \text{ otherwise}\} . \tag{3.1}$$

The selected control α^{k+1} is essentially a piecewise constant, "bang-bang" control. Although this procedure is technically convergent, the computed controls and responses are found, in practice, to contain large, high frequency oscillations, which arise principally because the optimal solution contains singular arcs where the control should lie on the interior of the constraint set and not at its boundaries. This algorithm is not suitable, therefore, and a modified approach is required.

We consider now the gradient projection algorithm, in which the new approximation to the control is chosen as $\alpha^{k+1} = \hat{P}(\alpha^k + \theta\nabla E(\alpha^k))$, where $\hat{P}$ is the L_2 projection -operator onto U_{ad}. The operator $\hat{P}$ is characterized by the property[13] $\langle v - \hat{P}v, \alpha - \hat{P}v\rangle \leqq 0$, $\forall \ \alpha \ \varepsilon \ U_{ad}$. For the selected control α^{k+1}, we have, therefore,

$$\langle\nabla E(\alpha^k), \alpha^{k+1} - \alpha^k\rangle \geqq \frac{1}{\theta} \|\alpha^{k+1} - \alpha^k\|_2^2 , \tag{3.2}$$

and it can be shown that for some choice of θ, $E(\alpha^{k+1}) \geqq E(\alpha^k)$ [8]. If the initial approximation $\alpha^0(t)$ is continuous, then the algorithm generates a sequence of continuous controls α^k for which the functions $E^k = E(\alpha^k)$ are monotonically non-decreasing, and, provided an optimal solution exists amongst the admissible controls, the process converges and the limiting control satisfies the necessary conditions. The complete algorithm is described as follows:

Algorithm

Step 1. Choose $\underline{\alpha}^0(t) \in U_{ad}$ $(\underline{\alpha}^0 \not\equiv 0)$, and $\theta \in (0,1]$.

Set $E^0 := 0$, $\nabla E^0 := 0$.

Step 2. <u>For</u> k: = 0, 1,2, ... <u>do</u>

Step 2.1. Set $\underline{\alpha}^{k+1} := \begin{cases} a_0 & \text{if } \alpha^k + \theta\nabla E^k \geqq a_0 \\ 0 & \text{if } \alpha^k + \theta\nabla E^k \leqq 0 \\ \alpha^k + \theta\nabla E^k & \text{otherwise} \end{cases}$.

Step 2.2. Solve state system (2.1)(2.2)(2.4) with $\underline{\alpha} := \underline{\alpha}^{k+1}$.

Step 2.3. Solve adjoint system (2.12)-(2.14) with $\underline{\alpha} := \underline{\alpha}^{k+1}$.

Step 2.4. Set $E^{k+1} := E(\underline{\alpha}^{k+1})$, $\nabla E^{k+1} := \nabla E(\underline{\alpha}^{k+1})$ and $s^{k+1} := \langle \nabla E(\alpha^k), \tilde{\alpha}^{k+1} - \alpha^k \rangle$, where $\tilde{\alpha}^{k+1}$ is given by (3.1).

Step 2.5. <u>If</u> s^{k+1} tol <u>then go to</u> Step 3.

Step 2.6. If $E^{k+1} \leqq E^k$ <u>then</u> set $\theta := \theta/2$ and <u>go to</u> Step 2.1.

Step 2.7. CONTINUE.

Step 3. Set $\underline{\alpha} := \underline{\alpha}^{k+1}$ and STOP.

3.2 Numerical Solution of State and Adjoint Systems

The numerical solution of the periodic-boundary value problem for the state system (2.1), (2.2), (2.4) in <u>Step 2.2</u> of the Algorithm is obtained by a discrete analogue of the iteration process (2.6) described in 2.2. With discrete initial data given at points $x_j = j\Delta x$, $j=0,1,\ldots,N$, for $t=0$ and $t=\Delta t$, the system is integrated forward by a finite difference method (using steps of size Δt) to obtain solutions at $t=T+\Delta t$. The integration is then repeated using these solutions as initial data. The process is continued until the difference between the initial and final solutions at $t=0$ and $t=T$ is within some tolerance. The solution of the periodic adjoint system (2.12)-(2.14) in <u>Step 2.3</u> is obtained by a similar iteration, using <u>backward</u> integration from discrete initial data given at $t=T$ and $t=T-\Delta t$.

In the case where the coefficients in the system equations are constant, the method of characteristics is used to construct finite difference approximations to the solutions. The characteristics of the system equations are straight lines in this case, and the equations can be integrated discretely by the trapezium method along the characteristics [14]. Since the boundary conditions are consistent, the scheme transfers solutions directly from outgoing to incoming characteristics at the boundaries, and the numerical procedure is stable.

In the variable coefficient case, the method of characteristics is unsatisfactory as the characteristic lines are curved, giving a non-uniform difference mesh which is inefficient for computation. In this case we use instead a modified version of the Leap-frog method [14] to integrate the state and adjoint systems on a uniform, staggered mesh. Straightforward application of the Leap-frog method leads to an unstable scheme due to the friction term $(-pu/h)$ in the momentum equation, but stability is restored by taking a time average of this term. Using a staggered mesh avoids certain difficulties in determining extra boundary conditions, but care must still be take in approximating the conditions (2.2) and (2.13) in order to maintain stability of the procedure.

The difference approximations for the state system are given by

$$b_j(\eta_j^{n+1} - \eta_j^{n-1}) = -\nu(A_{j+1}u_{j+1}^n - A_j u_j^n), \quad j = 0,1,\ldots,N-1,$$

$$u_j^{n+1} - u_j^{n-1} = -\nu g(\eta_j^n - \eta_{j-1}^n) - p\Delta t(u_j^{n+1} + u_j^{n-1})/h_j, \; j=1,2,\ldots,N-1, \tag{3.3}$$

with boundary approximations

$$u_0^n = \alpha^n[f^n - \tfrac{1}{2}(\eta_0^{n+1} + \eta_0^{n-1})], \quad u_N^n = 0. \tag{3.4}$$

The difference approximations for the adjoint system are similarly given by

$$b_j(\lambda_j^{n+1} - \lambda_j^{n-1}) = -\nu g(\mu_{j+1}^n - \mu_j^n), \quad j = 0,1,\ldots,N-1,$$

$$\mu_j^{n+1} - \mu_j^{n-1} = -\nu A_j(\lambda_j^n - \lambda_{j-1}^n) + p\Delta t(\mu_j^{n+1} + \mu_j^{n-1})/h_j, \; j=1,2,\ldots,N-1, \tag{3.5}$$

with boundary conditions

$$g\mu^n - \alpha^n A_0 \tfrac{1}{2}(\lambda_0^{n+1} + \lambda_0^{n-1}) = 2\alpha^n(f^n - \tfrac{1}{2}(\eta_0^{n+1} + \eta_0^{n-1})), \quad \mu_N^n = 0. \tag{3.6}$$

Here α^n approximates α at point $t=t_n$, η_j^n, λ_j^n approximate $\eta(x,t)$, $\lambda(x,t)$ at points $x=x_{j+\frac{1}{2}}$, $t=t_n$, and u_j^n, μ_j^n approximate $u(x,t)$, $\mu(x,t)$ at points $x=x_j$, $t=t_n$. The values of the system parameters are given by $b_j = b(x_{j+\frac{1}{2}})$, $A_j = A(x_j)$, $h_j = h(x_j)$, and $f^n = f(t_n)$. The difference mesh is defined such that $x_j = j\Delta x$, $j=0,1,2,\ldots,N$, where $\Delta x = 1/N$. The time levels are defined by $t_n = n\Delta t$, $n=0,1,\ldots$, where Δt is chosen such that the parameter $\nu = 2\Delta t/\Delta x$ satisfies

$$\nu^2 < \min_j \{b_j/gA_j\} \quad . \tag{3.7}$$

The difference schemes (3.3)-(3.4) and (3.5)-(3.6) can be written in completely explicit form. Stability for values of the parameter ν up to Courant-Friedrichs-Lewy limit of $\nu > \min\{1/c(x)\} \equiv \min\{\sqrt{b(x)/gA(x)}\}$, where $c(x)$ represents the local wave speed, is established by considering discrete "energy" norms analogous to the continuous norm (2.7) (with $k_\alpha(x) \equiv 0$). Convergence of the iterative processes for determining the periodic solutions of the state and adjoint equations is similarly proved for the discrete problems for the same parameter values. A detailed analysis of the stability and convergence of the numerical procedures is given by Birkett [2].

The remaining steps, Step 2.1 and Step 2.4, in the Algorithm are discretized in a natural way, as for the flat basin model [3], using the trapezium quadrature rule to evaluate integrals. The discrete approximations are all second-order accurate, and as the data and the solutions are all assumed to be smooth, we expect the discrete solutions to the optimization problem to converge asymptotically, with order two, as the mesh size $\Delta x \to 0$, with $\nu = 2\Delta t/\Delta x$ constant.

4. GENERALIZATIONS

We now generalize the partial differential model to include dual control of both turbines and sluices with non-linear head-flow relations, and to simulate ebb tide generation only.

For the dual control problem two independent bounded control functions $\alpha_1(t)$, $\alpha_2(t)$ are introduced, which determine the proportionate discharge across turbines and sluices, respectively. The velocity of the fluid across the barrier into the basin and the upstream boundary condition are now given by

$$u(0,t) = \alpha_1(t)P(\Delta H) + \alpha_2(t)R(\Delta H), \quad u(\ell,t) = 0, \tag{4.1}$$

where the discharges P,R are differentiable functions of the head difference, $\Delta H = f(t) - \eta(0,t)$, with derivatives $P' \geqq 0$, $R' \geqq 0$. The optimal control problem becomes

$$\max_{\underline{\alpha} \,\varepsilon\, U_{ad}} \quad E(\underline{\alpha}) = \int_0^T \underline{\alpha}_1 (t) P(\Delta H) \Delta H \, dt \tag{4.2}$$

subject to differential equations (2.1), periodic conditions (2.4) and modified boundary conditions (4.1). The admissible controls $\underline{\alpha} = [\alpha_1, \alpha_2]^T$ are again assumed to be "sufficiently smooth" for solutions to the system equations to exist, and are required to satisfy

$$0 \leqq \alpha_1(t) \leqq a_1, \quad 0 \leqq \alpha_2(t) \leqq a_2. \tag{4.3}$$

The results of §2 are easily extended to the modified problem. The gradient function $\nabla E(\underline{\alpha})$ now becomes

$$\underline{\nabla E}(\underline{\alpha}) = \begin{bmatrix} P(\Delta H)\Delta H + A(0)\lambda(0,t)P(\Delta H) \\ A(0)\lambda(0,t)R(\Delta H) \end{bmatrix} \tag{4.4}$$

where h, u satisfy the system equations (2.1), (2.4), (4.1) and λ, μ satisfy the adjoint equations (2.12), (2.14) and

$$g\mu(0,t) + A(0)\lambda(0,t)(\alpha_1 P' + \alpha_2 R') = \alpha_1(P'\Delta H + P), \quad \mu(\ell,t) = 0. \tag{4.5}$$

Necessary conditions for the optimum then require that $\alpha_i(t)$ takes values 0 or a_i depending on the sign of $\{\underline{\nabla E}(\underline{\alpha})\}_i$, $i=1,2$, except along singular arcs where the controls may take interior values on the constraint sets. The Algorithm of §3 is easily modified to solve the extended problem numerically.

For ebb generation only schemes, power generation takes place only when the head difference is negative. In this case the discharge function P has the general property $P(s) = 0$ if $s \geqq 0$. (In practice, since the function $P(s)$ is not smooth at $s = 0$, P is replaced by a smooth approximation which is as close to P as required).

5. RESULTS

The numerical procedure has been tested with various data, in order to examine the behaviour of the discrete algorithm. It is found that the iteration process converges reasonably rapidly and produces smooth solutions to the problem. The solutions are also found to converge as $\Delta x \to 0$ in all cases.

To illustrate the behaviour of the solutions we present results for a model approximating the Severn estuary. The data are given by

$b(x) = 1.84_{10^4} - 0.308\ x$ m., $h(x) = 13.0 - 1.719_{10^{-4}}\ x$ m.,

$A = bh\ m^2$, $p = 0.0025\ ms^{-1}$., $\ell = 5_{10^4}$m., $T = 44600$ s.,

$a_0 = 0.1877 = a_1$, $a_2 = 0.9406$.

We consider three models, each with different turbine and sluice discharge characteristics:-

Model 1 : $P(s) = s$, $R(s) = s$

Model 2 : $P(s) = 2.49 \tanh(0.55s)$ $R(s) = 1.42 \tanh(0.55s)$

Model 3 : $P(s) = \text{sgn}(s) 1.25\ (1 + \text{sgn}(s) \tanh(10(s - \text{sgn}(s) 2.27)))$

$R(s) = 2.84 \tanh\ (0.2s)$.

In the first model, the head flow relations are linear, and in the second model the relations are approximately linear around the origin, but have a restricted maximum discharge. The data for the third model approximate known plant characteristics, and as well as restricted maximum discharges, the turbine has a minimum cut off, that is, it does not operate for head differences below a given level. Representative solutions are shown in Figures 1 and 2 (for Model 1).

The maximum average power production (in GW) is shown in Table 1 for tidal amplitudes of 5m (spring) and 2.5m (neap). It can be seen that for two-way generation, the power output is significantly increased by the use of dual turbine and sluice controls over turbine control only schemes, in all cases. At low tidal amplitude the two-way dual control schemes give an improvement of about 30% over the ebb generation schemes, while at high tidal amplitude this improvement is less than 20% for models 2 and 3. This difference reflects the non-linearity in these two models. For large tidal amplitudes the power output of the non-linear schemes is considerably smaller than that of the linear scheme, due to the restriction on the maximum discharge. For low tidal amplitude the linear scheme gives a much better approximation to the non-linear schemes. We remark that in non-linear model 3 the turbines are essentially operational at all times, but flow through the turbines is not continuous, due to the minimum cut-off which acts as a natural switching control. The sluice control is then the critical factor in maximizing the power output.

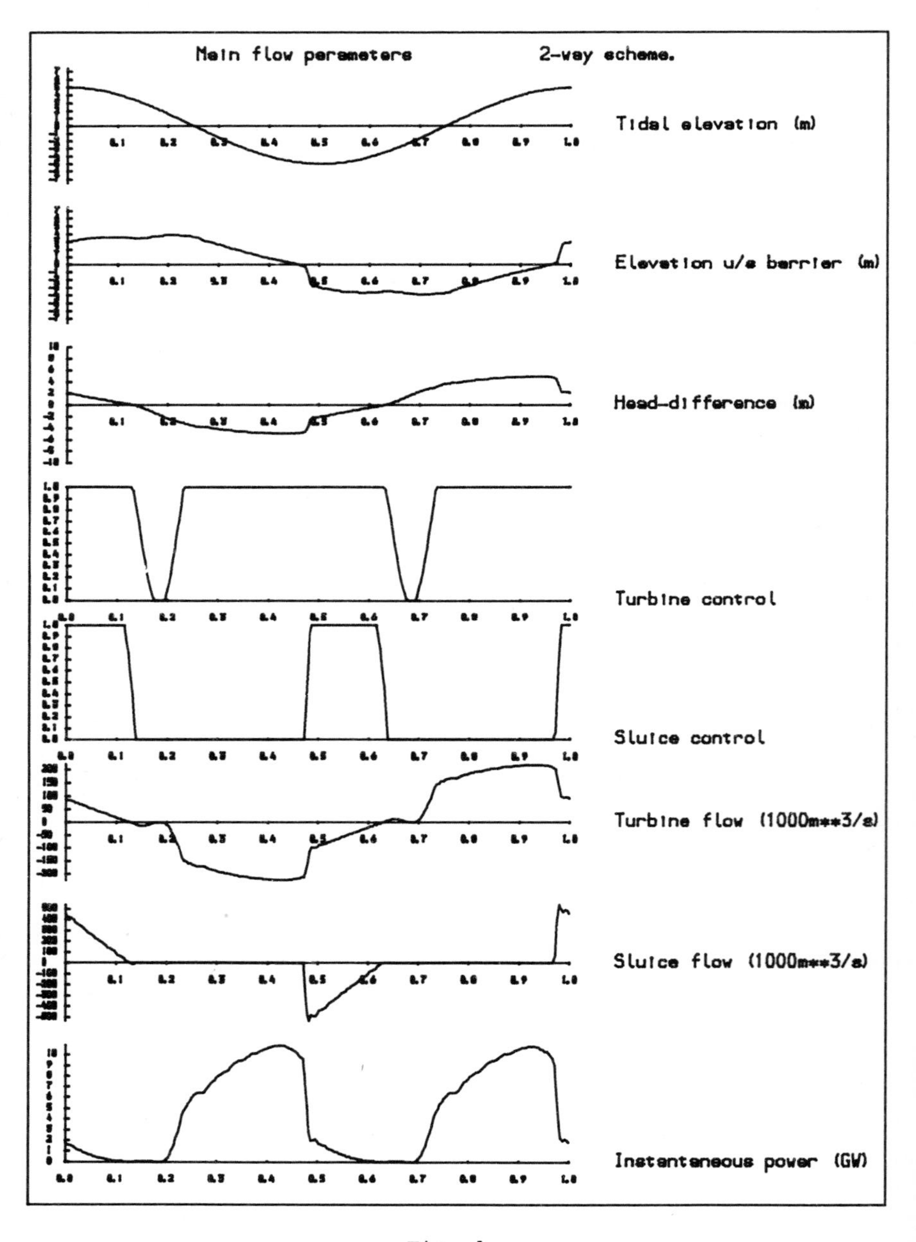
Main flow parameters
2-way scheme.
Tidal elevation (m)
Elevation u/s barrier (m)
Head-difference (m)
Turbine control
Sluice control
Turbine flow (1000m**3/s)
Sluice flow (1000m**3/s)
Instantaneous power (GW)

Fig. 1

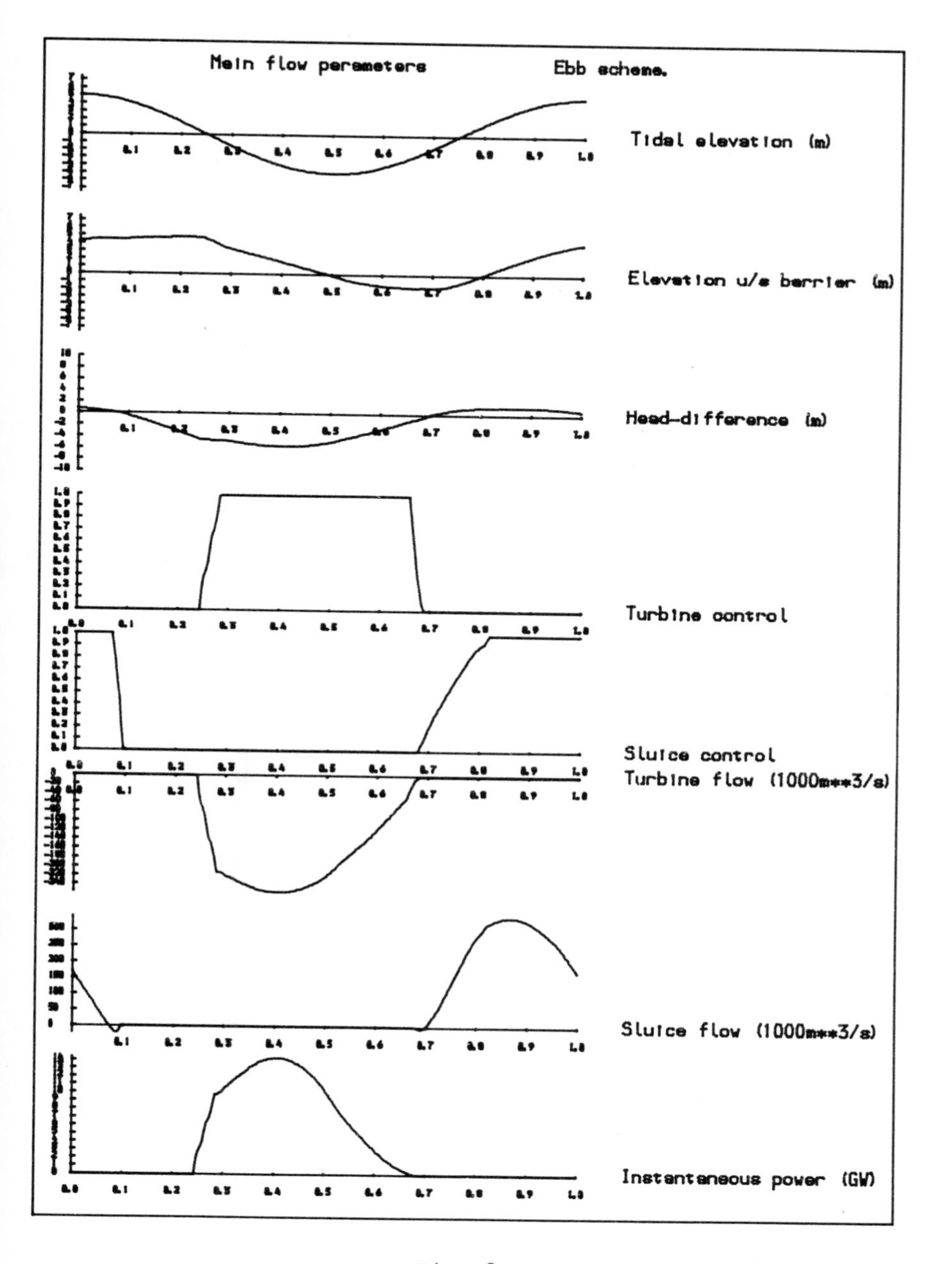
Main flow parameters
Ebb scheme.
Tidal elevation (m)
Elevation u/s barrier (m)
Head-difference (m)
Turbine control
Sluice control
Turbine flow (1000m**3/s)
Sluice flow (1000m**3/s)
Instantaneous power (GW)

Fig. 2

Table 1 Average Power Production (GW)

	Spring Tides (5m)			Neap Tides (2.5m)		
	Turbine	2-way/ Dual	Ebb/ Dual	Turbine	2-way/ Dual	Ebb/ Dual
Model 1	3.80	4.72	3.69	0.951	1.18	0.923
Model 2	2.89	3.22	2.71	0.975	1.11	0.847
Model 3	2.72	3.04	2.60	0.307	0.980	0.746

6. CONCLUSIONS

We examine here a general model of a tidal power generation scheme and develop an optimization technique for determining the maximum average energy output of the scheme. The linear channel flow equations with variable coefficients are used to model the dynamics in the tidal basin. The power generation problem is formulated as a problem in optimal control. It is shown that the system equations are well-posed (under certain assumptions) and necessary conditions for the optimal are given. A gradient projection algorithm is described for determining the optimal control strategy, and a numerical technique for solving the system and adjoint equations is given. Non-linear head-flow properties are incorporated and dual control of both turbines and sluices is treated. Both two-way and ebb generation schemes are simulated. Numerical results for various schemes are presented for a model approximating the River Severn.

We conclude that the optimal control approach to the tidal power problem is a feasible and attractive method for systematically computing flow control policies. The technique has now been extended to more realistic models in which the full estuary on both sides of the barrier is treated, and the assumption that the tidal elevation, unperturbed by flow across the barrier, is imposed directly on the seaward side of the barrier is removed [5]. The estimated power output is significantly reduced and the relative difference between the two-way and ebb schemes is also reduced. Further investigations of more realistic non-linear models and models taking into account the cost-effectiveness of the generated output are now being made with the support of CEGB and SERC.

7. REFERENCES

1. Berry, P.E. (1982) On the use of the describing function technique for estimating power output from a tidal barrage scheme. CEGB Marchwood Eng. Lab. Tech. Rpt. TPRD/M/1292/ N82 TF425.

2. Birkett, N.R.C. (1984) Optimal control of dynamic systems with switches. PH.D. Thesis, Univ. of Reading, Dept. of Maths.

3. Birkett, N.R.C. and Nichols, N.K. (1983) Optimal control problems in tidal power generation. Dept. of Maths., University of Reading, Num. Anal. Rpt. NA 8/83. (To appear in "Industrial Numerical Analysis - Case Histories", O.U.P.)

4. Birkett, N.R.C., Count, B.M., and Nichols, N.K. (1984) Optimal control problems in tidal power. Water power and Dam Construction, Jan. issue, pp. 37-42.

5. Birkett, N.R.C., and Nichols, N.K. (1983) The general linear problem of tidal power generation with non-linear head-flow relations. Dept. of Maths., Univ. of Reading, Num. Anal. Rpt. NA 3/83.

6. Count, B.M. (1980) Tidal power studies at M.E.L. CEGB Marchwood Eng. Lab. Tech. Rpt. MM/MECH/TF257.

7. Gelfand, I.M., and Fomin, S.V. (1963) Calculus of Variations. Prentice-Hall.

8. Gruver, W.A., and Sachs, E. (1980) Algorithmic methods in optimal control. Pitman.

9. H.M.S.O. (1981) Tidal power from the Severn estuary. Energy Paper No. 46.

10. Jeffreys, E.R. (1981) Dynamic models of tidal estuaries. Proc. of BHRA 2nd Int. Conf. on Wave and Tidal Energy.

11. Kreiss, H.O. (1979) Numerical Methods for Partial Differential Equations. Ed. S.V. Parter, Academic Press.

12. Kreyszig, E. (1978) Introductory Functional Analysis. Wiley.

13. Lions, J.L. (1971) Optimal Control of Systems Governed by Partial Differential Equations. Springer-Verlag.

14. Richtmyer, R.D. and Morton, K.W. (1967) Difference Methods for Initial Value Problems. Wiley Interscience (2nd ed.)

15. Stoker, J.E. (1957) Water Waves. Wiley Interscience.

16. Wilson, E.M. et al. (1981) Tidal energy computations and turbine specifications. Inst. of Civil Engineers Symp. on the Severn Barrage.

CHAPTER 4

NONLINEAR SYSTEMS

STATE OBSERVERS FOR A CLASS OF NONLINEAR SYSTEMS

E. Noldus

(University of Ghent, Belgium)

ABSTRACT

The construction of reduced order state observers for systems containing arbitrary nonlinearities is discussed. The basic principle of the proposed method can also be used to solve a variety of other design problems in automatic control.

1. INTRODUCTION

Most of the available theory on nonlinear state observers deals with some special types of systems of a well defined structure, for example bilinear systems[1-2]. This paper considers the design of state observers for a class of systems containing arbitrary nonlinearities. Existence criteria and a design method for a stabilizing observer feedback matrix will be derived, using Liapunov's direct method and some properties of the algebraic Riccati equation.

2. STABILITY ANALYSIS

Consider the system

$$\left.\begin{aligned} \dot{x} &= Ax + Bu + df(t\ ,\ e'x) \\ y &= Cx \end{aligned}\right\} \qquad (2.1)$$

with state vector $x \in R^n$, input $u \in R^m$ and output $y \in R^P$. $f(t\ ,\ \sigma)$ is a scalar function of time and of $\sigma \triangleq e'x$. If C has full rank $p < n$ we may assume that $C = [I_p\ ,\ 0]$, such that (2.1) can be partitioned as

$$\dot{x}_1 = A_{11}x_1 + A_{12}x_2 + B_1u + d_1f(t , e_1'x_1 + e_2'x_2) \tag{2.2.1}$$

$$\dot{x}_2 = A_{21}x_1 + A_{22}x_2 + B_2u + d_2f(t , e_1'x_1 + e_2'x_2) \tag{2.2.2}$$

The output $y = x_1$ can be measured directly. The standard structure of an observer for x_2 is

$$\dot{z} = A_{21}y + A_{22}z + B_2u + d_2f(t , e_1'y + e_2'z) +$$

$$+ H [\dot{y} - A_{11}y - A_{12}z - B_1u - d_1f(t , e_1'y + e_2'z)] \tag{2.3}$$

where z is the observation of x_2 and H is a constant matrix. The inconvenience of having to differentiate the output can be handled in the usual way by selecting a new observer state vector, $\hat{z} \triangleq z - Hy$. The observation error $\eta \triangleq z - x_2$, satisfies

$$\dot{\eta} = (A_{22} - HA_{12}) \eta + (d_2 - Hd_1)g(t , x , z)e_2'\eta \tag{2.4}$$

where

$$g(t , x , z) \triangleq \frac{f(t , e_1'y + e_2'z) - f(t , e_1'y + e_2'x_2)}{e_2'(z - x_2)}$$

If

$$0 \leqq \frac{\partial f(t , \sigma)}{\partial \sigma} \leqq k , \text{ all } t , \sigma$$

then also

$$0 \leqq g(t , x , z) \leqq k, \text{ all } t , x , z \tag{2.5}$$

The problem is to design H such that (2.4) is globally asymptotically stable for all g(t , x , z) which satisfy (2.5). Let

$$V(\eta) \triangleq \eta'P\eta , \quad P = P' > 0 \tag{2.6}$$

be a candidate for a Liapunov function for (2.4). Differentiating (2.6) along the solutions of (2.4) yields

$$\dot{V}(\eta , t , x , z) = \eta'[P(A_{22} - HA_{12}) + (A_{22} - HA_{12})'P]\eta +$$

$$+ \eta'[2P(d_2 - Hd_1) + e_2]g(t , x , z)e_2'\eta -$$

$$- \eta'e_2g(t , x , z)k^{-1}g(t , x , z)e_2'\eta - \lambda \qquad (2.7)$$

where

$$\lambda \triangleq \eta'e_2g(t , x , z) [1 - k^{-1}g(t , x , z)]e_2'\eta \geqq 0 , \quad \text{all } \eta , t , x , z$$

Suppose this expression can be written in the form

$$\dot{V} (\eta, t , x , z) = - [s'\eta - \nu g(t , x , z)e_2'\eta]^2 -$$

$$- \varepsilon \eta'PWP\eta - \lambda < 0 \qquad (2.8)$$

for $W = W' > 0$ and ε a small positive number. Then uniform global asymptotic stability of the error dynamics is ensured. Identification of the right hand sides of (2.7) and (2.8), eliminating s and ν and defining

$$H \triangleq P^{-1}\hat{H} \qquad (2.9)$$

results in the algebraic Riccati equation

$$P[\varepsilon W + d_2kd_2']P = P[-A_{22} + d_2k(\hat{H}d_1 - \tfrac{1}{2}e_2)'] +$$

$$+ [-A_{22} + d_2k(\hat{H}d_1 - \tfrac{1}{2}e_2)']'P +$$

$$+ \hat{H}A_{12} + A_{12}'\hat{H} - [\tfrac{1}{2}e_2 - \hat{H}d_1]k[\tfrac{1}{2}e_2 - \hat{H}d_1]' \qquad (2.10)$$

It must therefore be established whether there exists a matrix $\hat{H}$ such that (2.10) possesses a real solution $P = P' > 0$.

3. THE CASE $d_1 = 0$

If $d_1 = 0$ eq. (2.10) simplifies to

$$P[\varepsilon W + d_2kd_2']P = P[-A_{22} - \tfrac{1}{2}d_2ke_2'] + [-A_{22} - \tfrac{1}{2}d_2ke_2']'P + $$
$$+ \hat{H}A_{12} + A_{12}'H' - \tfrac{1}{4}e_2ke_2' \qquad (3.1)$$

Suppose that for a given $\hat{H} = \hat{H}_2$ and $\varepsilon = \varepsilon_2 > 0$, (3.1) has a solution

$$P = P_2 = P_2' > 0$$

Then P_2 also satisfies

$$P[(\varepsilon_2 - \varepsilon_1)W + d_2kd_2']P = P[-A_{22} - \tfrac{1}{2}d_2ke_2'] +$$
$$+ [-A_{22} - \tfrac{1}{2}d_2ke_2']'P + \hat{H}_2A_{12} + A_{12}'H_2' -$$
$$- \varepsilon_1P_2WP_2 - \tfrac{1}{4}e_2ke_2' \qquad (3.2)$$

where we choose $0 < \varepsilon_1 < \varepsilon_2$. Next select

$$\hat{H} = A_{12}'R \ , \ R = R' > 0$$

Then (3.1) becomes

$$P[\varepsilon W + d_2kd_2']P = P[-A_{22} - \tfrac{1}{2}d_2ke_2'] + [-A_{22} - \tfrac{1}{2}d_2ke_2']'P +$$
$$+ 2A_{12}'RA_{12} - \tfrac{1}{4}e_2ke_2' \qquad (3.3)$$

Comparing (3.2) and (3.3) and using lemma 2 in [3] shows that (3.3) has a maximal solution

$$P_1 \geqq P_2 > 0$$

if $\varepsilon = \varepsilon_2 - \varepsilon_1$ and

$$2A_{12}'RA_{12} - \tfrac{1}{4}e_2ke_2' \geqq \hat{H}_2A_{12} + A_{12}'\hat{H}_2' - \varepsilon_1P_2WP_2 - \tfrac{1}{4}e_2ke_2'$$

The latter expression can be written as

$$[A_{12}' - \hat{H}_2(2R)^{-1}]2R[A_{12}' - \hat{H}_2(2R)^{-1}]' + \varepsilon_1P_2WP_2 - \hat{H}_2(2R)^{-1}\hat{H}'_2 \geqq 0$$

which is satisfied for any sufficiently large positive definite R. For example, choose R of the form

$$R = \frac{1}{\theta} \hat{R}$$

Then theorem 1 follows immediately:

Theorem 1 : if there exists an $\hat{H}$ such that eq. (2.10) has a positive definite solution, then there is also a positive definite solution for

$$\hat{H} = \frac{1}{\theta} A_{12}'\hat{R}$$

where $\hat{R} = \hat{R}' > 0$ and otherwise arbitrary, and θ and ε are positive and sufficiently small.

If $e_2 = A_{12}'b$ for some vector b , (3.3) becomes

$$P[\varepsilon W + d_2 k d_2']P = P[-A_{22} - \frac{1}{2} d_2 kb'A_{12}] +$$

$$+ [-A_{22} - \frac{1}{2} d_2 kb'A_{12}]'P + A_{12}'[2R - \frac{1}{4} bkb']A_{12}$$

Choose $R > \frac{1}{8} bkb'$. Then there is a solution $P^+ > 0$ if the pair $[-A_{22} - \frac{1}{2} d_2 kb'A_{12} , A_{12}]$ is observable, or equivalently if the pair (A , C) is observable. Hence we have

Theorem 2 : If $e_2 = A_{12}'b$ and (A , C) is observable then a state observer can always be found.

4. DESIGN PROCEDURE

Let P^+ be a maximal solution of (3.3). Putting $R = \frac{1}{\theta}\hat{R}$, differentiating with respect to θ and k, and some standard calculations yield

$$P^+_\theta = - \frac{2}{\theta^2} \int_0^{+\infty} e^{\Psi' t} A_{12}'\hat{R} A_{12} e^{\Psi t} dt \leqq 0$$

and

$$P^+_k = - \int_0^{+\infty} e^{\Psi' t} [P^+ d_2 + \frac{1}{2} e_2][P^+ d_2 + \frac{1}{2} e_2]' e^{\Psi t} dt \leqq 0$$

where

$$\Psi \triangleq - A_{22} - \frac{1}{2}d_2ke_2' - [\varepsilon W + d_2kd_2']P^+$$

is stable[4]. Also , P^+ increases as ε decreases (see lemma 3 in [3]). A design procedure based on these properties is :

1. Choose a small positive ε and any $R > 0$ and $W > 0$.
2. Solve (3.3) for $R = (1/\theta)\hat{R}$ and for decreasing θ until a $P^+ > 0$ is found.
3. If for $\theta \to 0$ no positive definite P^+ appears, then start again with a smaller k until $P^+ > 0$.
4. Calculate

$$H = \frac{1}{\theta}(P^+)^{-1}A_{12}'\hat{R}$$

The search method is exhaustive in the sense that for any k , an $\hat{H}$ such that (2.10) has a positive definite solution will be found, if one exists.

5. THE CASE $d_1 \neq 0$

If $d_1 \neq 0$ we may assume that $d_2 = 0$. Indeed if $d_2 \neq 0$ the system's state equation may be rewritten in terms of a new state component

$$x_3 \triangleq x_2 - d_2(d_1'd_1)^{-1}d_1'x_1$$

which satisfies an equation of the form

$$\dot{x}_3 = A_{31}x_1 + A_{32}x_3 + B_3u$$

The Riccati equation (2.10) becomes

$$\varepsilon PWP = - PA_{22} - A_{22}'P + \Phi \qquad (5.1)$$

with

$$\Phi \triangleq \hat{H}A_{12} + A_{12}'\hat{H}' - [\frac{1}{2}e_2 - \hat{H}d_1]k[\frac{1}{2}e_2 - \hat{H}d_1]'$$

Two subcases must be discerned :

<u>1. p = 1</u>

For a scalar output $d_1kd_1' = kd_1^2 > 0$ and Φ can be written as

$$\Phi = -[\hat{H} - (A_{12} + \tfrac{1}{2}d_1 ke_2')'(kd_1^2)^{-1}]kd_1^2 [\hat{H}' - (kd_1^2)^{-1}(A_{12} + \tfrac{1}{2}d_1 ke_2')] + (A_{12} + \tfrac{1}{2}d_1 ke_2')'(kd_1^2)^{-1}(A_{12} + \tfrac{1}{2}d_1 ke_2') - \tfrac{1}{4}e_2 ke_2'$$

Φ attains a maximum for

$$\hat{\hat{H}} = (A_{12} + \tfrac{1}{2}d_1 ke_2')'(kd_1^2)^{-1} \tag{5.2}$$

and (5.1) becomes

$$\varepsilon PWP = -PA_{22} - A_{22}'P + (A_{12} + \tfrac{1}{2}d_1 ke_2')'(kd_1^2)^{-1}(A_{12} + \tfrac{1}{2}d_1 ke_2') - \tfrac{1}{4}e_2 ke_2' \tag{5.3}$$

By lemma 2 in [3] we have

Theorem 3 : If there exists an H such that (5.1) has a positive definite solution then there is also a positive definite solution for H as given by (5.2).

Differentiating (5.3) with respect to k yields

$$P^+_k = -\frac{1}{k^2 d_1^2}\int_0^{+\infty} e^{\Psi' t} A_{12}'A_{12} e^{\Psi t}\, dt \leqq 0$$

where

$$\Psi \triangleq -A_{22} - \varepsilon WP^+$$

is stable. Hence the observer design consists in solving (5.3) for decreasing k until a positive definite P^+ is obtained. Then (2.9) and (5.2) define an observer feedback matrix which stabilizes the error dynamics for all nonlinearities with slopes restricted to the interval [o , k].

If $e_2 = A_{12}'\gamma$, for some scalar γ , then

$$\Phi_{max} = \mu A_{12}'A_{12}$$

where

$$\mu \triangleq \frac{1}{kd_1^2}\left(1 + \frac{1}{2}d_1 k\nu\right)^2 - \frac{1}{4}\nu^2 k > 0$$

for k sufficiently small. This implies

Theorem 4 : If $e_2 = A_{12}'\nu$ and (A , C) is observable, a state observer can always be found for k sufficiently small.

2.p > 1

By a transformation of the state component x_1 we can put $d_1 k d_1'$ in the form

$$d_1 k d_1' = \begin{pmatrix} \lambda_o & 0 \\ 0 & 0 \end{pmatrix}$$

where the scalar $\lambda_o \neq 0$. Now the design problem is considerably more difficult, and, for the sake of simplicity we shall only consider feedback amplifiers of the type

$$H = P^{-1}\hat{H}$$

$$= P^{-1}(A_{12} + \frac{1}{2}d_1 k e_2')'\tilde{H} \quad , \qquad (5.4)$$

which have the same structure as in the case p = 1. Substitution in (5.1) and a similar argument as before show that the best choice for H is

$$\tilde{H} = \begin{pmatrix} \lambda_o^{-1} & \\ & \frac{1}{\rho}\hat{R} \end{pmatrix} \quad , \quad \hat{R} = \hat{R}' > 0 \quad , \qquad (5.5)$$

with $\rho > 0$ and sufficiently small. Furthermore, for H given by (5.4) and (5.5), the maximal solution again satisfies

$$P^+_\rho \leqq 0 \quad , \quad P^+_k \leqq 0$$

The design therefore consists in solving the Riccati equation for decreasing ρ and k until a positive definite solution is found. Then (5.4) and (5.5) provide the observer feedback matrix. Also, theorem 4 remains valid.

6. EXAMPLES

Consider a system of the form (2.1) with

$$A = \begin{bmatrix} 1 & 0 & 1 \\ -1 & 0 & 0 \\ 0 & 1 & 0 \end{bmatrix}, \quad B = \begin{bmatrix} 1 \\ 0 \\ -1 \end{bmatrix}, \quad d = \begin{bmatrix} 0 \\ 1 \\ 0 \end{bmatrix}, \quad e = \begin{bmatrix} -1 \\ 1 \\ 0 \end{bmatrix}$$

$$C = \begin{bmatrix} 1 & 0 & 0 \end{bmatrix}$$

and $0 \leqq \frac{\partial f}{\partial \sigma}(t\,,\,\sigma) \leqq k$ for all t and σ. Choose $\hat{R} = 1$. Then for $\varepsilon W \underset{\rightarrow}{>} 0$ the maximal solution of eq. (3.3) approaches

$$P^{+} \rightarrow \begin{bmatrix} -\frac{1}{2} + \left(\frac{2}{k}\right)^{\frac{1}{2}} \left(\frac{2}{k\theta}\right)^{\frac{1}{4}} & -\left(\frac{2}{k\theta}\right)^{\frac{1}{2}} \\ -\left(\frac{2}{k\theta}\right)^{\frac{1}{2}} & \frac{2}{\theta}^{\frac{1}{2}} \left(\frac{2}{k\theta}\right)^{\frac{1}{4}} \end{bmatrix} > 0$$

if

$$k^{3}\theta < 8 \tag{6.1}$$

Hence an observer can be found for any $k > 0$. Its state equation is (2.3), where

$$H = \frac{1}{\theta}\,(P^{+})^{-1}A_{12}',$$

and θ satisfies (6.1).

Next consider a fourth order case with

$$A = \begin{bmatrix} -1 & 0 & 1 & 1 \\ 0 & -1 & 0 & 1 \\ 1 & 0 & 0 & -1 \\ 0 & 0 & 1 & 1 \end{bmatrix}, \quad B = \begin{bmatrix} 1 \\ 0 \\ -1 \\ 0 \end{bmatrix}, \quad d = \begin{bmatrix} 1 \\ 0 \\ 0 \\ 0 \end{bmatrix}, \quad e = \begin{bmatrix} -1 \\ 0 \\ 1 \\ 0 \end{bmatrix}$$

$$C = \begin{bmatrix} 1 & 0 & 0 & 0 \\ 0 & 1 & 0 & 0 \end{bmatrix}$$

and $f(t\ ,\ \sigma)$ as before. Since

$$A_{12} + \frac{1}{2}d_1 k e_2{}' = \begin{bmatrix} 1 + \frac{k}{2} & 1 \\ 0 & 1 \end{bmatrix}$$

is nonsingular, any H can be written in the form (5.4).
For $\hat{R} = 1$,

$$\tilde{H} = \begin{bmatrix} \frac{1}{k} & \\ & \frac{1}{\rho} \end{bmatrix}$$

and for $\varepsilon W \underset{\rightarrow}{>} 0$ the maximal solution of the Riccati equation approaches

$$P^+ \rightarrow \begin{bmatrix} \frac{1}{2} + \frac{1}{2k} + \frac{1}{\rho} & \frac{1}{2}(1 + \frac{1}{k}) \\ \frac{1}{2}(1 + \frac{1}{k}) & \frac{1}{2} + \frac{1}{k} + \frac{1}{\rho} \end{bmatrix}$$

H is given by (5.4) where $P = P^+$. The error dynamics can be stabilized for any k by selecting ρ such that $P^+ > 0$.

7. DISCUSSION AND CONCLUSIONS

There exist a variety of design problems in automatic control which can be solved using essentially the same principle as developed in the preceding sections. For example, one may think of the stabilization of (linear or nonlinear) time-delay systems using memoryless feedback [5], the design of state observers, or state feedback laws for bilinear systems [3], certain disturbance rejection problems [6] etc. These problems usually involve the implementation of an amplifier matrix H with the purpose of stabilizing some state - or error dynamics. Using a quadratic Liapunov function to prove stability then produces an equation of the general form:

$$PWP = A'P + PA + \Phi(P , H) \quad , \quad W = W' > 0 \tag{7.1}$$

which must have a solution (P , H) with $P = P' > 0$. Suppose that

$$\Phi(P , H) \leqq Q \text{ for all } P \text{ and } H$$

Then the following simple principle holds: If eq. (7.1) has a solution (P , H) with $P = P' > 0$, then there is also a solution which satisfies

$$\Phi(P , H) = Q \tag{7.2}$$

and

$$PWP = A'P + PA + Q \quad , \quad P = P' > 0 \tag{7.3}$$

If an amplifier matrix is selected which satisfies eq. (7.2), the design problem is reduced to analysing eq. (7.3), which is usually much simpler than eq. (7.1). The matrix Q may depend on a number of free parameters. Then the algorithms essentially require repeatedly solving a parameter dependent Riccati equation until a positive definite solution is found.

In the present paper this design principle was illustrated by synthesizing reduced order state observers for a class of systems containing an arbitrary nonlinear component. However most of the results may be extended to the case of several nonlinearities. Other extensions might deal with systems whose output equation contains a nonlinear function of the state. A disadvantage of the design is that the system's nonlinear component must be known exactly, as it appears in the observer's state equation.

8. REFERENCES

1. Hara, S., and Furuta, K., (1976) Minimal order state observers for bilinear systems, Int. J. Control 24 , 705-718.

2. Derese, I., and Noldus, E. (1981). Existence of bilinear state observers for bilinear systems, I.E.E.E. Trans. Auto. Control 26 , 590-592.

3. Derese, I., and Noldus, E. (1980). Design of linear feedback laws for bilinear systems, Int. J. Control 31 , 219-237.

4. Willems, J.C. (1971). Least squares stationary optimal control and the algebraic Riccati equation, I.E.E.E. Trans. Auto. Control 16 , 621-634.

5. Thowsen, A. (1981) Stabilization of a class of linear time-delay systems, Int. J. Systems Sci. 12 , 1485-1492.

6. Derese, I. and Noldus, E. (1980). Rejection of multiplicative disturbances, R.A.I.R.O. 14 , 233-253.

ALGORITHMS FOR PIECEWISE CONSTANT CONTROL OF NONLINEAR SYSTEMS

M.J. Chapman
(Department of Mathematics, Coventry (Lanchester) Polytechnic)

A.J. Pritchard
(Control Theory Centre, University of Warwick, Coventry)

and

S. Verrall
(Electrowatt Engineering Services, (UK) Ltd., W. Sussex)

ABSTRACT

The purpose of this paper is to demonstrate the use of a fixed point formulation in constructing piecewise constant controls to steer a single input nonlinear system to a desired state. Admissible controls are restricted to an n dimensional subspace of piecewise constant functions having n-1 preassigned switching times, where n represents the state space dimension. Numerical simulations for a Van der Pol equation are presented.

1. INTRODUCTION

In this paper we consider the nonlinear control system

$$\dot{x}(t) = f(x(t),u(t),t) \quad x(t_o) = x_o \quad x(t) \in \mathbb{R}^n \quad u(t) \in \mathbb{R} \qquad (1.1)$$

The object will be to derive control strategies that steer the state from x_o to some desired final state x_d over the time interval $[t_o,T]$. Let $\bar{u}(.)$ represent some nominal control and $\bar{x}(.)$ the corresponding state trajectory initiating at x_o. So

$$\dot{\bar{x}}(t) = f(\bar{x}(t), \bar{u}(t),t) \quad \bar{x}(t_o) = x_o \qquad (1.2)$$

Writing $x(t) = \bar{x}(t) + x'(t)$, $u(t) = \bar{u}(t) + u'(t)$ and assuming appropriate smoothness conditions on the nonlinearity f, we see that

$$\dot{x}'(t) = A(t)x'(t) + b(t)u'(t) + N(x'(t),u'(t),t) \quad x'(t_o) = 0 \qquad (1.3)$$

where $A(t) = (^{\partial f}/\partial x)(\bar{x}(t), \bar{u}(t), t)$, is an $(n \times n)$ time varying matrix, and $b(t) = (\partial f/\partial u)(\bar{x}(t), \bar{u}(t),t)$ a time varying n-vector. Notice that $A(t)$ and $b(t)$ will normally be time varying even if f itself does not depend explicitly on time. Integrating equation (1.3) yields the following expression for x'.

$$x'(t) = \int_{t_o}^{t} U(t,s)b(s)u'(s)ds + \int_{t_o}^{t} U(t,s)N(x'(s),u'(s),s)ds \tag{1.4}$$

where $U(.,.)$ represents the evolution operator generated by $A(.)$. Let $\mathcal{U}$ denote the space of admissible control functions and assume (1.1) has a well defined solution for each $u(.) \in \mathcal{U}$, then we may define a map $R: \mathcal{U} \to \mathbb{R}^n$ by $R(u(.)) = x(T)$. Our problem now becomes that of constructing a local pseudo-inverse for R around $\bar{x}(T)$. Let us assume that the nonlinear system (1.1) is controllable to some neighbourhood of the state $\bar{x}(T)$. In this case it is not unreasonable to conjecture the existence of an n dimensional space of admissible controls, $\mathcal{U}$, for which R becomes a bijective map from some neighbourhood of $\bar{u}(.)$, which we will assume lies in $\mathcal{U}$, to some neighbourhood of $\bar{x}(t)$. Therefore, providing x_d is not too far from $\bar{x}(T)$, we might expect to be able to invert R at x_d as desired. Two possible methods for restricting the control class are:

(i) To use optimality criteria

(ii) To use piecewise constant controls with (n-1) preassigned switching times.

In case (i) it is meant that admissible controls would steer the linearised equation from x_o to some final state $X_L(T)$ in an optimal fashion, e.g. minimising control energy. Such controls are unlikely to be optimal for the nonlinear system but the interest of this paper is only in finding 'steering' controls and not in optimality. In fact we shall restrict our attention to method (ii) alone in the following discussion. For the remainder of this section, we examine the linear control system.

$$\dot{x}'(t) = A(t)x'(t) + b(t)u'(t) \quad x'(t_o) = 0 \tag{1.5}$$

Let $t_o < t_1 < \ldots < t_n = T$ be a partition of $[t_o, T]$ and let $\chi_i = \chi_{(t_{i-1},t_i]}$, $i = 1 \ldots, n$ be characteristic functions, that is

$$\chi_i(t) = \begin{cases} 1 & t \in (t_{i-1},t_i] \\ 0 & \text{Otherwise} \end{cases}$$

The control class $\mathcal{U}$ is now taken as

$$\mathcal{U} = \{u(.) : (t_o, T] \to \mathbb{R} \text{ with } u = \sum_{i=1}^{n} \alpha_i \chi_i\} \tag{1.6}$$

Writing $u' = \sum_{i=1}^{n} \alpha_i' \chi_i$ and integrating (1.5) over $[0,T]$ gives

$$x'(T) = \sum_{i=1}^{n} \alpha_i' \int_{t_{i-1}}^{t_i} U(T,s)b(s)ds \triangleq F\alpha' \tag{1.7}$$

where $\alpha' = [\alpha_1', \quad , \alpha_n']^T$. Assuming the pair $(A(.), b(.))$ is controllable over $[t_o,T]$, it is possible to show there exists a partition for which F is invertible.[1] It seems reasonable to conjecture that, apart from certain degenerate cases, F will remain invertible for almost all choices of partition. Henceforth we assume this to be the case for our given partition.

Define $G(T,.) : \mathbb{R}^n \to \mathcal{U}$ through

$$G(T,t)x = (F^{-1}x)_i \text{ for } t \in (t_{i-1}, t_i] \tag{1.8}$$

Then the control $u' \in \mathcal{U}$ given by

$$u'(t) = G(T,t)x_d' \tag{1.9}$$

will steer the linear system from zero to x_d' over $[0,T]$. In the next section we show how the use of fixed point theorems may be applied to extend the above to the nonlinear setting of (1.1), (1.3).

2. A FIXED POINT PROBLEM

As in the previous section, let our control class $\mathcal{U}$ be given by (1.6).

Substituting $u' = \sum_{i=1}^{n} \alpha_i' \chi_i$ into (1.4) at time $t = T$ yields

$$x'(T) = F\alpha' + \int_{t_o}^{T} U(T,s)N(x'(s),u'(s),s)ds \tag{2.1}$$

with F as given by (1.7). Our main assumption will be that the map F is invertible. In particular this implies the linearised model is controllable. Defining G as before, (1.8), the control $u = \bar{u} + u'$ will steer the state of the nonlinear system (1.1) from x_o to x_d providing u' satisfies

$$u'(t) = G(T,t)[x_d - \bar{x}(T) - \int_{t_o}^{T} U(T,s)N(x'(s),u'(s),s) \, ds] \tag{2.2}$$

where x' satisfies (1.4)

$$x'(t) = \int_{t_0}^{t} U(t,s)[b(s)u'(s) + N(x'(s),u'(s), s)]\, ds \tag{2.3}$$

Let us consider the trajectory x' as lying in some appropriate Banach space X. We also place an appropriate Banach space structure on $\mathcal{U}$ and the product space $Z = X \times \mathcal{U}$. For $(x',u') \in Z$ we define

$$\phi_1(x',u') = \int_{t_0}^{\cdot} U(.,s)[b(s)u'(s) + N(x'(s),u'(s),s)]ds \tag{2.4}$$

$$\phi_2(x',u') = G(T,.)[x_d - \bar{x}(T) - \int_{t_0}^{T} U(T,s)N(x'(s),u'(s),s)\, ds\,] \tag{2.5}$$

and let $\phi(x',u') = (\phi_1(x',u'),\phi_2(x'u'))$. Our problem becomes that of finding a fixed point in Z for the map ϕ. Perhaps the most well known fixed point theorem is the Contraction Mapping Principle.

Theorem (Contraction Mapping Principle)

Let S be a closed subset of a Banach space, Z, and let ϕ be a contraction on S, that is

(i) $\phi(S) \subset S$

(ii) $||\phi(s) - \phi(\hat{s})|| \leq K||s - \hat{s}|| \quad 0 \leq K < 1$

Then there exists a unique fixed point s* of ϕ lying in S, that is $\phi(s^*) = s^*$. Moreover the sequence $s_{n+1} = \phi(s_n)$ converges to s* irrespective of the choice of starting value $s_0 \in S$.

The aim of the next sections will be to apply this result to the map ϕ as given by (2.4), (2.5).

3. THREE ITERATIVE ALGORITHMS

The aim of this section is to derive algorithms $\{(x_i(.),u_i(.))\}$, so that $u_n(.)$ converges to our desired piecewise constant control, u*(.) and $x_i(.)$ converges to the corresponding state trajectory, x*(.). Notice these are not the perturbed quantities x',u'. The reason for this is that it will normally be easier, and hopefully more accurate, to use calculations involving the nonlinear function, f, as opposed to N.

All the algorithms we consider here will involve an update on u defined through

$$u'_{i+1}(t) = G(T,t)[x_d - \bar{x}(T) - \int_{t_0}^{T} U(T,s)N(x'_i(s),u'_i(s),s)ds] \tag{3.1}$$

From (1.1)-(1.3) we see that $N(x',u',s) = f(x,u,s) - f(\bar{x},\bar{u},\bar{s}) - A(s)(x-\bar{x}) - b(s)u'$. If this is substituted into (3.1), using the relation

$$-\int_{t_0}^{T} U(T,s)A(s)\bar{x}(s)ds = \int_{t_0}^{T} (\partial/\partial s)(U(T,s))\bar{x}(s)ds$$

$$= [U(T,s)\bar{x}(s)]_{t_0}^{T} - \int_{t_0}^{T} U(T,s)f(\bar{x}(s),\bar{u}(s),s)ds$$

$$= \bar{x}(T) - U(T,t_0)x_0 - \int_{t_0}^{T} U(T,s)f(\bar{x}(s),\bar{u}(s),s)ds$$

we arrive at the formula

$$u'_{i+1}(t) = G(T,t)[x_d - U(T,t_0)x_0 - \int_{t_0}^{T} U(T,s)(f(x_i(s),u_i(s),s) - A(s)x_i(s) - b(s)u'_i(s))ds]$$

From the definition of G, (1.8), it is easy to show that

$$G(T,t)\int_{t_0}^{T} U(T,s)b(s)u_i'(s)ds = u'_i(t)$$

and so finally we have

$$u_{i+1}(t) = u_i(t) + G(T,t)[x_d - U(T,t_0)x_0 - \int_{t_0}^{T} U(T,s)(f(x_i(s),u_i(s),s) - A(s)x_i(s))ds] \tag{3.2}$$

Algorithm 1

From the contraction mapping principle we see that an algorithm based upon $s_{n+1} = \phi(s_n)$ will converge providing ϕ is contractive. From (2.4) we have

$$x'_{i+1}(t) = \int_{t_0}^{t} U(t,s)[b(s)u'_i(s) + N(x'_i(s),u'_i(s),s)]ds \tag{3.3}$$

Differentiating this and substituting for N leads to

$$\dot{x}_{i+1}(t) = A(t)(x_{i+1}(t) - x_i(t)) + f(x_i(t),u_i(t),t) \quad x_i(t_0) = x_0 \tag{3.4}$$

This implies $f(x_i(s),u_i(s),s) - A(s)x_i(s) = \dot{x}_{i+1}(s) - A(s)x_{i+1}(s)$, which may then be substituted into (3.2). After an integration by parts to remove the $\dot{x}_{i+1}$ term, (3.2) reduces to

$$u_{i+1}(t) = u_i(t) + G(T,t)[x_d - x_{i+1}(T)] \tag{3.5}$$

Equations (3.4), (3.5) together give algorithm 1. Starting values are given by $x_0(t) = \bar{x}(t)$, $u_0(t) = \bar{u}(t)$.

Algorithm 2

This time, in place of the explicit formula (3.3), we use the implicit relation

$$x'_{i+1}(t) = \int_{t_0}^{t} U(t,s)[b(s)u'_{i+1}(s) + N(x'_{i+1}(s),u'_{i+1}(s),s)]ds \tag{3.6}$$

Differentiation and substitution for N now leads to

$$\dot{x}_{i+1}(t) = f(x_{i+1}(t),u_{i+1}(t),t) \quad x_{i+1}(t_0) = x_0 \tag{3.7}$$

Replacing $f(x_i(s),u_i(s),s)$ by $\dot{x}_i(s)$ in (3.2) and integrating by parts gives

$$u_{i+1}(t) = u_i(t) + G(T,t)[x_d - x_i(T)] \tag{3.8}$$

It is important to note that, in this algorithm, each $x_i(.)$ satisfies the original nonlinear equation (1.1) with control $u_i(.)$. Hence the difference $x_d - x_i(T)$ provides us with a precise measure of the effectiveness of the control $u_i(.)$. This is not the case in algorithm 1, (or 3), and the iterative procedure must then be terminated only when it is deemed that the sequence $\{u_n(.)\}$ has converged.

Algorithm 3

There are many other possibilities for seeking a fixed point for the map ϕ as given by (2.4), (2.5), each having different convergence rates and convergence regions. One possibility is to substitute $\phi_2(x'(.),u'(.))$ for $u'(.)$ in (2.4) to produce

$$\phi'_1(x',u') = \int_{t_0}^{\cdot} U(.,s)b(s)G(T,s)[x_d-\bar{x}(T) - \int_{t_0}^{T} U(T,\rho)N(x'(\rho),u'(\rho),\rho)d\rho]ds + \int_{t_0}^{\cdot} U(.,s)N(x'(s),u'(s),s)ds \tag{3.9}$$

and redefine ϕ by $\phi(x',u') = (\phi_1'(x',u'),\ \phi_2(x',u'))$. Using the idea of algorithm 1 we obtain $x'_{i+1}(t) = \phi'_1(x'_i,u'_i)(t)$, which upon differentiation and substitution for N yields

$$\dot{x}_{i+1}(t) = A(t)(x_{i+1}(t) - x_i(t)) + b(t)(u_{i+1}(t) - u_i(t)) + f(x_i(t),u_i(t),t) \quad x_i(t_0) = x_0 \tag{3.10}$$

If the formula

$f(x_i(s),u_i(s),s) - A(s)x_i(s) = \dot{x}_{i+1}(s) - A(s)x_{i+1}(s) - b(s)(u'_{i+1}(s) - u'_i(s))$ is substituted into (3.2), it reduces after integration by parts to the expression $G(T,t)[x_d-x_{i+1}(T)] = 0$. This is of course of no use in finding an updating formula for the control, but serves to illustrate the fact that $x_i(T) = x_d$ at each iterate $i \geq 1$. ($x_0(T) = \bar{x}(T) \neq x_d$ in general). For this algorithm we leave (3.2) as our recursive formula for u. Since theoretically $x_i(T) = x_d$ we may use the difference $x_d - x_i(T)$ as a measure of errors occurred in integrating (3.10) and, as mentioned previously, terminate the iterative procedure when the control sequence is deemed to have converged.

4. CONVERGENCE RESULTS

In this section we discuss convergence properties of Algorithms 1-3 of the previous section. To shorten the treatise only outlines of the various proofs are presented. For simplicity let us examine first Algorithm 3.

Algorithm 3

$$x'_{i+1}(t) = \phi'_1 (x'_i,u'_i)(t), \qquad u'_{i+1}(t) = \phi_2(x'_i,u'_i)(t) \tag{4.1}$$

Under mild constraints such as the Lebesque square integrability of U(.,) and b(.) we can deduce, by means of the Schwartz inequality, the existence of constants K_1 and K_2 for which

$$||\phi'_1(x,u) - \phi'_1(\hat{x},\hat{u})||_X \leqslant K_1||N(x(.),u(.),.) - N(\hat{x}(.),\hat{u}(.),.)||_X \tag{4.2}$$

$$||\phi_2(x,u) - \phi_2(\hat{x},\hat{u})||_{\mathcal{U}} \leqslant K_2||N(x(.),u(.),.) - N(\hat{x}(.),\hat{u}(.),.)||_X \tag{4.3}$$

where X is taken here to equal $L_2[t_o,T; \mathbb{R}^n]$ and $\mathcal{U}$ is given the topology of $L_2[t_o,T; \mathbb{R}]$. Define now a norm on $Z = X \times \mathcal{U}$ by

$$||(x,u)||^2_Z = ||x||^2_X + ||u||^2_{\mathcal{U}} \tag{4.4}$$

For simplicity of notation we write z for the pair (x,u) and similarly write N(x,u,t) as N(z,t). We shall make use of the following mild assumption

$$||N(z(.),.) - N(\hat{z}(.),.)||_X \leqslant k(||z(.)||_Z,||\hat{z}(.)||_Z)||z(.) - \hat{z}(.)||_Z \tag{4.5}$$

with $k(.,.): \mathbb{R}^+ \times \mathbb{R}^+$ continuous, symmetric and satisfying $k(o,o) = 0$. Using (4.2), (4.3) we deduce

$$||\phi(z) - \phi(\hat{z})||^2_Z \leqslant (K_1{}^2 + K_2{}^2)k(||z||,||\hat{z}||)^2||z - \hat{z}||^2 \tag{4.6}$$

Hence for z, $\hat{z}$ lying in a sufficently small ball in Z, ϕ is indeed a contraction. More precisely we require the ball to have radius smaller than a where

$$K(a) \triangleq (K_1{}^2 + K_2{}^2)^{\frac{1}{2}} \sup_{\theta_1,\theta_2 \leqslant a} k(\theta_1,\theta_2) < 1 \tag{4.7}$$

In order to apply the contraction mapping principle we also need this ball to be invariant under ϕ. We have the existence of a constant K_3 for which

$$||\phi(z)||_Z{}^2 \leqslant K_3{}^2||x_d - \bar{x}(T)||^2 + (K_1{}^2 + K_2{}^2)k(||z||,0)^2||z||^2 \tag{4.8}$$

Therefore, for a satisfying (4.7), we have

$$||\phi(z)||_Z^2 \leq K_3^{\ 2}||x_d - \bar{x}(T)||^2 + K(a)a^2 \tag{4.9}$$

and hence, since $K(a) < 1$, $||\phi(z)|| \leq a$ for $||x_d - \bar{x}(T)||$ sufficiently small. This completes the required convergence proof. The region of convergence clearly depends on the constants K_1, K_2 and K_3 which in practice may well prove intractable, or at least very difficult to estimate. If the algorithm fails to converge either a new algorithm must be chosen or a better initial estimate $(\bar{u}(.),\bar{x}(.))$ found.

Algorithm 1

$$x'_{i+1}(t) = \phi_1(x_i',u_i')(t) \quad u'_{i+1}(t) = \phi_2(x_i',u_i')(t) \tag{4.10}$$

As in the above, we deduce the existence of constants M_1, M_2 and M_3 for which

$$||\phi_1(x,u) - \phi_1(\hat{x},\hat{u})||_X \leq M_1||u-\hat{u}||_{\mathcal{U}} + M_2||N(x(.),u(.),.) - N(\hat{x}(.),\hat{u}(.),.)||_X \tag{4.11}$$

$$||\phi_2(x,u) - \phi_2(\hat{x},\hat{u})||_{\mathcal{U}} \leq M_3||N(x(.),u(.),.) - N(\hat{x}(.),\hat{u}(.),.||_X \tag{4.12}$$

Due to the term $M_1||u-\hat{u}||_{\mathcal{U}}$ in (4.11), not appearing in (4.2), we now define our norm on $Z = X\times \mathcal{U}$ by

$$||(x,u)||^2_Z = \varepsilon||x||^2_X + ||u||^2_{\mathcal{U}} \tag{4.13}$$

where $\varepsilon > 0$ is a parameter free for choice. Again assuming relation (4.5), but now with the new cross product norm, we find

$$||\phi(z) - \phi(\hat{z})||^2_Z \leq [2\ \varepsilon\ M_1^2 + (2\ \varepsilon\ M_2^2+M_3^2)k(||z||,||\hat{z}\ ||)^2]\ ||z - \hat{z}||^2 \tag{4.14}$$

We see that the requirements that z, $\hat{z}$ lie in a sufficiently small ball together with a suitably chosen ε imply the desired contractive property for ϕ. Proceeding as before, it is then possible to deduce, for $||x_d - \bar{x}(T)||$ sufficiently small, the required convergence result for algorithm 1.

Algorithm 2

$$\begin{aligned} \dot{x}_i(t) &= f(x_i(t),u_i(t),t) \quad x_i(t_0) = x_0 \\ u_{i+1}(t) &= u_i(t) + G(T,t)[x_d - x_i(T)] \end{aligned} \tag{4.15}$$

Clearly in this case we need only show convergence for the control sequence $\{u_i\}$ since this will then imply $x_d - x_i(T) \to 0$ which is precisely our requirement as $x_i(.)$ satisfies the original nonlinear equation (1.1). From (3.1) we have

$$u'_{i+1}(t) = G(T,t)[x_d - \bar{x}(T) - \int_{t_0}^{T} U(T,s)N(x'_i(s),u'_i(s),s)ds]$$

$$\triangleq \hat{\phi}(x'_i,u'_i)$$

However given u'_i and hence u_i we can solve (4.15) for $\underline{x}_i$. In this manner we may write $x'_i = g(u'_i)$ say. Now define $\bar{\phi}(u) = \hat{\phi}(g(u),u)$. $\bar{\phi}$ can readily be shown to be contractive in a suitably chosen ball in $\mathcal{U}$, providing it is first shown that g satisfies a Lipschitz condition. Straightforward manipulation of (3.6) gives the desired Lipschitz property, thus completing our outlined proof of convergence.

Numerical Results

Numerical simulations were performed for the Van der Pol equation,

$$\ddot{z}(t) - 4(1 - z^2(t))\dot{z}(t) + z(t) = u(t) \quad z(0) = 0 \quad \dot{z}(0) = 0$$

Setting $x_1(t) = z(t)$, $x_2(t) = \dot{z}(t)$ and $x(t) = [x_1(t),x_2(t)]^T$ gives

$$\dot{x}(t) = \begin{bmatrix} \dot{x}_1(t) \\ \dot{x}_2(t) \end{bmatrix} = \begin{bmatrix} x_2(t) \\ 4(1-x_1^2(t))x_2(t) - x_1(t)+u(t) \end{bmatrix} = f(x(t),u(t))$$

with $x(0) = 0$.

The control $\bar{u}(.) \equiv 0$ was chosen for the nominal trajectory with the corresponding state function, $\bar{x}(.) \equiv 0$. Linearising about these trajectories yields

$$\dot{x}(t) = \begin{bmatrix} 0 & 1 \\ -1 & 4 \end{bmatrix} x(t) + \begin{bmatrix} 0 \\ 1 \end{bmatrix} u(t) + \begin{bmatrix} 0 \\ -4x_1{}^2(t)x_2(t) \end{bmatrix}$$

The admissible controls were restricted to the 2-dimensional control class, $\mathcal{U}$, of piecewise constant functions over the time interval $[0,1]$ with a single switch at $t = 0.5$. The problem considered was that of estimating the reachable states using controls generated via algorithms 1 and 2. For algorithm 3, it was found that the amount of computation needed to accurately calculate the desired control for several differing final states, x_d ruled out the feasibility of estimating the associated reachable set.

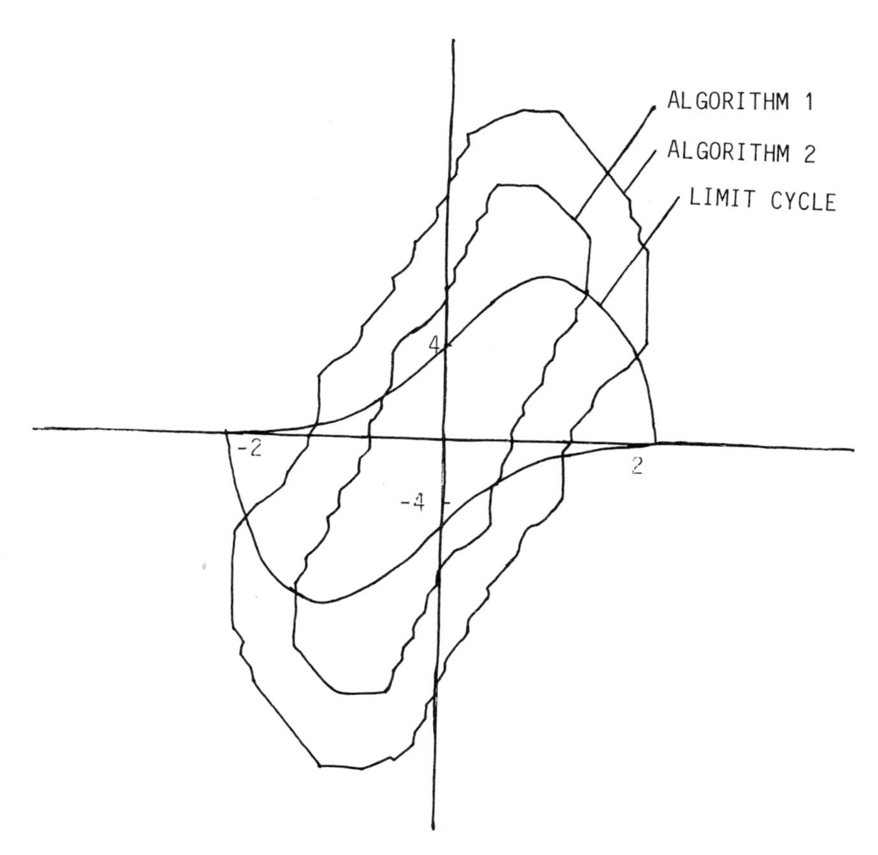

Fig. 1

In figure 1 a phase plot is presented for the estimated reachable sets, together with the limit cycle associated with the free Van der Pol equation. Equations (3.4) and (3.7) were integrated using an Euler method with time step h = 0.005. These plots were obtained by scanning along a grid in the phase plane with increments of 0.1 in the x_1 direction and 0.5 in the x_2 direction. A point was deemed unreachable if either

(i) An overflow condition was anticipated. (Instability)

OR (ii) The algorithm failed to converge to a desired accuracy within 30 iterations, where the criterion

$$||x_d - x_n(1)|| \leqslant 10^{-4}||x_d||$$

was used to judge successful convergence.

The fact that certain points within the limit cycle lie outside the reachable sets is due in part to the shortness of the time interval chosen.

5. REFERENCES

1. Furi, M., Nistri, P., Pera, M.P. and Zezza, P.L. (1980) 'Linear controllability by piecewise constant controls with preassigned switching times'. Preprint.

2. Furi, M., Nistri, P., Pera, M.P. and Zezza, P.L. (1980) 'Topological methods for the global controllability of nonlinear systems'. Preprint.

3. Kassara, K., and Eljai, A. (1983) 'Algorithmes pour la commande d'une classe de systèmes à paramètres répartis non linéaires'. Moroccan Journal of Control Computer Science and Signal Processing. Vol. 1, pp. 3-24.

STABILITY ANALYSIS OF STRONGLY NON-LINEAR INTERCONNECTED SYSTEMS

I. Zambettakis, F. Rotella and J.P. Richard
(Institut Industriel du Nord, France)

ABSTRACT

The choice of an original Lyapunov function induces an easy stability study for a large class of strongly non linear systems. Interconnected SI-SO non linear systems belong to this class.

0. INTRODUCTION

This paper concerns the stability analysis of dynamic systems by Lyapunov's direct method. Considering the free-motion state-space representation of a general continuous system, we propose a special decomposition of the evolution matrix coefficients. This configuration induces a particularly well-adapted Lyapunov candidate function that leads to sufficient stability conditions. Several important particular cases are indicated : the presented work thus appears as a generalization of classical criteria.

Two examples of application are then proposed. First, it is shown that the specific decomposition may be brought to bear on the modelling of the general class of single input - single output non-linear systems which are therefore capable of study using the presented theory. Taking this result into account, we extend it, in the last part, to an interconnection of such systems, this interconnection ensuring the existence of the decomposition.

1. SYSTEM DESCRIPTION

This study concerns non-linear continuous systems described by the following free-motion state-space representation :

$$\overset{\circ}{x}(t) = \frac{dx(t)}{dt} = A(x,t)\ x(t) \tag{1.1}$$

with
$$\begin{cases} t \in \mathcal{T} = [\,0\ ,\ +\infty\,[\\ x \in \mathbb{R}^q,\ x = (x_1,\ \ldots,\ x_q)^T \ : \text{state-vector} \\ A(x,t) = [\,\alpha_{ij}(x,t)\,],\ A(x,t) \in \mathbb{R}^{q\times q} \end{cases}$$

All $\alpha_{ij}(x,t)$ are bounded for any bounded value of x, and for any time t. The considered equilibrium is $x = 0$.

A stability study of equilibrium $x = 0$ by Lyapunov's direct method is proposed from the following special decomposition of the evolution matrix coefficients[1] :

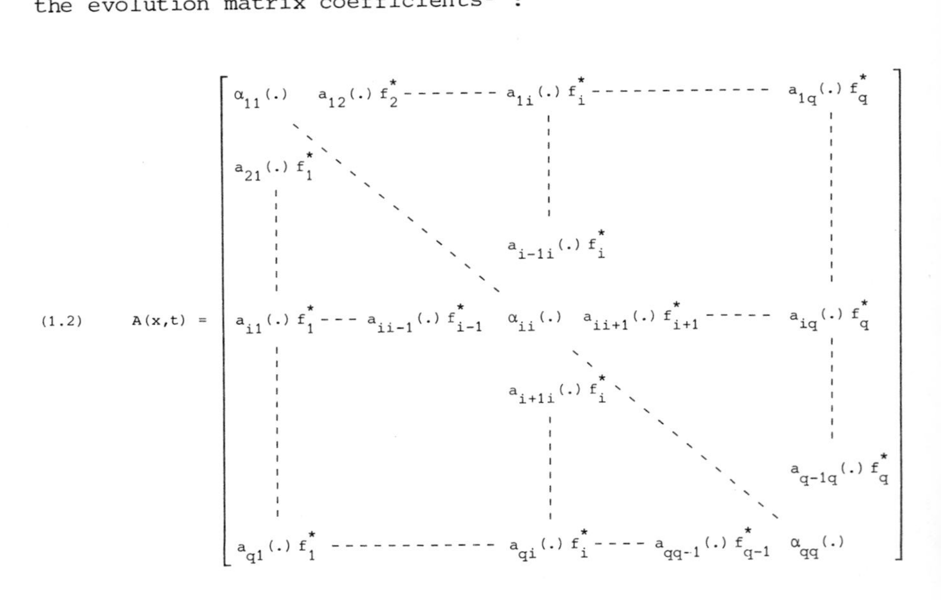

* $a_{ij}(.) = a_{ij}(x,t)$ $(i \neq j)$ are $q(q-1)$ scalar functions from $\mathbb{R}^q \times \mathcal{T}$ into $\mathbb{R}$.

* $f_i(x_i)$ $(i = 1, \ldots, q)$ are q scalar functions everywhere defined and integrable on $\mathbb{R}$, continuous and zero at the origin $x = 0$, and where instantaneous gains are defined by :

$$\text{(1.3)} \qquad \forall\, i \in \{1, \ldots, q\} \qquad f_i^* = \lim_{u_i \to x_i} \frac{f_i(u_i)}{u_i} \qquad \forall\, x \in \mathbb{R}^q$$

2. STABILITY CONDITIONS

2.1 Direct study

Previous works[2] have shown that the choice of an integral type Lyapunov function $W(x)$ defined in (2.1) :

$$W(x) = \sum_{i=1}^{q} \int_0^{x_i} f_i(u_i)\, du_i \tag{2.1}$$

leads to the following stability conditions :

Theorem 1 :

□ The equilibrium $x = 0$ of the system (1.1) (1.2) is stable if there exists a domain Ω of $\mathbb{R}^q$ including a neighbourhood of the origin $x = 0$ such that :

$$\forall\, x \in \Omega - \{0\},\ \forall\, t \in T,\ \forall\, i \in \{1, \ldots, q\}$$

a) $$\int_0^{x_i} f_i(u_i)\, du_i > 0 \qquad \forall\, x_i \neq 0$$

b) $$f_i^* \left[\alpha_{ij}(x,t) + f_i^* \sum_{\substack{j=1 \\ j \neq i}}^{q} \frac{|a_{ij}(x,t) + a_{ji}(x,t)|}{2} \right] \leqq 0$$

If conditions b) are strict, the equilibrium $x = 0$ is asymptotically stable. □

This theorem appears as a generalization of Rosenbrock criterion[3] on $A + A^T / 2$, which corresponds to the particular case where, for every i, $f_i^* = 1$.

Another important particular case must also be pointed out, the case of "pseudo-antisymmetrical" matrices[1] defined by (1.2) and (2.2) :

$$\forall\, (i,j) \in \{1, \ldots, q\}^2,\ i \neq j \qquad a_{ij}(.) = - a_{ji}(.) \tag{2.2}$$

When such a decomposition exists, we remark that these coefficients are not appearing anymore in the stability conditions, which are then particularly easy to check, with regard to a system that can present a lot of non-linearities.

2.2 Quadratic criterion

In this part, we suppose that the diagonal terms of $A(x,t)$ are decomposed into :

$$\forall\, i \in \{1, \ldots, q\},\ \alpha_{ii}(x,t) = a_{ii}(x,t)\ f_i^* + d_i(x,t) \tag{2.3}$$

where $a_{ii}(x,t)$ and $d_i(x,t)$ are 2q - scalar functions from $\mathbb{R}^q \times T$ to $\mathbb{R}$.

Under this hypothesis of decomposition, equation (1.1) becomes :

$$\mathring{x} = \tilde{A}(x,t)\ F(x) + \Delta x \tag{2.4}$$

where $\tilde{A}(x,t) = \left[\, a_{ij}(x,t) \,\right] \in \mathbb{R}^{q \times q}$

$$F(x) = \left[\, f_1(x_1), \ldots, f_q(x_q) \,\right]^T \in \mathbb{R}^q$$

$$\Delta = \operatorname{diag}\{d_i(x,t)\} \in \mathbb{R}^{q\times q} \qquad i \in \{1, \ldots, q\}$$

The choice of the Lyapunov function $W(x)$, defined in (2.1), then leads to the following stability criterion :

Theorem 2 :

□ The equilibrium $x = 0$ of system (2.4) is stable if there exists a domain Ω of $\mathbb{R}^q$ including a neighbourhood of the origin such that :

$$\forall\, x \in \Omega - \{0\},\ \forall\, t \in T$$

α) $\int_0^x F^T(u)\ du > 0$ where $du = (du_1, \ldots, du_q)^T \in \mathbb{R}^q$

β) $\tilde{A}(x,t) + \tilde{A}^T(x,t)$ is negative semi-definite[4]

γ) $\forall\, i \in \{1, \ldots, q\}$ $\quad d_i(x,t)\ f_i^* \leqq 0$

Moreover, if $\tilde{A} + \tilde{A}^T$ is negative definite, the equilibrium $x = 0$ is asymptotically stable. □

Remark : If matrix Δ is zero, the conditions γ) vanish.

This theorem can be interpreted as a generalization of another classical stability criterion proposed by Krasovskii[5]. This last criterion corresponds to the particular case of a non-decomposition : $\tilde{A}(x,t) = A(x,t)$.

So, the choice of a specific integral type Lyapunov function (2.1), linked with the decomposition (2.4), leads us to apply a classical stability criterion to a simplified model defined by the evolution matrix $\tilde{A}$.

Two applications of this theory are proposed in the following parts, where the aim is then to obtain a decomposition of type (2.4).

3. SINGLE INPUT - SINGLE OUTPUT (SI - SO) NON-LINEAR SYSTEMS

As a first example, let us now consider the large class of systems formed of the continuous, single input - single output Luri'e Postnikov systems.

These are composed of a linear part of transfer function $N(s)/D(s)$, controlled by a non-linear gain $f^*(\varepsilon)$, which depends on input - output difference (see Figure 1). The linear part is assumed non degenerate[6]

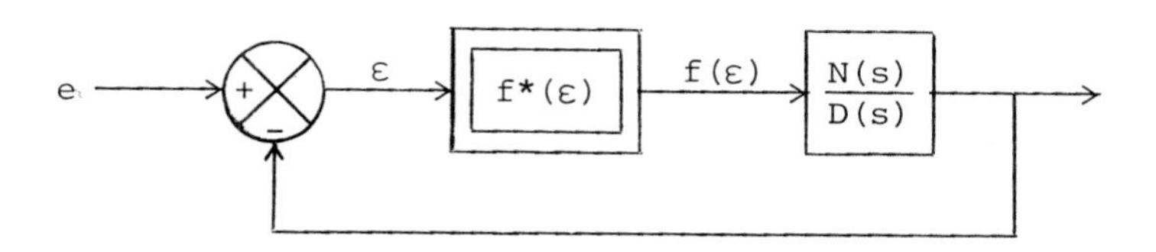

Fig. 1

with

$$\begin{cases} N(s) = \sum_{i=0}^{q-1} b_i\, s^i \qquad D(s) = \sum_{i=0}^{q-1} a_i\, s^i + s^q \\ \forall\, i \in \{0, \dots, q-1\},\ (a_i\, , b_i) \in \mathbb{R}^2\ ;\ b_{q-1}) \geqq 0 \\ \forall\, \varepsilon \in \mathbb{R} \qquad f^*(\varepsilon) = \lim_{\mu \to \varepsilon} \dfrac{f(\mu)}{\mu} \end{cases}$$

It has been shown[7] that this class of systems can be characterized in free-motion by a "thick arrow form matrix"[8]. This needs to choose a parameter (a) such that the polynomial $D(s) + a\,N(s)$ presents a real root s_o, and (q-1) others (real, complex, single or multiple). We denote :

$$D(s) + a\,N(s) = (s - s_o) \prod_{i=1}^{r} (s - \lambda_i)^{n_i} \prod_{i=1}^{t} \left[(s - \mu_i)^2 + \nu_i^2 \right]^{m_i} \tag{3.1}$$

where :

* $(a, s_o) \in \mathbb{R}^2$, $\sum_{i=1}^{r} n_i + 2 \sum_{i=1}^{t} m_i = q - 1$

* $\forall\ (i,k) \in \{1, \ldots, r\}^2$; $\lambda_i \in \mathbb{R}$, $n_i \in \mathbb{N}$, $\lambda_i \neq \lambda_k \Longleftrightarrow i \neq k$

* $\forall\ (i,k) \in \{1, \ldots, t\}^2$; $(\mu_i, \lambda_i) \in \mathbb{R}^2$, $m_i \in \mathbb{N}$,

$(\mu_i \neq \mu_k \text{ or } |\nu_i| \neq |\nu_k|) \Longleftrightarrow i \neq k$

Then, $A(x,t) = A(\varepsilon)$ is similar[4] to :

$$A(\varepsilon) = \begin{bmatrix} M_1 & & & & & & \alpha_1 . g(\varepsilon) \\ & M_2 & & & & & \alpha_2 . g(\varepsilon) \\ & & \ddots & & 0 & & \vdots \\ & & & M_r & & & \alpha_r . g(\varepsilon) \\ & & & & C_1 & & \beta_1 . g(\varepsilon) \\ & 0 & & & & \ddots & \vdots \\ & & & & & C_t & \beta_t . g(\varepsilon) \\ 1 & \cdots & & & & 1 & \gamma(\varepsilon) \end{bmatrix} \tag{3.2}$$

where :

* $\forall\ \varepsilon \in \mathbb{R}$, $g(\varepsilon) = f^*(\varepsilon) - a$, $\gamma(\varepsilon) = s_o - b_{q-1} . g(\varepsilon)$

* $\forall\ i \in \{1, \ldots, r\}$ $M_i \in \mathbb{R}^{n_i \times n_i}$, $\alpha_i \in \mathbb{R}^{n_i}$ constant vector

$$M_i = \begin{bmatrix} \lambda_i & 1 & & 0 \\ & \ddots & \ddots & \\ & & \ddots & 1 \\ & 0 & & \lambda_i \end{bmatrix}$$

* $\forall\, i \in \{1, \ldots, t\}$ $\quad C_i \in \mathbb{R}^{2m_i \times 2m_i}$, $\beta_i \in \mathbb{R}^{2m_i}$ constant vector

$$C_i = \begin{bmatrix} \mu_i & \nu_i & 1 & 0 & & & & 0 \\ -\nu_i & \mu_i & 0 & 1 & & & & \\ & & \ddots & & \ddots & & & \\ & & & \ddots & & \ddots & 1 & 0 \\ & & & & \ddots & & 0 & 1 \\ & 0 & & & & & \mu_i & \nu_i \\ & & & & & & -\nu_i & \mu_i \end{bmatrix}$$

The analytic expressions of the components of constant vectors α_i and β_i are given in [7]. Thus, in every case these SI - SO systems belong to the general class of processes studied in this paper (1.1), and theorem 1 or 2 can be applied.

Remark : Theorem 2 states that all the roots of $(D(s)+aN(s))\,/\,(s-s_o)$ are with a negative real part, and s_o is real negative.

Particular cases :

If the parameter (a) is chosen such that D(s)+aN(s) has one zero root ($s_o = 0$) the decomposition (2.4) for $A(\varepsilon)$ can be realised with matrix Δ equal to zero[9].

In the case where D(s)+aN(s) has (q-1) real negative separate roots, the method leads to define a reduced order model[10] for which linear conjecture holds. The stability of the equilibrium of this reduced model involves the same property for the equilibrium of the original system.

4. NON-LINEAR INTERCONNECTED SI - SO SYSTEMS

In the last part, we propose to extend the above results to the case of interconnected systems. Let us consider n SI - SO nonlinear subsystems (described in part III), interconnected in the following way (see Figure 2).

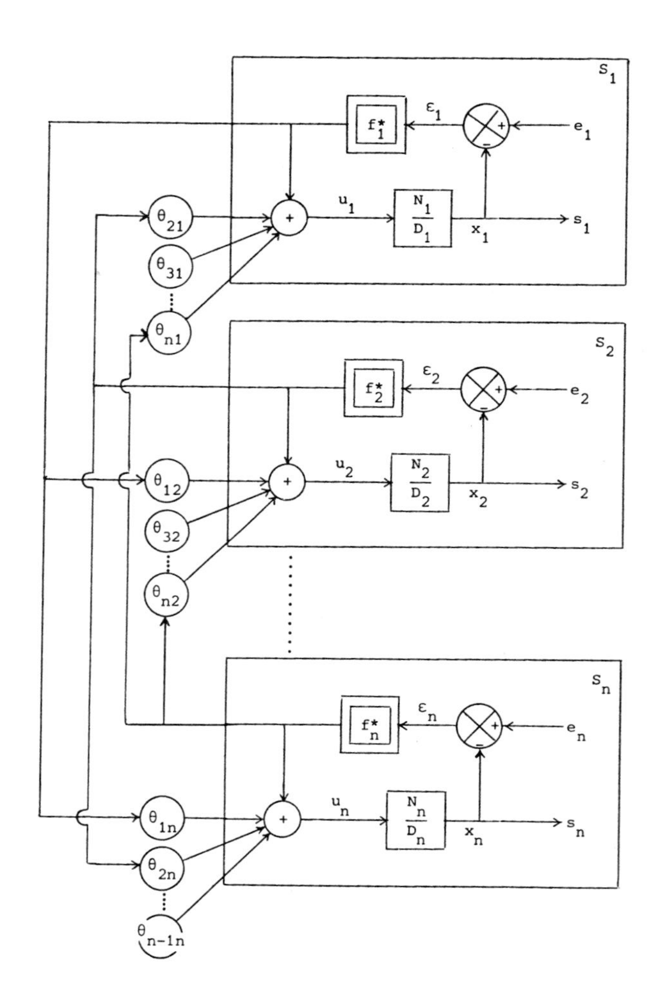

Fig. 2

Each SI - SO system S_i is characterized by :

* a transfer function $F_i(s) = N_i(s) / D_i(s)$, with :

$$N_i(s) = \sum_{j=0}^{q_i-1} b_{ij} s^j \; ; \; D_i(s) = s^{q_i} + \sum_{j=1}^{q_i-1} a_{ij} s^j$$

$(a_{ij}, b_{ij}) \in \mathbb{R}^{2q_i}$ q_i is the order of subsystem S_i,

* a non-linear feedback $f_i(\varepsilon_i)$ whose instantaneous gain is defined by :

$$\forall\ \varepsilon_i \in \mathbb{R} \qquad f_i^* = \lim_{\mu_i \to \varepsilon_i} \frac{f_i(\mu_i)}{\mu_i}$$

The n SI - SO systems S_i are linearly interconnected by means of n(n-1) constant coefficients θ_{ji} $(i \neq j)$. So, the system represented in Figure 2 can be described by the following set of equations :

$$\begin{cases} \forall\ i = 1, \ldots, n \\ \mathring{x}_i = A_i\, x_i + B_i\, u_i & A_i \in \mathbb{R}^{q_i \times q_i},\ B_i \in \mathbb{R}^{q_i} \\ s_i = C_i^T\, x_i & C_i \in \mathbb{R}^{q_i} \\ \varepsilon_i = e_i - s_i \\ u_i = f_i^*\, \varepsilon_i + \sum_{\substack{j=1 \\ j \neq i}}^{n} \theta_{ji}\, f_j^*\, \varepsilon_j & \theta_{ji} \in \mathbb{R} \\ x_i \in \mathbb{R}^{q_i}, \text{ state-vector of subsystem } S_i \end{cases} \tag{4.1}$$

The aim of the study is the stability analysis, so we consider the free-motion ($e_i = 0$; $i = 1, \ldots, n$). A matrix representation of the studied system is then obtained from (4.1) :

$$\begin{cases} \forall\ i = 1, \ldots, n \\ \mathring{x}_i = M_i\, x_i - \sum_{j=1}^{n} T_{ji}\, x_j \\ M_i = A_i - B_i\, f_i^*\, C_i^T \\ \underset{(j \neq i)}{T_{ji}} = B_i\, \theta_{ji}\, f_j^*\, C_j^T \end{cases} \tag{4.2}$$

where M_i is the evolution matrix of the deconnected subsystem S_i, and the n(n-1) matrices T_{ji} characterize the interconnections. We choose the state-vector x_i of subsystem S_i such that M_i is an arrow form matrix of general type (3.2) (ε_i being the last component of state-vector x_i). The whole considered state-vector is then :

$$X = \left[x_1^T, x_2^T, \ldots, x_n^T \right] \in \mathbb{R}^q \qquad q = \sum_{i=1}^{n} q_i$$

This choice of state-space basis implies that each interconnection matrix T_{ji} has its non-zero coefficients regrouped in its last column, and proportional to the same non-linear term f_j^*.

Moreover the diagonal blocks of each matrix M_i are constituted from (q_i-1) non zero roots of polynomial $D_i(s)$ ($a = 0$ in equation (3.1)).

Therefore all the non-linear coefficients of M_i are regrouped in its last column too. Except for the diagonal term, they are proportional to the same non-linearity f_i^*. Finally the whole system can be represented by the following matrix equation (4.3):

$$(4.3) \quad \left\{ \begin{array}{l} \mathring{X} = M\,X \\[1ex] X = \left[x_1^T, x_2^T, \ldots, x_n^T \right]^T \in \mathbb{R}^q \\[2ex] M = \left[\begin{array}{cc|cc|c|cc} \begin{smallmatrix} \diagdown & 0 \\ 0 & \diagdown \\ 1 - & 1 \end{smallmatrix} & \begin{array}{c} .f_1^* \\ \gamma_1^* \end{array} & 0 & .f_2^* & \cdots\cdots & 0 & .f_n^* \\ \hline 0 & .f_1^* & \begin{smallmatrix} \diagdown & 0 \\ 0 & \diagdown \\ 1 - & 1 \end{smallmatrix} & \begin{array}{c} .f_2^* \\ \gamma_2^* \end{array} & \cdots\cdots & 0 & .f_n^* \\ \hline \vdots & & \vdots & & & \vdots & \\ \hline 0 & .f_1^* & 0 & .f_2^* & \cdots\cdots & \begin{smallmatrix} \diagdown & 0 \\ 0 & \diagdown \\ 1 - & 1 \end{smallmatrix} & \begin{array}{c} .f_n^* \\ \gamma_n^* \end{array} \end{array} \right] \end{array} \right.$$

So, this system belongs to the general class of studied processes (1.2), and its evolution matrix can also be decomposed according to (2.4) with :

* $F^* = \text{diag}\,\{\underbrace{1, \ldots, 1, f_1^*}_{q_1}, \underbrace{1, \ldots, 1, f_2^*}_{q_2}, \underbrace{1, \ldots, 1, f_n^*}_{q_n}\}$

  $F^* \in \mathbb{R}^{q\times q}$

* $\tilde{A}$ and Δ are constant matrices.

Theorem 1 and 2 can then be applied to obtain stability conditions.

5. CONCLUSION

The purpose of this paper was to present a new method to choose a candidate Lyapunov function in the general case of strongly non-linear systems. The proposition of this specific function, linked with a special decomposition of the state-space representation, leads to stability conditions which can be easily checked.

The way of decomposing the matrix is very general and can be considered for many systems. The use of the arrow form matrix particularly allows such a configuration for interconnected single input - single output non-linear systems to be obtained.

6. REFERENCES

1. Zambettakis, I., (1983) "Contribution à l'étude de systèmes à non-linéarités multiples. Application aux systèmes électromécaniques".
   Thèse de Docteur-Ingénieur, Lille, n° 326.

2. Rotella, F., Richard, J.P. and Zambettakis, I., (1983) "Stability of non-linear continuous systems".
   Congrès IASTED Robotics and Automation, Lugano, Switzerland.

3. Rosenbrock, H.H., (1965) "A method of investigating stability".
   Conférence IFAC, Bâle.

4. Gantmacher, F.R., (1966) "Théorie des matrices".
   Collection Universitaire de Mathématiques, Dunod.

5. Krasovski, N.N., (1963) "Certain problem of the theory of stability of motion".
   Stanford, California.

6. Popov, V.M., (1973) "L'hyperstabilité des systèmes automatiques".
   Dunod.

7. Rotella, F., (1983) "Détermination de nouvelles représentations d'état adaptées à l'analyse et à la synthèse de systèmes continus non linéaires".
Thèse de Docteur-Ingénieur, Lille, n° 327.

8. Benrejeb, M., (1980) "Sur l'analyse et la synthèse de processus complexes hiérarchisés. Application aux systèmes singulièrement perturbés".
Thèse de Doctorat ès Sciences Physiques, Lille.

9. Rotella, F., Zambettakis, I. and Richard, J.P., (1982) "Application de la méthode du lieu des racines à la synthèse de systèmes non linéaires".
Congrès MECO 83, Tunis.

10. Zambettakis, I., Richard, J.P. and Rotella, F., (1984) "Simplification of models for stability analysis of large scale systems".
Chapter 6 of "Multivariable Control. New concepts and tools", D. Reidel Publishing Company. Editor : S.G. Tzafestas.

LINEAR STRUCTURES IN BILINEAR SYSTEMS

D. H. Owens
(Department of Mathematics, University of Sheffield)

P. E. Crouch
(Department of Engineering, University of Warwick)

ABSTRACT

Cone containment properties of bilinear systems are studied in terms of positivity structures of an induced linear system and related to the system inverse transfer function and pole-zero plot.

1. INTRODUCTION

In this paper we study the bilinear system in R^n

$$\underline{\dot{x}}(t) = A\underline{x}(t) + \sum_{i=1}^{m} u_i(t)\; N_i\underline{x}(t) \tag{1.1}$$

with bounded, measurable controls and identify conditions on $A, N_1, N_2, \ldots, N_m$ and the initial state $\underline{x}_o$ under which solutions $\underline{x}(t)$ lie in a proper cone in R^n with vertex $\{0\}$. The results have a strong connection with the classical result that the linear system $\underline{\dot{x}}(t) = A\underline{x}(t)$, $\underline{x}(o) = \underline{x}_o$, in R^n has positive solutions $\underline{x}(t) \geq 0$, $t \geq 0$, (in the sense that $x_i(t) \geq 0$, $t \geq 0$, $1 \leq i \leq n$) if $A = D+E$ where D is diagonal, and E has zero diagonal entries and positive off-diagonal entries. This is proved by writing

$$\underline{x}(t) = e^{Dt}\underline{x}_o + \int_o^t e^{D(t-s)} E\; \underline{x}(s)\; ds \quad , \; t \geq 0 \tag{1.2}$$

and noting that the solution of any integral equation

$$\underline{X}(t) = \underline{W}_o(t) + \int_o^t W_1(t,s)\; \underline{X}(s) ds \quad , \quad t \geq 0 \tag{1.3}$$

has a unique positive solution $\underline{X}(t) \geq 0$ if $\underline{W}_o(t) \geq 0$ and $W_1(t,s) \geq 0$ for all $t \geq 0$, $s \geq 0$.

Lie algebraic techniques have been extensively used (see, for example, [1]) in the study of bilinear controllability problems, but it appears that more straightforward techniques have been neglected. In particular, little use has been made of the natural relationship of the bilinear system (1.1) to the induced linear system

$$\dot{\underline{x}}(t) = A\underline{x}(t) + B\underline{v}(t) \quad , \quad \underline{y}(t) = C\underline{x}(t) \tag{1.4}$$

where

$$B = [B_1, B_2, \ldots, B_m] \quad , \quad C = \begin{pmatrix} C_1 \\ C_2 \\ \vdots \\ C_m \end{pmatrix} \tag{1.5}$$

are constructed by full-rank factorization of N_i of the form $N_i = B_i C_i$ with $B_i(n \times k_i)$, $C_i(k_i \times n)$, $1 \leq i \leq m$. Such a factorization has been shown by P d'Alessendro et al [2] and Crouch [3] to relate strongly to realization theory for bilinear systems.

No attempt is made to be exhaustive but the results obtained indicate the possibilities inherent in the approach and will hopefully prompt further work in the area of relating bilinear phenomena to properties of the underlying linear system and its transfer function.

2. SUFFICIENT CONDITIONS FOR CONE CONTAINMENT

Let the linear system (1.4) be induced by the bilinear system (1.1) by factorization of N_i in the manner indicated above, and introduce the matrices

$$\underline{H}_o(t,k) \triangleq Ce^{(A-kBC)t}\underline{x}_o \tag{2.1}$$

$$H(t,k) \triangleq Ce^{(A-kBC)t}B \tag{2.2}$$

parameterized by the real scalar k. Let

$$\Lambda(u,k) \triangleq \text{block diag}\,\{I_{k_j}(u_j+k)\}_{1 \leq j \leq m} \tag{2.3}$$

and write (1.1) in the form

$$\underline{\dot{x}}(t) = (A - k \sum_{i=1}^{m} B_i C_i)\underline{x}(t) + \sum_{i=1}^{m} (u_i(t)+k)B_i C_i \underline{x}(t)$$

$$\underline{x}(o) = \underline{x}_o \qquad (2.4)$$

to obtain the fundamental identity

$$C\underline{x}(t) = \underline{H}_o(t,k) + \int_o^t H(t-s,k)\Lambda(u(s),k)C\underline{x}(s)ds \qquad (2.5)$$

Theorem 1: All solutions $\underline{x}(t)$ corresponding to the initial state x_o of the bilinear system (1.1) lie in the cone $\{\underline{x} : \underline{x} \in R^n,\ C\underline{x} \geq 0\}$ if there exists a real number k* such that, for all $k \geq k^*$ and $t \geq 0$,

$$\underline{H}_o(t,k) \geq 0 \quad , \quad H(t,k) \geq 0 \qquad (2.6)$$

Proof: Fix a time T>0 and controls u(t) on $[0,T]$. There exists $k' \geq 0$ such that $\Lambda(u(t),k')$ has only positive entries on $[0,T]$. Now take $k \geq \max(k',k^*)$ and note that (2.5) is a special case of (1.3) indicating that $C\underline{x}(t) \geq 0$, $t \in [0,T]$. Since T and u(t) were arbitrary, we obtain the desired result.

Corollary 1.1: The bilinear system is not point controllable on $R^n-\{0\}$ if there exists a real k* such that $H(t,k) \geq 0$ for all $k \geq k^*$, for all $t \geq 0$.

Proof: Let $\underline{\alpha} \geq 0$ and $\underline{x}_o = B\underline{\alpha}$, then $\underline{H}_o(t,k) = H(t,k)\underline{\alpha} \geq 0$, $k \geq k^*$, $t \geq 0$ and the reachable set is contained in the proper cone $\{\underline{x} : C\underline{x} \geq 0\}$.

Theorem 1 has an asymptotic structure that, at first sight, is difficult to confirm. In the remainder of the paper therefore we concentrate on the single-input case (m = 1) with rank N = 1 and N = BC where B and C^T are elements of R^n. In such a situation, the induced linear system is single-input/single-output with transfer function

$$g(s) = C(sI - A)^{-1}B \qquad (2.7)$$

and hence

$$\mathcal{L}\{H(t,k)\} = \frac{g(s)}{1+kg(s)} \qquad (2.8)$$

This 'feedback interpretation' is used in the following section to investigate conditions for the validity of (2.6).

3. TRANSFER FUNCTION ANALYSIS AND CONE CONTAINMENT

Elementary classical root-locus arguments and inspection of (2.8) indicate that:

Proposition 1: A necessary condition for (2.6) to hold is that $CB \neq 0$.

'Proof': The positivity condition on $H(t,k)$ requires that the pole-zero excess of $g(s)$ is unity to prevent oscillation of the dominant modes. This requirement is equivalent to $CB \neq 0$.

We will therefore assume for the remainder of the paper that $CB \neq 0$ and, without loss of generality, that

$$CB = 1 \tag{3.1}$$

as the input can always be scaled to produce this condition.

The inverse system plays an important role in the analysis so we write

$$g^{-1}(s) = s + \alpha + h(s) \tag{3.2}$$

where $h(s)$ is strictly proper and uniquely defined with impulse response

$$h(t) \triangleq \mathcal{L}^{-1}\{h(s)\} \tag{3.3}$$

Substitution into (2.8) yields

$$\mathcal{L}\{H(t,k)\} = \frac{g_k(s)}{1 + g_k(s)h(s)} \tag{3.4}$$

(after a little manipulation) where

$$g_k(s) = \frac{1}{s + \alpha + k} \tag{3.5}$$

Before stating the main result of this section we need the following technical lemma:

Lemma 1: Let

$$\eta_\ell(t,k) \triangleq \frac{(-\mu(k))^{\ell+1}}{\ell!} t^\ell e^{\mu t} \tag{3.6}$$

where $\mu(k)$ is continuous and satisfies $\lim_{k\to+\infty} k^{-1}\mu(k) = -1$. Then

$$\lim_{k\to+\infty} \int_0^T \eta_\ell(t,k)f(t)dt = \begin{cases} 0 & , \quad T = 0 \\ f(o) & , \quad T > 0 \end{cases} \tag{3.7}$$

for any continuous function $f(t)$ on $[0,T]$.

(Remark: $\eta_\ell(t,k)$ is, in effect, a representation of the unit delta function).

<u>Proof</u>: It is clear that $\eta_\ell(t,k) \geq 0$, $t \geq 0$, k large and that it converges uniformly to zero on any interval $[\delta,+\infty[$ with $\delta>0$. Also

$$\begin{aligned} \int_0^\delta \eta_\ell(t,k)dt &= \frac{(-\mu)^{\ell+1}}{\ell!} \int_0^\delta t^\ell e^{\mu t}\, dt \\ &= \frac{(-\mu)^{\ell+1}}{\ell!} \frac{d^\ell}{d\mu^\ell} \{\int_0^\delta e^{\mu t}\, dt\} \\ &= \frac{(-\mu)^{\ell+1}}{\ell!} \frac{d^\ell}{d\mu^\ell} \{\mu^{-1}(e^{\mu\delta} - 1)\} \end{aligned} \tag{3.8}$$

Noting however that, for $0 \leq j \leq \ell$

$$\begin{aligned} &\left(\frac{d^j}{d\mu^j}(\mu^{-1})\right)\left(\frac{d^{\ell-j}}{d\mu^{\ell-j}}(e^{\mu\delta} - 1)\right) \\ &\qquad = \frac{(-1)^j\, j!}{\mu^{j+1}}\, \delta^{\ell-j}\, (e^{\mu\delta} - \delta_{j\ell}) \end{aligned} \tag{3.9}$$

and that $\lim_{k\to\infty} k^{-1}\mu(k) = -1$ by assumption, we obtain

$$\lim_{k\to+\infty} \int_0^\delta \eta_\ell(k,t)dt = 1 \tag{3.10}$$

directly from (3.8). Now let $T>0$ be arbitrary (the case of $T = 0$ is trivial) and denote the norm of f in $L_p(0,T)$ by $\|f\|_p$, $1 \leq p \leq \infty$. It is clear that

$$\begin{aligned} \lim_{k\to\infty} \left|\int_\delta^T \eta_\ell(k,t)f(t)dt\right| &\leq \|f\|_\infty (T-\delta) \lim_{k\to\infty} \sup_{\delta\leq t\leq T} |\eta_\ell(t,k)| \\ &= 0 \end{aligned} \tag{3.11}$$

and hence that we need only prove (3.7) for an arbitrary $\delta>0$. By the continuity of f choose δ such that $f(o)-\varepsilon \leq f(t) \leq f(o)+\varepsilon$ for $t \in [0,\delta]$ with ε arbitrarily small. It is trivially shown that, for all large k,

$$(f(o)-\varepsilon) \int_o^\delta \eta_\ell(t,k)dt \leq \int_o^\delta \eta_\ell(t,k)f(t) \leq (f(o)+\varepsilon) \int_o^\delta \eta_\ell(t,k)dt \tag{3.12}$$

or, using (3.10) and (3.11),

$$f(o)-\varepsilon \leq \lim_{k\to\infty} \int_o^T \eta_\ell(t,k)f(t)dt \leq f(o)+\varepsilon \tag{3.13}$$

which proves the lemma as ε is arbitrary.

Theorem 2: A necessary and sufficient condition for the existence of k* such that $H(t,k) \geq 0$ for $k > k^*$ and $t \geq 0$ is that

$$h(t) \leq 0 \quad , \quad t \geq 0 \tag{3.14}$$

Moreover, under this condition, if $\lim_{t\to+\infty} e^{-\beta t}h(t) = 0$ we can choose any
$k^* > ||e^{-\beta t}h||_1 - (\alpha+\beta)$.

Proof: The proof makes use of the feedback structure (3.4) by defining an operator L_k by the relation

$$L_k f \triangleq g_k * h * f \tag{3.15}$$

where * denotes convolution, and g_k is interpreted as $\mathcal{L}^{-1}\{g_k(s)\} = e^{-(k+\alpha)t}$. The domain of definition presents a minor problem as we will be using boundedness conditions to prove the result. If we concentrate on the case of $k+\alpha > 0$ and h(t) stable, L_k is a mapping of $C(0,\infty)$ into itself. The first condition presents no problem as we are interested in the case of $k\to+\infty$ but the stability of h(t) requires the assumption that linear system (1.4) is minimum phase. We will continue with this assumption as $H(t,k) \geq 0$, $t \geq 0$, $k \geq k^*$ iff $e^{-\beta t}H(t,k) \geq 0$, $t \geq 0$, $k \geq k^*$ and standard shift theorems for Laplace transforms indicate that this transformation is equivalent to replacing h(t) by $e^{-\beta t}h(t)$. It is clear that a suitable choice of β will ensure stability of $e^{-\beta t}h(t)$!

To estimate the norm $||L_k||_\infty$ of L_k in $C(0,\infty)$ note that

$$\|L_k f\|_\infty = \sup_{t \geq o} |(h*(g_k*f))(t)|$$
$$\leq \|h\|_1 \sup_{t \geq o} |\int_o^t e^{-(k+\alpha)t'} f(t-t')dt'|$$
$$\leq \|h\|_1 (\int_o^\infty e^{-(k+\alpha)t} dt) \|f\|_\infty$$
$$= \frac{\|h\|_1}{k+\alpha} \|f\|_\infty \qquad (3.16)$$

whence

$$\|L_k\|_\infty \leq \|h\|_1 (k+\alpha)^{-1} \qquad (3.17)$$

and L_k is a contraction for $k > \|h_1\| - \alpha$. Under these conditions we can write $H(t,k)$ as a uniformly convergent power series in k of the form

$$H(t,k) = g_k(t) - (L_k g_k)(t) + (L_k^2 g_k)(t) - \ldots$$
$$= g_k(t) - (L_k g_k)(t) + (L_k^2 g_k)(t) + O(k^{-3}) \qquad (3.18)$$

as $\|L_k^j g_k\|_\infty \leq \|L_k\|_\infty^j \|g_k\|_\infty = \|L_k\|_\infty^j = O(k^{-j})$. A necessary condition for $H(t,k) \geq 0$ for $t \geq 0$ and large k is that, for each fixed $t > 0$,

$$0 \leq \lim_{k\to\infty} (\alpha+k)^2 H(t,k)$$
$$= \lim_{k\to\infty} \{(\alpha+k)^2 g_k(t) - (\alpha+k)^2 (L_k g_k)(t) + (\alpha+k)^2 (L_k^2 g_k)(t)\}$$
$$= -h(t) \qquad (3.19)$$

as

$$\lim_{k\to\infty} (\alpha+k)^2 g_k(t) = \lim_{k\to\infty} (\alpha+k)^2 e^{-(k+\alpha)t} = 0 , \qquad (3.20)$$

$$\lim_{k\to\infty} (\alpha+k)^2 (L_k g_k)(t) = \lim_{k\to\infty} (\alpha+k)^2 ((g_k * g_k)*h)(t)$$
$$= \lim_{k\to\infty} (\alpha+k)^2 \int_o^t t' e^{-(k+\alpha)t'} h(t-t')dt'$$

$$= \lim_{k\to\infty} \int_0^t \eta_1(t',k)h(t-t')dt'$$

$$= h(t) \qquad (3.21)$$

by lemma 1 with $f(t') = h(t-t')$ and $\mu(k) = -(k+\alpha)$, and

$$\lim_{k\to\infty} (\alpha+k)^2(L_k^2 g_k)(t) = \lim_{k\to\infty} (\alpha+k)^2((g_k * g_k * g_k)*(h*h))(t)$$

$$= \lim_{k\to\infty} (\alpha+k)^2 \int_0^t \frac{(t')^2}{2!} e^{-(k+\alpha)t'} (h*h)(t-t')dt'$$

$$= \lim_{k\to\infty} \frac{(-1)}{(k+\alpha)} \int_0^t \eta_2(t',k)(h*h)(t-t')dt'$$

$$= \lim_{k\to\infty} \frac{(-1)}{(k+\alpha)} (h*h)(t) = 0 \qquad (3.22)$$

by lemma 1 again with $f(t') = (h*h)(t-t')$ and $\mu(k) = -(k+\alpha)$. (3.19) indicates that we need $h(t) \le 0$ for $t > 0$ and hence for $t \ge 0$ by continuity.

Conversely, if $k > k^* > ||h||_1 - \alpha$ and $h(t) \le 0$, $t \ge 0$, then it is trivially verified that $(-1)^\ell (L_k^\ell g_k)(t) \ge 0$, $t \ge 0$, $\ell \ge 0$ and hence, using the series expansion (3.18) $H(t,k) \ge 0$, $t \ge 0$. This proves sufficiency of the conditions.

(Note: In the case of $CB \neq 0$, (3.14) can be replaced by $(CB)h(t) \le 0$, $t \le 0$, where h is obtained from the decomposition $g^{-1}(s) = s(CB)^{-1} + \alpha + h(s))$.

Theorem 2 provides a simple simulation-based method for checking the conditions of theorem 1 by evaluation of a minimal realization $S(A_o, B_o, C_o)$ of $h(s)$ (generically in R^{n-1}) and noting that $h(t)$ can be obtained as the solution of

$$\dot{\underline{z}}(t) = A_o \underline{z}(t) \quad , \quad \underline{z}(o) = B_o \quad , \quad h(t) = C_o \underline{z}(t) \qquad (3.23)$$

If $h(t) \le 0$, $t \ge 0$, then we conclude from theorems 1 and 2 that

---

<u>Theorem 3</u>: All solutions of the bilinear system

$$\dot{\underline{x}}(t) = (A + u(t)BC)\underline{x}(t) \qquad (3.24)$$

from the initial conditions $\underline{x}(o) = B\alpha$, $\alpha \geq 0$, lie in the cone $\{\underline{x} : C\underline{x} \geq 0\}$ if $h(t) \leq 0$, $t \geq 0$.

Corollary 3.1: If $h(t) \leq 0$, $t \geq 0$, the bilinear system (3.24) is not point controllable in $R^n - \{0\}$.

An alternative approach is to use pole-zero analysis of $h(s)$ to deduce that $h(t) \leq 0$ for $t \geq 0$. The following results provide simple sufficient conditions for this to be true:

Proposition 2: If $h(s)$ has the pole-zero structure

$$h(s) = \frac{k_o(s-\mu_1) \ldots (s-\mu_{n-2})}{(s-z_1)(s-z_2)\ldots(s-z_{n-1})} \tag{3.25}$$

with $k_o \leq 0$ and real $\{\mu_j\}$ and $\{z_j\}$ satisfying the interlacing condition

$$z_1 < \mu_1 < z_2 < \mu_2 < \ldots < \mu_{n-2} < z_{n-1} \tag{3.26}$$

then $h(t) \leq 0$, $t \geq 0$.

Proof: An elementary application of partial fraction methodology to prove that

$$h(t) = \sum_{j=1}^{n-1} \hat{R}_j e^{z_j t} \quad , \quad \hat{R}_j < 0 \ , \ 1 \leq j \leq n-1 \tag{3.27}$$

Proposition 3: If $g(s)$ has the structure

$$g(s) = \frac{(s-z_1)(s-z_2)\ldots(s-z_{n-1})}{(s-p_1)(s-p_2)\ldots(s-p_n)} \tag{3.28}$$

where the poles and zeros of g satisfy the interlacing condition

$$p_1 < z_1 < p_2 < z_2 < \ldots < z_{n-1} < p_n \tag{3.29}$$

then $h(t) \leq 0$, $t \geq 0$.

Proof: If $g(s)$ is interlaced then elementary root-locus arguments indicate that $\mathcal{L}(H(t,k))$ is interlaced for all k. Pole-residue analysis then yields the observation that $H(t,k) \geq 0$, $t \geq 0$, for all k which is equivalent to $h(t) \leq 0$ by theorem 2.

4. CONES OF INITIAL CONDITIONS

If the conditions of theorem 2 hold, it is natural to ask the question - for what class of initial conditions $\underline{x}_o$ is $H_o(t,k) \geq 0$, $t \geq 0$, $k \geq k^*$. An answer is provided by the following results. Firstly, write

$$G_o(s) = C(sI-A)^{-1}\underline{x}_o \tag{4.1}$$

and note that

$$\mathcal{L}\{H_o(k,t)\} \equiv \frac{G_o(s)}{1+kG(s)} \tag{4.2}$$

As $CB \neq 0$ we can write, with h' strictly proper,

$$G^{-1}(s)G_o(s) = h_o + h'(s) \tag{4.3}$$

to yield the identity

$$H_o(k,t) = h_o H(k,t) + (H*h')(t) \tag{4.4}$$

Lemma 2: $H_o(t,k) \geq 0$, $t \geq 0$, $k \geq k^*$ if (a) $H(k,t) \geq 0$, $t \geq 0$, $k \geq k^*$ (b) $h_o \geq 0$ and (c) $h'(t) \geq 0$, $t \geq 0$

We will also need the following result:

Lemma 3: There exists a basis in R^n denoted $\{\underline{t}_1, \underline{t}_2, \ldots, \underline{t}_n\}$ such that

$$C(sI_n-A)^{-1}\underline{t}_j = s^{j-1}/|sI_n-A| \tag{4.5}$$

Proof: Transform S(A,B,C) to observable canonical form and let $\{\underline{t}_1, \underline{t}_2, \ldots, \underline{t}_n\}$ be the corresponding basis.

Now write

$$\underline{x}_o = \sum_{j=1}^{n} \varepsilon_j \underline{t}_j \tag{4.6}$$

and note that $G_o(s) = \sum_{j=1}^{n} \varepsilon_j s^{j-1}/|sI_n-A|$ to obtain

$$G^{-1}(s)G_o(s) = \frac{1}{z(s)} \sum_{j=1}^{n} \varepsilon_j s^{j-1} = \varepsilon_n + \sum_{j=1}^{n-1} (\varepsilon_j - \varepsilon_n b_j) \frac{s^{j-1}}{z(s)} \tag{4.7}$$

where $z(s) = \sum_{j=1}^{n} b_j s^{j-1}$ is the zero polynomial of S(A,B,C) and $b_n = 1$.

Theorem 3: Suppose that the conditions of theorem 2 hold. Then all trajectories originating from the initial condition (4.6) lie in $\{\underline{x} : C\underline{x} \geq 0\}$ for all $t \geq 0$ if $\varepsilon_n \geq 0$ and

$$\sum_{j=1}^{n-1} (\varepsilon_j - b_j \varepsilon_n) \frac{d^{j-1}\psi(t)}{dt^{j-1}} \geq 0 \quad , \qquad t \geq 0 \tag{4.8}$$

where $\psi(t) = \mathcal{L}^{-1}\{1/z(s)\}$.

Proof: Use (4.7) in lemma 2 with $h_o = \varepsilon_n$ and $h'(s) = \sum_{j=1}^{n-1} \ldots$
$\ldots (\varepsilon_j - b_j)s^{j-1}/z(s)$ noting that $\mathcal{L}\{\overset{(j-1)}{\psi}(t)\} = s^{j-1}/z(s)$.

---

5. CONCLUSIONS

The paper has demonstrated that a number of systems theoretical problems for a class of homogeneous bilinear systems in R^n can be fruitfully resolved by investigating simple properties of an induced linear system and checked in practice by linear simulation techniques in R^{n-1} or pole-zero analysis of a related transfer function. The results use a judicious mix of asymptotic analysis and functional analysis that are reminiscent of root-locus theory and suggest that more complex problems may possibly be resolved by similar techniques. In particular, it may be possible to extend the cone containment results to cones other than $\{\underline{x} : C\underline{x} \geq 0\}$ and to extend the transfer function analysis to the multi-input case. Work continues in this area.

6. REFERENCES

1. Jurdjevic, V. and Sussmann, H.J. (1972). 'Control systems in Lie Groups', Jrnl. Diff. Eqns., 12, 313-329.

2. d'Alessendro, P., Isidori, A. and Ruberti, A. (1974). 'Realization and structure theory of bilinear systems', SIAM Jrnl. Control, Vol.12, 3, 517-535.

3. Crouch, P.E. (1980). 'On the relation between internal and external symmetries of bilinear systems', Control Theory Centre Report No.80, University of Warwick.

CHAPTER 5

UNCERTAINTY AND ROBUSTNESS

PROPERTIES OF ADAPTIVE DISCONTINUOUS MODEL-FOLLOWING CONTROL SYSTEMS

A.S.I. Zinober

(Department of Applied and Computational Mathematics, University of Sheffield)

ABTRACT

Model-following control with a discontinuous variable structure controller is described and the properties arising in the case of output model-following are studied using the concept of linear equivalent control. The robust controller does not require any global convergence conditions and is well suited to the control of a class of uncertain time-varying systems.

1. INTRODUCTION

The technique of Model Reference Adaptive Control (MRAC) using Lyapunov and hyperstability theory has been widely discussed in the literature[1]. The actual plant is required to follow the dynamic behaviour of a specified model plant. MRAC relies upon the global convergence of parameter identification schemes or certain related auxiliary signals. The global convergence requirement raises serious problems in the case of uncertain time-varying plants and much research has been carried out in this field[1,2].

Here we shall study model-following control using the variable structure design approach[4,5] which will be introduced in Section 3. The robust self-adaptive scheme does not require any convergence criteria to hold. Properties of the output model-following problem will be studied in Section 4 and related to the linear state feedback approach of Shaked[6]. Implementation and links with recent work on the control of uncertain systems will then be outlined.

2. MODEL-FOLLOWING CONTROL SYSTEMS

We shall consider the system with the plant described by

$$\dot{\underline{x}} = A\,\underline{x} + B\,\underline{u} \tag{2.1}$$
$$\underline{y} = C\,\underline{x}$$

where $\underline{x} \varepsilon R^n$, $\underline{u} \varepsilon R^m$, $y \varepsilon R^\ell$ and the pairs (A,B) and (C,A) are respectively controllable and observable. The plant output $\underline{y}$ is required to follow $\underline{z}$ the output of a constant model plant given by

$$\dot{\underline{w}} = M\,\underline{w} + N\,\underline{r} \tag{2.2}$$
$$\underline{z} = L\,\underline{w}$$

with $\underline{w} \varepsilon R^p$, $\underline{r} \varepsilon R^r$ and $\underline{z} \varepsilon R^\ell$. The model is assumed to be stable and completely controllable and observable. The matrices B, C, N and L have full rank and $p \leqslant n$. We shall first study the case when all the matrices are constant and later consider the case when the plant matrices A and B are time-varying.

The matching conditions required for perfect model-following to be possible have been studied for a number of different cases[6-8]. It will be assumed that the necessary structural matching conditions are satisfied.

3. VARIABLE STRUCTURE SYSTEMS

We shall first outline the properties when all the plant states are required to follow the model states, i.e. $n = p = \ell$ and $C = L = I$. Many authors[9-13] have investigated model-following control systems using variable structure (VS) controllers which have desirable properties such as parameter invariance and disturbance rejection. Define the error state to be $\underline{e} = \underline{w} - \underline{x}$. Variable structure systems are characterised by control discontinuous on m switching hyperplanes, v_i, in the state space. Suppose that the control components have the form

$$u_i = f_i(\underline{x},\underline{e},\underline{r}) + g_i(\underline{x},\underline{e},\underline{r})\,\mathrm{sgn}(v_i) \qquad i = 1,2,\ldots m$$
$$\underline{v} = (v_1\; v_2\; \ldots\; v_m)^T = G\,\underline{e} \tag{3.1}$$

where G is a constant $m \times n$ matrix. The model-following error system

$$\dot{\underline{e}} = M\,\underline{e} + (M - A)\,\underline{x} + N\,\underline{r} - B\,\underline{u} \tag{3.2}$$

is obtained from (2.1) and (2.2). The design philosophy is similar to that of the multivariable VS system[1,2]. The switching of the control on the switching hyperplanes yields the desirable properties of total (or selective) invariance to system parameter variations and disturbances[14], as well as precise closed-loop eigenvalue placement.

Sliding or chatter motion occurs, if at a point on a switching surface, the directions of motion along the state trajectories on either side of the surface are not directed away from the surface. Filippov[15] has defined rigorously the solution of differential equations with discontinuous right hand sides as we have here in the neighbourhood of the switching surface. During the sliding mode on the intersection of the switching hyperplanes the state remains on the switching surface S and satisfies the equations

$$\underline{v} = \underline{0} \quad \text{and} \quad \dot{\underline{v}} = \underline{0}. \tag{3.3}$$

The equations governing the system dynamics in the sliding mode may be obtained by substituting an equivalent control $\underline{U}$ for the original control $\underline{u}$. For GB non-singular it can easily be shown[12] that

$$\underline{U} = -(GB)^{-1} G\{M\,\underline{e} + (M-A)\underline{x} + N\underline{r}\} \tag{3.4}$$

and

$$\dot{\underline{e}} = [I - B(GB)^{-1}G]\; M\,\underline{e} = F\,\underline{e} \tag{3.5}$$

providing the perfect model-following properties hold.

Thus the system behaviour on the switching surface may be obtained by applying the discontinuous control or the linear equivalent control $\underline{U}$. The former yields the desired dynamics in the presence of uncertainty in the plant parameters and disturbances, and is chosen so that the state 'reaches' the switching surface S from all initial conditions and subsequently slides on S to the error state origin. The linear control achieves the desired result only in the case of known parameter values from initial conditions satisfying $\underline{v} = \underline{0}$; a particular case being zero error at t = 0.

The design of discontinuous control laws, which ensure the reaching of S and a subsequent stable sliding mode, has been studied by numerous authors including Utkin[1], Young[10] and Ryan[16] and will not be discussed at this stage.

The choice of the matrix G (3.1) yields the desired error transient during the sliding mode and El-Ghezawi et al[17] have developed a geometric design approach. The closed-loop system eigenvalues are the transmission zeros of the related multivariable linear control system with the triple (M,N,G)[18] (see also 9, 10, 18 and 20). Introducing the similarity transformation

$$\bar{\underline{x}} = T\,\underline{x} \tag{3.6}$$

with

$$T = \begin{bmatrix} C \\ P \end{bmatrix} \quad \text{and} \quad T^{-1} = [V_1 \;\; W_1] \tag{3.7}$$

where P is chosen so that I is nonsingular with $CV_1 = I_m$, $PW_1 = I_{n-m}$, $PV_1 = 0$ and $CW_1 = 0$. The matrices V_1 and W_1 can be taken to be generalised inverses of C and P such that $PC^+ = 0$ and $CP^+ = 0$. Then

$$\dot{\bar{\underline{x}}} = \begin{bmatrix} 0 & 0 \\ PFC^+ & PFP^+ \end{bmatrix} \bar{\underline{x}} \tag{3.8}$$

and

$$\underline{y} = [I_m \ \ 0]\ \bar{\underline{x}} \tag{3.9}$$

which is the unobservability decomposition. The (n-m) eigenvalues of PFP^+ are the system zeros and the remaining m eigenvalues of the closed-loop system are zero-valued. If parameter variations and plant disturbances lie in the range space of B, total invariance properties hold[10,17].

4. OUTPUT MODEL-FOLLOWING

4.1 The System Equations

Here the output of the plant is required to follow the output of a model plant so that the error tends to zero as $t \to \infty$. The output is an ℓ-vector with $\ell < n$ and $p \leq n$. Using the error formulation with the error between the model and plant outputs given by $\underline{e} = \underline{z} - \underline{y}$, the error equation is

$$\begin{aligned} \dot{\underline{e}} &= \dot{\underline{z}} - \dot{\underline{y}} \\ &= L(M\underline{w} + N\underline{r}) - C(A\underline{x} + B\underline{u}) + E(\underline{z} - \underline{y}) - E(\underline{z} - \underline{y}) \\ &= E\,\underline{e} + \{(LM - EL)\underline{w} + LN\underline{r} + (EC - AC)\underline{x} - CB\underline{u}\} \end{aligned} \tag{4.1}$$

where E is an $\ell \times \ell$ diagonal matrix which has eigenvalues with negative real parts. The terms with matrix E will be shown below to give additional eigenvalue assignment.

Perfect model-following can be achieved if the terms in the brackets {} above can be made equal to zero. This requires the structural property

$$\begin{aligned} \text{rank } (CB) &= \text{rank } (LM-EL \quad CB) \\ &= \text{rank } (LN \quad CB) \\ &= \text{rank } (EC-CA \quad CB) \end{aligned} \tag{4.2}$$

Shaked[6] has shown that for these conditions to hold CB must be full rank.

4.2 Equivalent Control

When CB is square and full rank, the conditions are satisfied for all A, L, M, N and E, so a suitable control $\underline{u}$ exists and has the form

$$\underline{u} = (CB)^{-1}\,[(EC - CA)\underline{x} + (LM - EL)\underline{w} + LN\underline{r}] \tag{4.3}$$

Then (4.1) becomes

$$\dot{\underline{e}} = E\,\underline{e} \tag{4.4}$$

which is asymptotically stable and ensures that $\underline{e} \to \underline{0}$ as $t \to \infty$. The control $\underline{u}$ (4.3) can be expressed in the form

$$\underline{u} = (CB)^{-1} CA\underline{x} + (CB)^{-1} EC\underline{x} + (CB)^{-1} \{(LM-EL)\underline{w} + LN\underline{r}\}$$
$$= \underline{u}_1 + \underline{u}_2 + \underline{u}_3 \qquad (4.5)$$

The term $\underline{u}_1 = (CB)^{-1} CA\underline{x}$ is the equivalent control arising in a variable structure system with $\underline{v}=C\underline{x}$. We require the error $\underline{e}$ to be zero, and therefore $\dot{\underline{e}}=\underline{0}$ as well. Equating $\underline{v}$ with $\underline{e}$ shows the analogue with the sliding mode when both $\underline{v}$ and its derivative $\dot{\underline{v}}$ are identically zero (3.3).

4.3 Stability, Eigenvalue Assignment and Order Reduction

With the control (4.5) the plant states satisfy

$$\dot{\underline{x}} = [\{I - B(CB)^{-1}C\} A \underline{x} + B(CB)^{-1}EC\underline{x}+B(CB)^{-1}[(LM-EL)\underline{w} + LN\underline{r}] \qquad (4.6)$$

which is stable if the matrix

$$H = F + G_1 \qquad (4.7)$$

is stable where

$$F = \{I-B(CB)^{-1}C\}A \qquad (4.8)$$

and

$$G_1 = B(CB)^{-1}EC . \qquad (4.9)$$

The stability can readily be assessed by using the transformation (3.6). Then

$$T H T^{-1} = \begin{bmatrix} CHC^+ & CHP^+ \\ PHC^+ & PHP^+ \end{bmatrix} \qquad (4.10)$$

Noting that F is the matrix arising from the equivalent control $\underline{u}_1$ and that $CF=0$, $CC^+ = I_m$ and $CP^+ = 0$, (4.10) simplies to

$$z = T H T^{-1} = \begin{bmatrix} E & 0 \\ PHC^+ & PFP^+ \end{bmatrix} \qquad (4.11)$$

The eigenvalues of Z are determined by E and $Q=PFP^+$. The eigenvalues of Q are the system zeros which cannot be altered by state feedback. The equivalent control $\underline{u}_1$ automatically assigns zero values to ℓ of the eigenvalues of the matrix F. The feedback term $\underline{u}_2$ in (4.5) reassigns the values of the these eigenvalues to the sp(E). So stability is ensured by having chosen E to have eigenvalues with negative real parts. The equivalent control automatically cancels all of the plant zeros so we require that the plant has only minimum phase zeros to ensure a stable closed-loop system. The plant is driven into the unobservable sub-space since

$$\underline{y} = C\underline{x} = CT^{-1}\underline{x} = [I_\ell \ 0]\underline{\bar{x}} \quad (4.12)$$

and the effective plant order has been reduced by ℓ. The term $\underline{u}_3$ provides the set points of the system as specified by $\underline{w}$ and $\underline{r}$ in the model plant.

Full state feedback is required so an observer may be employed to reconstruct unmeasured states. A dynamic observer for the model reduction case has been described by Shaked and Karcanias[21]. Full state feedback control for a constant system with zero initial conditions has been proposed by Shaked[6]. It has the form

$$\underline{u}_s = KW^+ \underline{x} + K_r \underline{r} \quad (4.13)$$

where K is a set of input zero directions and W^+ is a generalised inverse of the associated state zero directions. The system order is effectively reduced by cancelling plant zeros. Variation of the system parameter values may lead to the reappearance of cancelled dipoles owing to imperfect pole-zero cancellation and the design should attempt to keep the residues small[21]. The equivalent control $\underline{u}_1$ and the first term of $\underline{u}_s$ have the same structure since[18]

$$K = -(CB)^{-1} \ CAW$$

and

$$KW+ = -(CB)^{-1} \ CA$$

but, since W^+ is not uniquely defined, the control terms are not necessarily identical.

The equivalent control $\underline{u}_1$ can be implemented as a discontinuous control and one then achieves the invariant properties of variable structure systems. The two remaining terms in (4.5) can be obtained from the known matrices M, N, L and E.

4.4 Decoupling

The control (4.5) decouples the system by yielding a diagonal and non-singular $\ell \times \ell$ transfer function matrix, G(s), with the result that each plant output variable is affected only by the associated model output variable. The structure of equivalent control can be shown using decoupling theory to yield $G(s)=(sI)^{-1}$. Additional feedback arising from the matrix E will be shown below to give decoupling with eigenvalue assignment.

Treating $B_e = B(CB)^{-1}$ as the effective plant input matrix, equations (4.6) - (4.12) and the transformation (3.6) give

$$G(s) = CT^{-1} [sI - \underline{Z}]^{-1} TB_e \quad (4.14)$$

Further simplification[22] yields

$$G(s) = [sI - E]^{-1}$$

so the inclusion of the term $\underline{u}_2$ in (4.5) introduces greater design freedom in the specification of the decoupled system transfer matrix G(s). The effective plant 'input' is from (4.5) $\{(LM-EL)\underline{w} + LN\underline{r}\}$ and it can be readily shown[22] that $\underline{y}(s) = \underline{z}(s)$ as required in the model-following problem. For $\ell \neq m$ see reference 22.

5. IMPLEMENTATION

Consider the time-varying uncertain plant satisfying

$$\dot{\underline{x}}(t) = A(\underline{x}(t),t)\ \underline{x}(t) + B(\underline{x}(t),t),t)\ \underline{u}(t)$$

where $A = A_c + A_v$ and $B = B_c + B_v$. The subscripts c and v refer to nominal (constant) and additive (varying) matrices respectively.

Let us first state some of the features of the linear feedback implementation of the controller (4.5). As has already been noted in Section 3 the robust invariance properties of VS controllers do not apply. With suitable choice of the matrix E stable control may be achieved provided (i) A_v does not introduce non-minimum phase zeros and (ii) the eigenvalues of A_v have sufficiently large negative real parts. (The results stated apply strictly only to the case of time-invariant matrices.) El-Ghezawi[22] has shown that, if the columns of A_v lie in the range space of B, the system zeros are not affected by A_v. This condition is identical to the invariance condition of VS systems. The decoupling properties are maintained if the columns of A_v lie in the null space of C. Care needs to be exercised to ensure that the closed-loop system eigenvalues, corresponding to the system zeros, have negative real parts. This may be achieved by choosing E to have eigenvalues sufficiently far into the left half of the complex plane. Similarly, the effect of variations B_v have been considered in reference 10 and 22, and robust properties exist provided certain conditions are satisfied.

VS implementation of the controller should yield the desirable robust properties of invariance to the appropriate classes of (time-varying) parameter variations and disturbances. When there are more outputs than control inputs, i.e. $\ell > m$, the switching hyperplanes matrix G is an $m \times \ell$ full rank matrix. This case is perhaps the most commonly occurring situation. Other cases, some of which do not yield favourable VS control, are considered in reference 22. The matrix D = GCB is of order $m \times m$ and is chosen to be nonsingular to ensure the uniqueness of the control in the sliding mode. Using the equivalent control approach yields the system equations (4.6) with the terms $(CB)^{-1}$ replaced by D^{-1}. GC squares the outputs down and new zeros are introduced which need to be placed in the left half of the complex plane. Further details are given in reference 22. Whether G has enough degrees of freedom to ensure this condition and also simultaneously suitably assigns the eigenvalues for all classes of system, is an unresolved question.

6. CONCLUDING REMARKS

The equivalent control technique of Variable Structure Systems theory has been used to determine a state feedback controller for the output model-following control problem. The VS implementation of the control strategy yields robust control which is insenitive to a class of parameter variations and disturbances. It should be noted that the feedback controller is fixed and the adaptive properties are achieved through the switching logic associated with the switching hyperplanes. The controller effectively reduces the order of the closed-loop system, specifies the eigenvalues of the system and error transients, and achieves decoupling between the plant output variables and the 'driving inputs' associated with each model output. Certain additive plant disturbances can also be handled[12].

Replacing the discontinuities $\mathrm{sgn}(v_i)$ (3.1) by $\underline{v}/\|\underline{v}\|$, enforces sliding to be initiated on the intersection of the switching hyperplanes and obviates the need for the rather complicated hierarchial control method[5,16] which ensures reaching and sliding on each switching hyperplane successively. Recent work[13,23] has shown that the undesirable chatter associated with the sliding mode can be removed by using control laws which are continuous in the neighbourhood of the switching surface S. Examples include saturating amplifiers and functions of the form $\underline{v}/(\|\underline{v}\| + \delta)$ where δ is a suitably small positive constant. The state point of the system during the 'sliding' mode remains in a neighbourhood of S. There are also close links with work on the control of uncertain systems[24-26] and the hyperstability design technique[27].

7. REFERENCES

1. Landau, I.D. (1979) 'Adaptive control: The model reference approach', M. Dekker Inc, New York.
2. Rohrs, C., Valavani, L., Athans, M. and Stein, G. (1982) 'Robustness of adaptive control algorithms in the presence of unmodelled dynamics', Proc IEEE CDC Conference, 3-11.
3. Cook, P.A. and Chen, Z.J. (1983) 'Robustness of continuous-time model reference adaptive control systems', CSC Report No.572, University of Manchester Institute of Science and Technology.

4. Utkin, V.I. (1971) 'Equations of sliding mode in discontinuous systems - I', Automat and Remote Control, 21, 1897-1907
5. Utkin, V.I. (1974) 'Sliding modes and their application in variable structure systems', MIR (Moscow) (English translation 1978).
6. Shaked, U. (1977) 'Design of general model-following control systems', Int J Control, 25, 213-238.
7. Erzberger, H. (1968) 'Analysis and design of model-following control systems by state space techniques', Proc JACC, Ann Arbor, 572-581.
8. Chen, Y.T. (1973) 'Perfect model-following with real model', Proc JACC, 1973, 287-293.
9. Young, K-K.D. (1977) 'Asymptotic stability of model reference systems with variable structure control', IEE Trans Automat Control, AC-22, 279-281.
10. Young, K-K.D. (1978) 'Design of variable structure model-following control systems', IEEE Trans Automat Control, AC-23, 1079-1085.
11. Zinober, A.S.I. (1981) 'Controller design using the theory of variable structure systems', in 'Self-tuning and adaptive control', ed Billings, S.A. and Harris, C.J., Peter Peregrinus Ltd, London, Chapter 9, 206-229.
12. Zinober, A.S.I., El-Ghezawi, O.M.E. and Billings, S.A. (1982) 'Multivariable variable structure adaptive model - following control systems', Proc IEE, 129D, 1-6.
13. Ambrosino, G., Celentano, G. and Garofalo, F. (1984) 'Variable structure model reference adaptive control systems', Int. J. Control, 39, No.6, pp 1339-1349.
14. Draženović, B. (1969) 'The invariance conditions in variable structure systems', Automatica, 5, 287-295.

15. Filippov, A.G. (1960) 'Application of the theory of differential equations with discontinuous right hand sides to non-linear problems in automatic control', Proc First IFAC Congress, Moscow, 2, 923-927.
16. Ryan, E.P. (1983) 'A variable structure approach to feedback regulation of a class of uncertain dynamic systems', Int.J. Control, 38, 1121-1134.
17. El-Ghezawi, O.M.E., Zinober, A.S.I. and Billings, S.A.(1983) 'Analysis and design of variable structure control systems using a geometric approach', Int.J. Control, 38, 657-671.
18. El-Ghezawi, O.M.E., Billings, S.A. and Zinober, A.S.I. (1983) 'Variable-structure systems and system zeros', Proc, IEE, 130D, 1-5.
19. Young, K-K.D., Kokotović, P.V. and Utkin, V.I. (1977) 'A singular perturbation analysis of high-gain feedback systems', IEEE Trans Automat Control, AC-22, 931-938.
20. MacFarlane, A.G.J. and Karcanias, N. (1976) 'Poles and zeros of linear multivariable systems: a survey of the algebraic

and complex-variable theory', Int. J. Control, 24, 33-74.

21. Shaked, U. and Karcanias, N. (1976) 'The use of zeros and zero directions in model reduction', Int. J. Control, 23, 113-135.
22. El-Ghezawi, O.M.E. (1982) 'Adaptive control using variable structure systems', PhD Thesis, University of Sheffield.
23. Slotine, J.J. and Sastry, S.S. (1983) 'Tracking control of non-linear systems using sliding surfaces with application to robot manipulators', Int. J. Control, 38, 465-492.
24. Gutman, S. and Palmor, Z. (1982) 'Properties of min-max controllers in uncertain dynamic systems', SIAM J Control Optimization, 20, 850-861.
25. Corless, M.J. and Leitmann, G. (1981) 'Continuous state feedback guaranteeing uniform ultimate boundedness for uncertain dynamical systems', IEEE Trans. Automat Control, AC-26, 1139-1144.
26. Ryan, E.P. and Corless, M. (1984) 'Ultimate boundedness and asymptotic stability of a class of uncertain dynamical systems via continuous and discontinuous feedback control', submitted for publication.
27. Balestrino, A., De Maria, G. and Zinober, A.S.I. (1984) 'Nonlinear adaptive model following control', Automatica, 20, 559-568.

TRACKING IN THE PRESENCE OF BOUNDED UNCERTAINTIES

M. Corless and G. Leitmann
(Dept of Mechanical Engineering, University of California-Berkeley)
and
E.P. Ryan
(School of Mathematics, University of Bath)

ABSTRACT

The tracking problem for uncertain dynamical systems is considered. Based only on knowledge of functional properties and bounds relating to the uncertainties, a class of feedback controls is developed which, under a realistic assumption of feasibility of the motion to be tracked, guarantees tracking with arbitrary prescribed accuracy. The case of infeasible motion is also studied.

1. INTRODUCTION

Mathematical modelling of a physical dynamical process generally induces some degree of imprecision or uncertainty. Based only on knowledge of functional properties and bounds relating to the uncertain elements in the model, a deterministic theory of feedback control, guaranteeing certain behaviour of the uncertain system, has been presented in Refs.[1-10]. Here, under similar hypotheses, this theory is extended to a problem of tracking in the presence of uncertainties. A class of continuous state feedback controls is proposed which, given a desired state motion $\bar{x}(\cdot)$, guarantees that $\bar{x}(\cdot)$ can be tracked to within calculable accuracy; moreover, under the realistic assumption of feasibility of $\bar{x}(\cdot)$, tracking with arbitrarily small error can be achieved.

2. SYSTEM DESCRIPTION AND PROBLEM STATEMENT

Consider an uncertain dynamical system modelled by

$$\dot{x}(t) = f(x(t),t) + \Delta f(x(t),t,r(t)) + [B + \Delta B(x(t),t,r(t))]u(t)$$
$$x(t_o) = x^o \tag{2.1}$$

with $t \in \mathbf{R}$, state $x(t) \in \mathbf{R}^n$, and control $u(t) \in \mathbf{R}^m$. Uncertainty in the system description is modelled by an unknown Lebesgue measurable function $r(\cdot): \mathbf{R} \to \mathcal{R}$, where the uncertainty bounding set $\mathcal{R} \subset \mathbf{R}^p$ is known and compact. $B \in \mathbf{R}^{n \times m}$, $f(\cdot): \mathbf{R}^n \times \mathbf{R} \to \mathbf{R}^n$, $\Delta f(\cdot): \mathbf{R}^n \times \mathbf{R} \times \mathcal{R} \to \mathbf{R}^n$, and $\Delta B(\cdot): \mathbf{R}^n \times \mathbf{R} \times \mathcal{R} \to \mathbf{R}^{n \times m}$ are assumed known. Given an absolutely continuous function $\overline{x}(\cdot): \mathbf{R} \to \mathbf{R}^n$ (the desired state motion), the problem treated here may be state as follows. Construct a class $\mathcal{C}$ of Carathéodory (see Appendix) feedback control functions $\phi(\cdot): \mathbf{R}^n \times \mathbf{R} \to \mathbf{R}^m$ such that, given any neighbourhood $\mathcal{B}$ of the origin in $\mathbf{R}^n$, there exists $\phi(\cdot) \in \mathcal{C}$ such that, for *any* Lebesgue measurable function $r(\cdot): \mathbf{R} \to \mathcal{R}$, the feedback-controlled system

$$\dot{x}(t) = f(x(t),t) + \Delta f(x(t),t,r(t)) + [B + \Delta B(x(t),t,r(t))]\phi(x(t),$$
$$x(t_o) = x^o \tag{2.2}$$

has existence and continuation of solutions and $\mathcal{B}$-tracks $\overline{x}(\cdot)$, in the sense of the following definitions.

Defn 2.1: System (2.2) has existence and continuation of solutions iff, for each $(x^o,t_o) \in \mathbf{R}^n \times \mathbf{R}$, there exists a solution $x(\cdot): [t_o,t_1) \to \mathbf{R}^n$ of (2.2) (i.e. an absolutely continuous function satisfying (2.2) a.e. on $[t_o,t_1)$) and every solution of (2.2) can be extended into a solution defined on $[t_o,\infty)$.

Defn 2.2: With existence and continuation of solutions, system (2.2) $\mathcal{B}$-tracks $\overline{x}(\cdot)$ (tracks $\overline{x}(\cdot)$ to within $\mathcal{B}$) iff:-
(i) Uniform boundedness: Given any $\delta \geq 0$, $\exists\ d(\delta) \geq 0$ such that, if $x(\cdot): [t_o,\infty) \to \mathbf{R}^n$ is any solution of (2.2) with $\|x(t_o) - \overline{x}(t_o)\| \leq \delta$, then $\|x(t) - \overline{x}(t)\| \leq d(\delta)\quad \forall\ t \in [t_o,\infty)$.
(ii) There exists a neighbourhood $\mathcal{B}_o$ of the origin in $\mathbf{R}^n$ such that, if $x(\cdot): [t_o,\infty) \to \mathbf{R}^n$ is any solution of (2.2) with $x(t_o) - \overline{x}(t_o) \in \mathcal{B}_o$, then $x(t) - \overline{x}(t) \in \mathcal{B}\quad \forall\ t \in [t_o,\infty)$.
(iii) Uniform ultimate boundedness within $\mathcal{B}$: For each $\delta \geq 0$, $\exists\ T(\delta) \geq 0$ such that, if $x(\cdot): [t_o,\infty) \to \mathbf{R}^n$ is any solution of (2.2) with $\|x(t_o) - \overline{x}(t_o)\| \leq \delta$, then $x(t) - \overline{x}(t) \in \mathcal{B}\ \ \forall t \geq t_o + T(\delta)$.

Remark 2.1: If, for *any* neighbourhood $\mathcal{B}$ of the origin in $\mathbf{R}^n$, system (2.2) $\mathcal{B}$-tracks $\overline{x}(\cdot)$, then (2.2) is globally uniformly asymptotically stable about $\overline{x}(\cdot)$.

In order to achieve our objective, we make the following assumptions:

A1. Linearizability and stabilizability: There exist a matrix $A \in \mathbf{R}^{n \times n}$ and a Carathéodory function $g(\cdot): \mathbf{R}^n \times \mathbf{R} \to \mathbf{R}^m$ such that for all $(x,t) \in \mathbf{R}^n \times \mathbf{R}$,

$$f(x,t) = A\,x + B\,g(x,t) \tag{2.3}$$

and (A,B) is a stabilizable pair.

A2. Matching of uncertainties: There exist Carathéodory function $h(\cdot): \mathbf{R}^n \times \mathbf{R} \times \mathcal{R} \to \mathbf{R}^m$ and $E(\cdot): \mathbf{R}^n \times \mathbf{R} \times \mathcal{R} \to \mathbf{R}^{m \times m}$ and a continuous function $\beta(\cdot): \mathbf{R}^n \times \mathbf{R} \to \mathbf{R}_+$ such that, for all $(x,t) \in \mathbf{R}^n \times \mathbf{R}$ and for all $r \in \mathcal{R}$,

$$\Delta f(x,t,r) = B\,h(x,t,r) \tag{2.4}$$

$$\Delta B(x,t,r) = B\,E(x,t,r) \tag{2.5}$$

$$\|E(x,t,r)\| \leq \beta(x,t) < 1\ . \tag{2.6}$$

A3. Feasibility of $\bar{x}(\cdot)$: There exists a function $\theta^f(\cdot): \mathbb{R} \to \mathbb{R}^m$, which is Lebesgue integrable on bounded intervals, such that a.e. on $\mathbb{R}$,

$$\dot{\bar{x}}(t) = A\bar{x}(t) + B\theta^f(t) \ . \tag{2.7}$$

3. PROPOSED CONTROLLERS

First, note that, if $(A, g(\cdot))$ is any pair which assures satisfaction of A1, then there exists $K \in \mathbb{R}^{m\times n}$ such that

$$\bar{A} = A + BK \tag{3.1}$$

is asymptotically stable (i.e. with spectrum $\sigma(\bar{A}) \subset \mathbb{C}^-$, the open left half complex plane); hence, given any $Q \in PD^n$ †, there exists a unique solution $P \in PD^n$ of the matrix Lyapunov equation

$$P\bar{A} + \bar{A}^T P + Q = 0 \ . \tag{3.2}$$

Now, let $(A, g(\cdot))$ be any pair which assures satisfaction of A1 and selecting any corresponding K and P as above, the members of a proposed class $\mathcal{C}$ of controls are functions $\phi(\cdot): \mathbb{R}^n \times \mathbb{R} \to \mathbb{R}^m$ given by

$$\phi(x,t) = \psi(x,t) + p_\varepsilon(x,t), \tag{3.3}$$

where

$$\psi(x,t) = K[x - \bar{x}(t)] - g(x,t) + \theta^f(t), \tag{3.4}$$

$\theta^f(\cdot)$ is any function which assures A3 and, for any $\varepsilon > 0$, $p_\varepsilon(\cdot): \mathbb{R}^n \times \mathbb{R} \to \mathbb{R}^m$ is any Carathéodory function such that, for all $(x,t) \in \mathbb{R}^n \times \mathbb{R}$,

$$\|p_\varepsilon(x,t)\| \leq \rho(x,t) \tag{3.5}$$

and

$$p_\varepsilon(x,t) = -\rho(x,t)\|\eta(x,t)\|^{-1}\eta(x,t), \text{ if } \|\eta(x,t)\| > \varepsilon \tag{3.6}$$

where

$$\eta(x,t) = B^T P[x - \bar{x}(t)]\rho(x,t) \tag{3.7}$$

and $\rho(x,t)$ satisfies

$$\rho(x,t) \geq \rho_0(x,t) = \max_{r\in\mathcal{R}} [1-\|E(x,t,r)\|]^{-1}\|h(x,t,r) + E(x,t,r)\psi(x,t)\| \tag{3.8}$$

$E(\cdot)$ and $h(\cdot)$ being any pair which assures A2. As a particular example of a function satisfying the above requirements, consider $p_\varepsilon(\cdot)$ given by

$$p_\varepsilon(x,t) = \begin{cases} -\rho_0(x,t)\|\eta(x,t)\|^{-1}\eta(x,t), & \text{if } \|\eta(x,t)\| > \varepsilon \\ -\rho_0(x,t)\,\varepsilon^{-1}\eta(x,t), & \text{if } \|\eta(x,t)\| \leq \varepsilon \end{cases} \tag{3.9}$$

with

$$\eta(x,t) = B^T P[x - \bar{x}(t)]\rho_0(x,t) \ . \tag{3.10}$$

---

† We use PD^n to denote the set of positive-definite symmetric members of $\mathbb{R}^{n\times n}$.

Remark 3.1: The controllers proposed here are similar in structure to those of Refs. 3,4,7,8.

4. PROPERTIES OF SYSTEMS WITH PROPOSED CONTROLLERS

Theorem 4.1: Consider a system described by (2.1) and an absolutely continuous function $\bar{x}(\cdot): \mathbf{R} \to \mathbf{R}^n$ which together satisfy A1, A2 and A3. Suppose that $\phi(\cdot): \mathbf{R}^n \times \mathbf{R} \to \mathbf{R}^m$ is a member of the corresponding class $\mathcal{C}$ of controls of §3. Then, for *any* Lebesgue measurable function $r(\cdot): \mathbf{R} \to \mathcal{R}$, system (2.2) has existence and continuation of solutions and tracks $\bar{x}(\cdot)$ to within any neighbourhood† $\mathcal{B}$ of

$$\mathcal{E}^f_\varepsilon = \left\{ z \in \mathbf{R}^n : \; z^T P z \le 4\varepsilon\lambda_1^{-1} \right\}$$

where

$$\lambda_1 = \lambda_{min}(P^{-1}Q) > 0 \text{ ¶}$$

Proof: See Appendix.

Noting that $z \in \mathcal{E}^f_\varepsilon \Rightarrow \|z\|^2 \le 4\varepsilon(\lambda_1\lambda_2)^{-1}$, where $\lambda_2 = \lambda_{min}(P) > 0$, it follows that, given any neighbourhood $\mathcal{B}$ of the origin, there exists $\varepsilon > 0$ such that $\mathcal{B}$ is also a neighbourhood of $\mathcal{E}^f_\varepsilon$; hence, we may deduce the following corollary.

Corollary 4.1: Consider a system described by (2.1) and an absolutely continuous function $\bar{x}(\cdot): \mathbf{R} \to \mathbf{R}^n$ which together satisfy A1, A2 and A3. Then, given *any* neighbourhood $\mathcal{B}$ of the origin in $\mathbf{R}^n$, there exists a feedback control function $\phi(\cdot) \in \mathcal{C}$ such that, for *any* Lebesgue measurable function $r(\cdot): \mathbf{R} \to \mathcal{R}$, system (2.2) has existence and continuation of solutions and tracks $\bar{x}(\cdot)$ to within $\mathcal{B}$.

Thus, $\mathcal{C}$ has the desired properties stated in §2.

5. RELAXATION OF THE FEASIBILITY ASSUMPTION

We now consider the consequences of A3 (feasibility of $\bar{x}(\cdot)$) *not* necessarily being satisfied. For this purpose, we shall suppose that $\text{rank}(B) = m$.

For each $\Gamma \in PD^n$, let $\theta^\Gamma(\cdot): \mathbf{R} \to \mathbf{R}^m$ and $\xi^\Gamma(\cdot): \mathbf{R} \to \mathbf{R}^n$ denote functions which satisfy the following a.e.

$$\theta^\Gamma(t) = [B^T \Gamma B]^{-1} B^T \Gamma [\dot{\bar{x}}(t) - A\bar{x}(t)], \tag{5.1}$$

$$\xi^\Gamma(t) = \dot{\bar{x}}(t) - A\bar{x}(t) - B\theta^\Gamma(t) . \tag{5.2}$$

---

† If $V \subset \mathbf{R}^n$, then a neighbourhood of V is any set which contains an open set containing V.

¶ For $M \in \mathbf{R}^{n\times n}$, $\lambda_{min}(M)$ denotes the minimum real eigenvalue of M.

Thus, a.e.

$$\dot{\bar{x}}(t) = A\bar{x}(t) + B\theta^{\Gamma}(t) + \xi^{\Gamma}(t), \tag{5.3}$$

$$\xi^{\Gamma}(t) = [I - B(B^T\Gamma B)^{-1}B^T\Gamma][\dot{\bar{x}}(t) - A\bar{x}(t)] . \tag{5.4}$$

Moreover, a.e.

$$\|\xi^{\Gamma}(t)\|_{\Gamma} = \|\dot{\bar{x}}(t) - A\bar{x}(t) - B\theta^{\Gamma}(t)\|_{\Gamma}$$
$$= \min_{\zeta \in \mathbf{R}^m} \|\dot{\bar{x}}(t) - A\bar{x}(t) - B\zeta\|_{\Gamma} \tag{5.5}$$

where

$$\|y\|_{\Gamma} = (y^T\Gamma y)^{\frac{1}{2}}, \tag{5.6}$$

i.e. $B\theta^{\Gamma}(t)$ is the nearest (in the sense of $\|\cdot\|_{\Gamma}$) vector in the range space of B to $\dot{\bar{x}}(t) - A\bar{x}(t)$.

It can be readily shown that, if A3 is satisfied, then for any $\Gamma \in PD^n$, $\theta^{\Gamma}(t) = \theta^f(t)$ and $\xi^{\Gamma}(t) = 0$ a.e., where $\theta^f(\cdot)$ assures A3. Conversely, from (5.3) it follows that if $\xi^{\Gamma}(t) = 0$ a.e. for some $\Gamma \in PD^n$, then A3 is satisfied by letting $\theta^f(t) = \theta^{\Gamma}(t)$ a.e.. Thus, a necessary and sufficient condition for A3 to hold is the existence of $\Gamma \in PD^n$ such that, a.e.,

$$\xi^{\Gamma}(t) = [I - B(B^T\Gamma B)^{-1}B^T\Gamma][\dot{\bar{x}}(t) - A\bar{x}(t)] = 0 .$$

We now introduce a relaxed version of Assumption A3.

A3.1: $\xi^{\Gamma}(\cdot)$ is essentially bounded for some $\Gamma \in PD^n$, i.e. there exists $b \in \mathbf{R}_+$ such that $\|\xi^{\Gamma}(t)\| \leq b$ a.e..

Remark 5.1: If $\Gamma_1, \Gamma_2 \in PD^n$, then, utilizing (5.3) and (5.4),

$$\xi^{\Gamma_2}(t) = [I - B(B^T\Gamma_2 B)^{-1}B^T\Gamma_2][\dot{\bar{x}}(t) - A\bar{x}(t)]$$
$$= [I - B(B^T\Gamma_2 B)^{-1}B^T\Gamma_2][B\,\theta^{\Gamma_1}(t) + \xi^{\Gamma_1}(t)]$$
$$= [I - B(B^T\Gamma_2 B)^{-1}B^T\Gamma_2]\,\xi^{\Gamma_1}(t)$$

hence, if A3.1 is satisfied, then $\xi^{\Gamma}(\cdot)$ is essentially bounded for any $\Gamma \in PD^n$.

The proposed class C of controls is now defined as in §3 with $\theta^f(\cdot)$ replaced by $\theta^P(\cdot)$, where P is the solution of (3.2) used in the construction of $p_{\varepsilon}(\cdot)$. Thus, the members $\phi(\cdot)$ of C are given by (3.3) where now

$$\psi(x,t) = K[x - \bar{x}(t)] - g(x,t) + \theta^P(t) \tag{5.7}$$

and $p_{\varepsilon}(\cdot)$ is any Carathéodory function which satisfies (3.5) - (3.8) for all $(x,t) \in \mathbf{R}^n \times \mathbf{R}$.

We now have the following theorem.

Theorem 5.1: Consider a system described by (2.1) and an absolutely continuous function $\bar{x}(\cdot): \mathbf{R} \to \mathbf{R}^n$ which together satisfy A1, A2 and A3.1. Suppose that $\phi(\cdot)$ is a member of the proposed class C of controls. Then, for *any* Lebesgue measurable function $r(\cdot): \mathbf{R} \to R$, system (2.2) has existence and continuation of solutions and tracks $\overline{x}(\cdot)$ to within any neighbourhood $\mathcal{B}$ of

$$\mathcal{E}_\varepsilon = \left\{ z \in \mathbb{R}^n: \quad z^T P z \le k_\varepsilon^2 \right\} \tag{5.8}$$

where

$$k_\varepsilon = c\lambda_1^{-1} + \sqrt{c^2\lambda_1^{-2} + 4\varepsilon\lambda_1^{-1}} \ ; \quad \lambda_1 = \lambda_{min}(P^{-1}Q) \tag{5.9}$$

and the constant c is such that

$$\| \xi^P(t) \|_P \le c \quad \text{a.e.} \tag{5.10}$$

Proof: See Appendix.

If we introduce the set

$$\mathcal{E}_o = \left\{ z \in \mathbb{R}^n: \quad z^T P z \le k_o^2 \right\} \text{ where } k_o = 2c\lambda_1^{-1} \tag{5.11}$$

then noting that, given any neighbourhood $\mathcal{B}$ of $\mathcal{E}_o$, there exists $\varepsilon > 0$ such that $\mathcal{B}$ is also a neighbourhood of $\mathcal{E}_\varepsilon$, we have the following corollary.

Corollary 5.1: Consider a system described by (2.1) and an absolutely continuous function $\overline{x}(\cdot): \mathbb{R} \to \mathbb{R}^n$ which together satisfy A1, A2 and A3.1. Then, given any neighbourhood $\mathcal{B}$ of $\mathcal{E}_o$, there exists $\phi(\cdot) \in \mathcal{C}$ such that, for any Lebesgue measurable function $r(\cdot): \mathbb{R} \to \mathcal{R}$, system (2.2) has existence and continuation of solutions and tracks $\overline{x}(\cdot)$ to within $\mathcal{B}$.

Remark 5.1: If A3 holds, then $\xi^P(t) = 0$ a.e.; hence, we may take $c = 0$. Thus, Theorem 4.1 is a consequence of Theorem 5.1. Also, Corollary 4.1 follows from Corollary 5.1.

6. APPLICATION TO ROBOTIC TRACKING

Consider the two-degrees-of-freedom manipulator (for horizontal planar tracking) illustrated in Figure 1. The arm can be extended or retracted and rotated about a vertical axis.

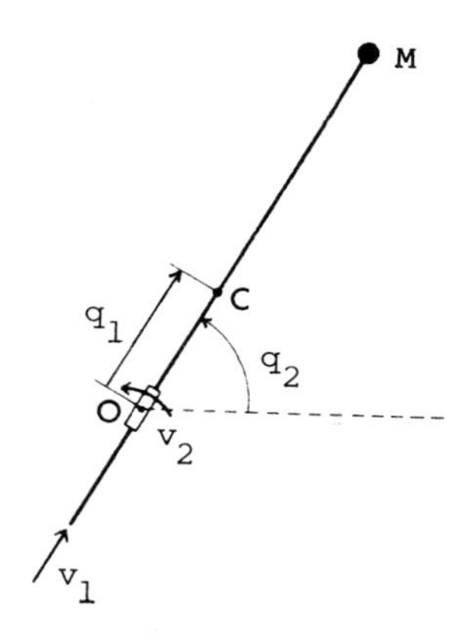

Fig. 1. Two-degrees-of-freedom manipulator

Loosely speaking, the problem to be considered is that of specifying the force v_1 and the moment v_2 to ensure that the load follows or tracks a prescribed desired motion. Treating the load as a point mass and letting (q_1,q_2) denote the position of the mass centre C of the arm in polar coordinates, the equations of motion are

$$(\mu + M)\ddot{q}_1 - R(q_1,M)\dot{q}_2^2 = v_1 \;; \quad I(q_1,M)\ddot{q}_2 + 2R(q_1,M)\dot{q}_1\dot{q}_2 = v_2$$

where (6.1)

$$R(q_1,M) = \mu q_1 + M(q_1 + a) \;; \quad I(q_1,M) = J_1 + J_2 + \mu q_1^2 + M(q_1 + a)^2 \tag{6.2}$$

a is the distance from C to the load, μ and M are the arm and load masses respectively, J_1 is the moment of inertia of the rotation mechanism about the vertical axis through O, and J_2 is the moment of inertia of the arm about the vertical axis through C. Let us suppose that all parameters are precisely known with the exception of the constant load mass M which is subject to known bounds, viz.

$$0 \le \underline{M} \le M \le \overline{M}, \tag{6.3}$$

where $\underline{M}$ and $\overline{M}$ are known constants. Let $(s(\cdot),\theta(\cdot))$ determine the desired motion of the load in polar coordinates, where $s(\cdot): \mathbb{R} \to \mathbb{R}_+$ and $\theta(\cdot): \mathbb{R} \to \mathbb{R}$ are known piecewise C^2 functions. To obtain a problem statement in the form considered in §2, we introduce the state vector

$$x = (x_1 \;\; x_2 \;\; x_3 \;\; x_4)^T = (q_1 \;\; q_2 \;\; \dot{q}_1 \;\; \dot{q}_2)^T \tag{6.4}$$

which results in

$$\begin{bmatrix} \dot{x}_1 \\ \dot{x}_2 \\ \dot{x}_3 \\ \dot{x}_4 \end{bmatrix} = \begin{bmatrix} x_3 \\ x_4 \\ (\mu + M)^{-1}R(x_1,M)x_4^2 + (\mu + M)^{-1} v_1 \\ -(I(x_1,M))^{-1}\, 2R(x_1,M)x_3x_4 + (I(x_1,M))^{-1} v_2 \end{bmatrix} \tag{6.5}$$

Adopting a nominal load mass M^o given by

$$M^o = \tfrac{1}{2}(\underline{M} + \overline{M}) \tag{6.6}$$

and letting

$$u = (u_1 \;\; u_2)^T = \Big((\mu + M^o)^{-1} v_1 \quad (I(x_1,M^o))^{-1} v_2\Big)^T \tag{6.7}$$

we obtain

$$\begin{bmatrix} \dot{x}_1 \\ \dot{x}_2 \\ \dot{x}_3 \\ \dot{x}_4 \end{bmatrix} = \begin{bmatrix} x_3 \\ x_4 \\ (\mu + M)^{-1} R(x_1,M)\, x_4^2 + (\mu + M)^{-1}(\mu + M^o)\, u_1 \\ -(I(x_1,M))^{-1}\, 2R(x_1,M)x_3x_4 + (I(x_1,M))^{-1} I(x_1,M^o)\, u_2 \end{bmatrix} \tag{6.8}$$

If we now define

$$\bar{x} = (\bar{x}_1 \quad \bar{x}_2 \quad \bar{x}_3 \quad \bar{x}_4)^T = ((s-a) \quad \theta \quad \dot{s} \quad \dot{\theta})^T \tag{6.9}$$

then we may state our problem as that of finding a collection $\mathcal{C}$ of feedback control functions $\phi(\cdot): \mathbb{R}^4 \times \mathbb{R} \to \mathbb{R}^2$ which assures that, given any neighbourhood $\mathcal{B}$ of the origin, there exists $\phi(\cdot) \in \mathcal{C}$ such that (6.8) (with $u(t) = \phi(x(t),t)$) has existence and continuation of solutions and $\mathcal{B}$ - tracks $\bar{x}$. Now, defining

$$\mathcal{R} = [\underline{M}, \overline{M}] \tag{6.10}$$

$$\left.\begin{aligned} f(x,t) &= (x_3 \quad x_4 \quad f_3(x,t) \quad f_4(x,t))^T \\ f_3(x,t) &= (\mu + M^o)^{-1} R(x_1, M^o)\, x_4^2 \\ f_4(x,t) &= -(I(x_1, M^o))^{-1}\, 2\, R(x_1, M^o) x_3 x_4 \end{aligned}\right\} \tag{6.11}$$

$$\left.\begin{aligned} \Delta f(x,t,M) &= (0 \quad 0 \quad \Delta f_3(x,t,M) \quad \Delta f_4(x,t,M))^T \\ \Delta f_3(x,t,M) &= (\mu + M)^{-1} (\mu + M^o)^{-1} (M - M^o) \mu\, a\, x_4^2 \\ \Delta f_4(x,t,M) &= (I(x_1, M))^{-1} (I(x_1, M^o))^{-1} \times \\ &\qquad 2(M - M^o)(\mu a x_1 - J_1 - J_2)(x_1 + a) x_3 x_4 \end{aligned}\right\} \tag{6.12}$$

$$B = \begin{bmatrix} 0^{2\times 2} \\ I^{2\times 2} \end{bmatrix}, \quad 0^{2\times 2} = \begin{bmatrix} 0 & 0 \\ 0 & 0 \end{bmatrix}, \quad I^{2\times 2} = \begin{bmatrix} 1 & 0 \\ 0 & 1 \end{bmatrix} \tag{6.13}$$

$$\left.\begin{aligned} \Delta B(x,t,M) &= \begin{bmatrix} & 0^{2\times 2} & \\ E_1(x,t,M) & & 0 \\ 0 & & E_2(x,t,M) \end{bmatrix} \\ E_1(x,t,M) &= -(\mu + M)^{-1} (M - M^o) \\ E_2(x,t,M) &= -(I(x_1, M))^{-1} (M - M^o)(x_1 + a)^2 \end{aligned}\right\} \tag{6.14}$$

system (6.8) can be re-written in the form (2.1), viz.

$$\dot{x}(t) = f(x(t),t) + \Delta f(x(t),t,M) + [B + \Delta B(x(t),t,M)]\, u(t) \tag{6.15}$$

To demonstrate that assumption A1 holds, let

$$A = \begin{bmatrix} 0^{2\times 2} & I^{2\times 2} \\ 0^{2\times 2} & 0^{2\times 2} \end{bmatrix} \text{ and } g(x,t) = (f_3(x,t) \quad f_4(x,t))^T \tag{6.16}$$

Assumption A2 is shown to hold by taking

$$h(x,t,M) = (\Delta f_3(x,t,M) \quad \Delta f_4(x,t,M))^T \tag{6.17}$$

and

$$E(x,t,M) = \begin{bmatrix} E_1(x,t,M) & 0 \\ 0 & E_2(x,t,M) \end{bmatrix} \tag{6.18}$$

for which it is readily shown that, for all $(x,t,M) \in \mathbb{R}^4 \times \mathbb{R} \times \mathcal{R}$,

$$\|E(x,t,M)\| \leq \beta = \tfrac{1}{2}[\underline{M}(J_1+J_2+\mu a^2)+\mu(J_1+J_2)]^{-1} \times (\overline{M}-\underline{M})(J_1+J_2+\mu a^2) \tag{6.19}$$

Thus, if $(\overline{M}-\underline{M})$ is sufficiently small, then $\beta < 1$ and A2 holds. Finally, assumption A3 is shown to hold by letting

$$\theta^f(t) = \begin{bmatrix} \ddot{s}(t) \\ \ddot{\theta}(t) \end{bmatrix} \quad \text{a.e..} \tag{6.20}$$

Selecting

$$K = [-\lambda^2 I^{2\times 2} \quad 2\lambda I^{2\times 2}], \tag{6.21}$$

where $\lambda < 0$, then

$$\overline{A} = A + BK = \begin{bmatrix} 0^{2\times 2} & I^{2\times 2} \\ -\lambda^2 I^{2\times 2} & 2\lambda I^{2\times 2} \end{bmatrix} \tag{6.22}$$

is asymptotically stable. Taking

$$Q = \begin{bmatrix} \lambda^2 I^{2\times 2} & 0^{2\times 2} \\ 0^{2\times 2} & I^{2\times 2} \end{bmatrix} \tag{6.23}$$

and solving (3.2) for P yields

$$P = \tfrac{1}{2}\begin{bmatrix} -3\lambda I^{2\times 2} & I^{2\times 2} \\ I^{2\times 2} & -\lambda^{-1} I^{2\times 2} \end{bmatrix} \tag{6.24}$$

The proposed class $\mathcal{C}$ of control functions is now given by (3.3) and (3.4), where

$$p_\varepsilon(x,t) = \begin{cases} -\rho(x,t)\|\eta(x,t)\|^{-1}\eta(x,t) & \text{if } \|\eta(x,t)\| > \varepsilon \\ -\rho(x,t)\,\varepsilon^{-1}\eta(x,t) & \text{if } \|\eta(x,t)\| \leq \varepsilon \end{cases} \tag{6.25}$$

with

$$\eta(x,t) = \rho(x,t)B^T P[x-\overline{x}(t)] \tag{6.26}$$

and, as in Ref. 8,

$$\rho(x,t) = (1-\beta)^{-1}[\rho_1(x,t)+\beta\|K(x-\overline{x}(t))\|+\gamma]$$

$$\rho_1(x,t) = \tfrac{1}{2}(\overline{M}-\underline{M})\Big[(\mu+\underline{M})^{-1}|\ddot{s}(t)-(x_1+a)x_4^2| + [J_1+J_2+\mu x_1^2+\underline{M}(x_1+a)^2]^{-1}|\ddot{\theta}(t)(x_1+a)^2+2(x_1+a)x_3x_4|\Big] \tag{6.27}$$

where $\gamma \geq 0$.

6.1 Numerical Simulation

For purposes of illustration, the following numerical values are taken: $\mu = 100\text{ kg} = \overline{M}$, $\underline{M} = 0$, $J_1 = J_2 = 100\text{ kg m}^2$, and $a = 1\text{ m}$. The control objective is to force the (uncertain) load to follow a straight line path AB in the Cartesian (y_1, y_2)-plane, from its initial rest position A, with coordinates (y_1^A, y_2^A), to prescribed final rest position B, with coordinates (y_1^B, y_2^B), in a prescribed time T. A C^2 function pair $(y_1(\cdot), y_2(\cdot))$ (Cartesian coordinate functions), which characterizes such a path, is given by

$$y_i(t) = \begin{cases} y_i^A & ;\ t < t_o \\ y_i^A + \nu_i (t-t_o)^3 & ;\ t_o \le t \le t_o + T/4 \\ y_i^A + \nu_i[(t-t_o)^3 - 2(t-t_o-T/4)^3] & ;\ t_o + T/4 < t \le t_o + 3T/4 \\ y_i^A + \nu_i[(t-t_o)^3 - 2(t-t_o-T/4)^3 + 2(t-t_o-3T/4)^3] & ; \\ & \quad t_o + 3T/4 < t \le t_o + T \\ y_i^B & ;\ t > t_o + T \end{cases} \tag{6.28}$$

for $i = 1,2$, where $\nu_i = 16(y_i^B - y_i^A)/(3T^3)$. For the path parameter values $T = 5\text{sec}$, $(y_1^A, y_2^A) = (2,0)\,\text{m}$, $(y_1^B, y_2^B) = (1,2)\,\text{m}$, and the controller parameter values $\varepsilon = 0.005$, $\gamma = 0.01$, $\lambda = -1$, the solid and dotted lines (virtually indistinguishable from the prescribed path AB) of Figure 2 represent the paths of the feedback controlled system in the (y_1, y_2) plane under maximum ($M = 100\text{kg}$) and minimum ($M = 0$) actual load conditions, respectively; for comparative purposes, also shown are the paths, under the same two extreme load conditions, generated by the nominal (open-loop) control which would generate the desired path AB under the nominal load condition $M = M^o$. Finally, the evolution of controls (together with the nominal control based on nominal load $M = M^o$) is shown in Figures 3 and 4.

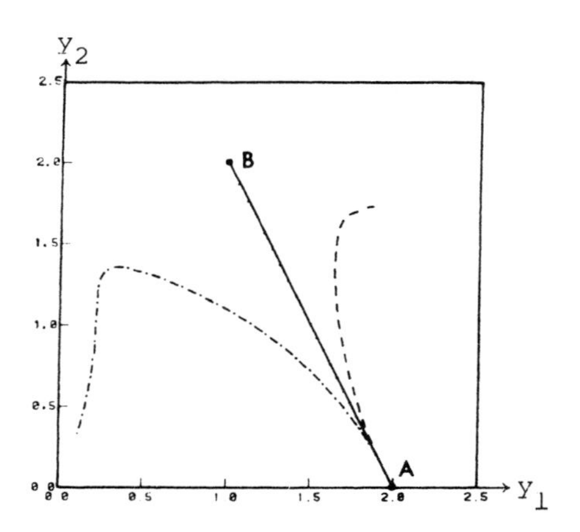

———— $M = \overline{M}$ } feedback control; ········· $M = \underline{M}$ } feedback control

– – – – – $M = \overline{M}$ } nominal control; –·–·–·– $M = \underline{M}$ } nominal control

Fig.2. Load paths

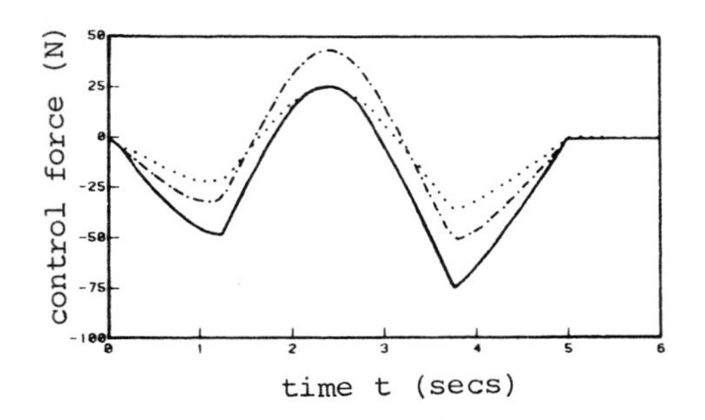

———————— $M = \bar{M}$ } feedback control

· · · · · · · · · · $M = \underline{M}$ }

—·—·—·— nominal control

Fig.3. Control force histories

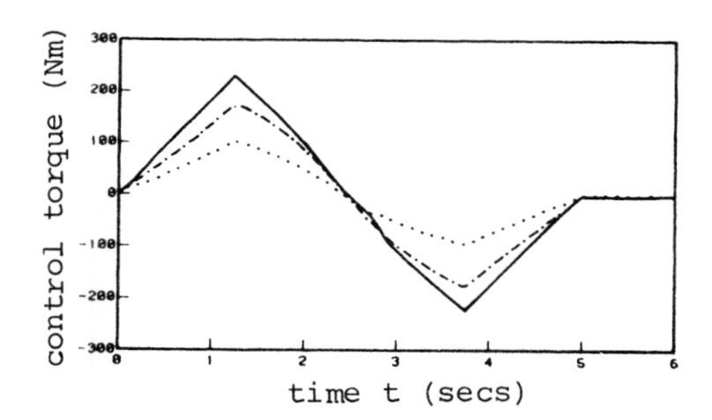

———————— $M = \bar{M}$ } feedback control

· · · · · · · · · · $M = \underline{M}$ }

—·—·—·— nominal control

Fig.4. Control torque histories

7. APPENDIX

In the following definition, D is a subset of $\mathbb{R}^n$ or $\mathbb{R}^n \times \mathbb{R}^p$.

Defn 7.1: A function $f(\cdot): D \times \mathbb{R} \to \mathbb{R}^m$ is *Carathéodory* iff: for each $t \in \mathbb{R}$, $f(\cdot,t): D \to \mathbb{R}^m$ is continuous; for each $z \in D$, $f(z,\cdot): \mathbb{R} \to \mathbb{R}^m$ is Lebesgue measurable; and for each compact subset U of $D \times \mathbb{R}$, there exists a Lebesgue integrable function $m_U(\cdot)$ such that $\|f(z,t)\| \le m_U(t)$ for all $(z,t) \in U$.

Proof of Theorem 4.1: First, we introduce a new state z(t), given by $z(t) = x(t) - \overline{x}(t)$, then, utilizing (2.2), (3.3) and (3.4), and assumptions A1, A2, A3, we obtain (for the sake of brevity, arguments are sometimes omitted below),

$$\begin{aligned} \dot{z} &= \dot{x} - \dot{\overline{x}} = f + \Delta f + [B + \Delta B]\phi - \dot{\overline{x}} \\ &= Ax + Bg + Bh + [B + BE]\phi - A\overline{x} - B\theta^f \\ &= Az - B[-g + \theta^f] + Bh + B[I + E]\phi \\ &= \overline{A}z - B[Kz - g + \theta^f] + Bh + B[I + E][\psi + p_\varepsilon] \\ &= \overline{A}z + B[h + E\psi] + B[I + E]p_\varepsilon \end{aligned} \tag{7.1}$$

Thus, we can prove Theorem 4.1 by demonstrating that (7.1) has existence and continuation of solutions and $\mathcal{B}$-tracks the zero function, where $\mathcal{B}$ is any neighbourhood of $\mathcal{E}_\varepsilon^f$. To this end, we note that (7.1) may be written as

$$\dot{z}(t) = F(z(t),t) \tag{7.2}$$

where the function $F(\cdot): \mathbb{R}^n \times \mathbb{R} \to \mathbb{R}^n$ is given by

$$\begin{aligned} F(z,t) = \overline{A}z &+ B[h(z + \overline{x}(t),t,r(t)) + E(z + \overline{x}(t),t,r(t))\psi(z + \overline{x}(t),t)] \\ &+ B[I + E(z + \overline{x}(t),t,r(t))]\, p_\varepsilon(z + \overline{x}(t),t) \end{aligned} \tag{7.3}$$

As a consequence of the Carathéodory-type assumptions on the functions of which $F(\cdot)$ is composed, one can readily show that $F(\cdot)$ is a Carathéodory function; hence (7.1) has existence of solutions (Refs. 11,12). To demonstrate continuation of solutions and $\mathcal{B}$-tracking, let $z(\cdot): [t_o,t_1) \to \mathbb{R}^n$, $t_1 > t_o$, be any solution of (7.1) and let $\tilde{V}(\cdot): [t_o,t_1) \to \mathbb{R}_+$ be given by

$$\tilde{V}(t) = (V \circ z)(t) = V(z(t)); \quad V(z) = z^T P z \tag{7.4}$$

Utilizing (7.1) and (3.2), we obtain, a.e. on $[t_o,t_1)$,

$$\begin{aligned} \dot{\tilde{V}} &= 2z^T P\dot{z} = 2z^T P[\overline{A}z + B(h + E\psi) + B(I + E)p_\varepsilon] \\ &= -z^T Qz + 2(B^T Pz)^T(h + E\psi) + 2(B^T Pz)^T(I + E)p_\varepsilon \\ &\le -z^T Qz + 2\|B^T Pz\|\,\|h + E\psi\| + 2\|B^T Pz\|\,\|E\|\,\|p_\varepsilon\| + 2(B^T Pz)^T p_\varepsilon \end{aligned}$$

whence

$$\dot{\tilde{V}} \le -z^T Qz + 2\|B^T Pz\|[\|h + E\psi\| + \|E\|\,\|p_\varepsilon\|] + 2(B^T Pz)^T p_\varepsilon \tag{7.5}$$

Considering the first term on the right of (7.5) and noting that

$$z^T Q z \geq \lambda_1 z^T P z \quad \forall\, z \in \mathbb{R}^n, \text{ where } \lambda_1 = \lambda_{min}(P^{-1} Q) \tag{7.6}$$

we have

$$-z(t)^T Q z(t) \leq -\lambda_1 \tilde{V}(t) \tag{7.7}$$

Considering the second term on the right of (7.5), it follows from (3.8) that

$$(1 - \|E\|)^{-1} \|h + E\psi\| \leq \rho \ ;$$

hence, utilizing (3.5),

$$\|h + E\psi\| + \|E\| \|p_\varepsilon\| \leq (1 - \|E\|)\rho + \|E\|\rho = \rho \tag{7.8}$$

Substituting (7.7) and (7.8) into (7.5) yields

$$\dot{\tilde{V}} \leq -\lambda_1 \tilde{V} + 2\|B^T P z\|\rho + 2(B^T P z)^T p_\varepsilon \ . \tag{7.9}$$

Now, if $\|\eta(x(t),t)\| \,(= \|B^T P z(t)\|\rho(x(t),t)\,) > \varepsilon$, then $p_\varepsilon = -\rho\|B^T P z\|^{-1} B^T P z$ and thus

$$\dot{\tilde{V}} \leq -\lambda_1 \tilde{V} + 2\|B^T P z\|\rho - 2\|B^T P z\|\rho = -\lambda_1 \tilde{V} \ .$$

On the other hand, if $\|\eta(x(t),t)\| \leq \varepsilon$, then

$$\dot{\tilde{V}} \leq -\lambda_1 \tilde{V} + 2\|B^T P z\|\rho + 2\|B^T P z\|\rho \leq -\lambda_1 \tilde{V} + 4\varepsilon \ .$$

Hence, a.e. on $[t_o, t_1)$,

$$\dot{\tilde{V}}(t) \leq -\lambda_1 \tilde{V}(t) + 4\varepsilon \ . \tag{7.10}$$

Since (7.4) and (7.10) hold for any solution $z(\cdot)$ of (7.1), one may readily show using standard arguments that (7.1) has continuation of solutions and $\mathcal{B}$-tracks the zero function, where $\mathcal{B}$ is any neighbourhood of

$$\mathcal{E}_\varepsilon^f = \left\{ z \in \mathbb{R}^n : \ z^T P z \leq 4\lambda_1^{-1}\varepsilon \right\} \ . \qquad \blacksquare$$

Proof of Theorem 5.1: The proof of Theorem 5.1 proceeds in the same manner as that of Theorem 4.1. However, in place of (7.1), one has

$$\dot{z} = \bar{A} z + B[h + E\psi] + B[I + E]p_\varepsilon - \xi^P, \tag{7.11}$$

from which one obtains

$$\dot{\tilde{V}}(t) \leq -\lambda_1 \tilde{V}(t) + 4\varepsilon - 2\, z^T(t) P \xi^P(t) \ . \tag{7.12}$$

But,

$$\begin{aligned} -z^T(t) P \xi^P(t) &\leq \|z(t)\|_P \|\xi^P(t)\|_P \\ &\leq c\|z(t)\|_P = c\,\tilde{V}^{\frac{1}{2}}(t) \end{aligned} \tag{7.13}$$

whence

$$\begin{aligned} \dot{\tilde{V}}(t) &\leq -\lambda_1 \tilde{V}(t) + 2c\tilde{V}^{\frac{1}{2}}(t) + 4\varepsilon \\ &= -\lambda_1 [\tilde{V}^{\frac{1}{2}}(t) + \ell_\varepsilon][\tilde{V}^{\frac{1}{2}}(t) - k_\varepsilon] \end{aligned} \tag{7.14}$$

where

$$\ell_\varepsilon = -c\lambda_1^{-1} + \sqrt{c^2\lambda_1^{-2} + 4\varepsilon\lambda_1^{-1}} \geq 0$$

and

$$k_\varepsilon = c\lambda_1^{-1} + \sqrt{c^2\lambda_1^{-2} + 4\varepsilon\lambda_1^{-1}} \geq 0 .$$

Since (7.4) and (7.14) hold for any solution of (7.11), one may readily show, using standard arguments, that (7.11) has continuation of solutions and tracks the zero function to within any neighbourhood of

$$E_\varepsilon = \left\{ z \in \mathbb{R}^n : \ \|z\|_P \leq k_\varepsilon \right\}.$$

■

8. REFERENCES

1. Gutman, S., and Leitmann, G. (1976) "Stabilizing feedback control for dynamical systems with bounded uncertainty", *Proc. IEEE Conf. Decision & Control.*

2. Gutman, S. (1979) "Uncertain dynamical systems - Lyapunov min-max approach", *IEEE Trans. Automatic Control,* AC-24, 437.

3. Leitmann, G. (1980) "Deterministic control of uncertain systems", *Acta Astronautica,* 7, 1457.

4. Corless, M., and Leitmann, G. (1981) "Continuous state feedback guaranteeing uniform ultimate boundedness for uncertain dynamic systems", *IEEE Trans. Automatic Control,* AC-26, 1139.

5. Barmish, B.R., and Leitmann, G. (1982) "On ultimate boundedness control of uncertain systems in the absence of matching conditions", *IEEE Trans. Automatic Control,* AC-27, 153.

6. Barmish, B.R., Corless, M., and Leitmann, G. (1983) "A new class of stabilizing controllers for uncertain dynamical systems", *SIAM J. Control & Optimiz.,* 21, 246.

7. Ryan, E.P., and Corless, M. (1984) "Ultimate boundedness and asymptotic stability of a class of uncertain dynamical systems via continuous and discontinuous feedback control", *IMA J. Math. Control & Information,* to appear.

8. Ryan, E.P., Leitmann, G., and Corless, M. "Practical stabilizability of uncertain dynamical systems: application to robotic tracking", *J. Optimiz. Theory & Applications,* to appear.

9. Slotine, J.J., and Sastry, S.S. (1983) "Tracking control of non-linear systems using sliding surfaces with application to robot manipulators", *Int. J. Control,* 38, 465.

10. Ryan, E.P. (1983) "A variable structure approach to feedback regulation of uncertain dynamical systems", *Int. J. Control,* 38, 1121.

11. Coddington, E.A., and Levinson, N. (1955) *Theory of Ordinary Differential Equations*, McGraw-Hill, New York.

12. Hale, J.K. (1980) *Ordinary Differential Equations*, Krieger.

ACKNOWLEDGEMENTS

This paper is based on research supported by the U.S. National Science Foundation, the U.S. Air Force Office of Scientific Research, and the U.K. Science & Engineering Research Council.

MODEL REFERENCE ADAPTIVE CONTROL FOR NONLINEAR PLANTS: AN APPLICATION TO POSITIONAL CONTROL OF MANIPULATORS

G. Ambrosino, G. Celentano and F. Garofalo
(Dipartimento di Informatica e Sistemistica, University of Naples, Italy)

ABSTRACT

In this paper we present a controller design procedure for positional control of industrial robots in the presence of parameter variations. The procedure is based on a methodology of feedback control, recently developed by the authors in[1], that makes use of a reference model to specify the design objectives and enables the designer to prescribe arbitrarily the rate of convergence of the model-plant error and its set of ultimate boundedness. We show that, taking advantage of the particular structure of the dynamical model of the robot and selecting a suitable reference model, the design procedure becomes simple and straightforward.

1. INTRODUCTION

The dynamic control of an industrial robot involves the determination of the input torques at each joint which enable the hand to pass through a prescribed sequence of corner points. If there are no path constraints and the work space is obstacle free, the control problem becomes a standard set-point regulation problem.

Conventional control methods do not work well for these problems because of both the nonlinear and coupled nature of the equations describing the dynamics of the manipulator[2,3], and the large system parameter variations mainly due to the variety of objects gripped by the manipulator, especially in flexible manufacturing environments.

If a detailed model of the whole system and information on load characteristics are available, it is possible to use nonlinear compensation techniques as in[4], or on-line computational schemes as in[5]. When only bounds on parameter variations are known, the most efficient approach for the solution is the so called sliding mode control[6]. The salient feature of this approach is that a 'sliding' state trajectory on a switching surface is enforced by means of discontinuous feedback controls. When in sliding mode the controlled system remains insensitive to parameter variations and disturbances[6]. This insensitivity property makes this technique useful and appealing in solving the manipulator control problem.

Control schemes based on the above-mentioned approach have been proposed by Young[7] and Balestrino et al.[8] The former uses the theory of variable structure systems[6] to solve directly both positional and tracking control problems. The latter utilizes a reference model to specify the desired dynamic behaviour of the controlled plant and uses the hyperstability theory to synthesize discontinuous adaptive control signals guaranteeing the tracking between model and plant outputs.

Unfortunately, in both cases, the price paid to obtain good performance and robustness to parameter variations is given by input torques chattering at very high frequency; these torques, in practice, are difficult or impossible to obtain with the hydraulic, pneumatic or electrical motors driving each joint of the manipulator.

In this paper we eliminate this drawback by using a model reference adaptive control scheme which makes use of continuous approximations of the previously mentioned discontinuous adaptation laws as already proposed in [1,9,10]. This control scheme guarantees that the model-plant error remains bounded and tends exponentially to an arbitrarily small neighbourhood of the origin. Furthermore, we suggest a procedure that enables the designer to increase arbitrarily the rate of convergence of the error simply by solving a pole placement problem.

In order to apply this adaptive control scheme to robots, the manipulator dynamical model, derived from classical Newtonian mechanics, has been suitably rearranged. The resulting form of the model makes the design procedure straightforward.

2. BACKGROUND

Consider the dynamic nonlinear plant

$$\dot{x} = a(x,t)+A(x,t)x+B(x,t)u+d(t) \tag{2.1}$$

where $x \in R^n$ is the state, $u \in R^m$ is the input, d is a bounded input disturbance, a(.,.) is a norm-bounded, nonlinear, time-varying vector, and A(.,.), B(.,.) are matrices of appropriate dimensions.

We assume that there exists a matrix $A_o \in R^{n\times n}$, a full rank matrix $B_o \in R^{n\times r}$, and two matrices $R_o(x,t) \in R^{r\times m}$, $H(x,t) \in R^{m\times r}$ such that:
$a(x,t) \in \mathcal{R}(B_o)$, $d(t) \in \mathcal{R}(B_o)$, $\mathcal{R}(A(x,t)-A_o) \subseteq \mathcal{R}(B_o)$, $B(x,t)=B_o R_o(x,t)$,

$R_o'(x,t)=R_o(x,t)H(x,t)$ is positive definite (p.d.) and bounded in norm, the pair (A_o, B_o) is controllable, $A(x,t)-A_o$ is bounded in norm.

Moreover we assume that the knowledge of the plant regards only the values of n and m, the matrices A_o, B_o and H(x,t), lower bounds of the norm $R_o'(x,t)$ and upper bounds of d(t), a(x,t) and $A(x,t)-A_o$.

The control objectives are specified in terms of the following reference model

$$\dot{\hat{x}} = \hat{A}\hat{x}+\hat{B}\hat{u} \tag{2.2}$$

where $\hat{x} \in R^n$ is the state, $\hat{u} \in R^m$ is the reference input, $\hat{A}$ is asymptotically stable and such that the pair $(\hat{A}, B_o)$ is controllable, and finally the pair $(\hat{A}, \hat{B})$ satisfies the so-called matching conditions

$$\mathcal{R}(\hat{B}) \subseteq \mathcal{R}(B_o), \quad \mathcal{R}(A(x,t)-\hat{A}) \subseteq \mathcal{R}(B_o). \tag{2.3}$$

A practical control problem can be that of designing a controller which results in bounded error between the state of the plant and that of the model. Moreover, it would be appreciable to have information on the rate of convergence of the error and on the size of its set of ultimate boundedness.

The following theorems address the problem of construction of such a controller.

Theorem 1. Consider system (2.1) and the reference model (2.2). Using a control law of the form

$$u = K_x x + K_{\hat{x}}\hat{x} + K_{\hat{u}}\hat{u} - H(x,t)\left(h_x(\varepsilon,x) + h_{\hat{x}}(\varepsilon,\hat{x}) + h_{\hat{u}}(\varepsilon,\hat{u}) + h_o(\varepsilon)\right) \quad (2.4)$$

where

$$\varepsilon = x - \hat{x} \quad (2.5)$$

$$h_z(\varepsilon,z) = h_z \frac{v}{\|v\|+\delta}\|z\| \,, \quad \delta>0, \quad z=x,\hat{x},\hat{u} \quad (2.6a)$$

$$h_o(\varepsilon) = h_o \frac{v}{\|v\|+\delta} \quad (2.6b)$$

$$v = B_o^T P\varepsilon \quad (2.7)$$

$$P: \quad E^T P + PE = -Q \,, \quad Q \text{ p.d.} \quad (2.8)$$

$$E = \hat{A} + B_o K_e \,, \quad K_e \text{ being an arbitrary matrix such that } E \text{ is asymptotically stable} \quad (2.9)$$

$$h_w > \{\max\left(\lambda_M(G_w),\ \lambda_M(G_w + (B_o^T P B_o)^{-1} B_o^T PE)\right) + \delta/\gamma\}/\lambda_m(R_o') \,, \quad \gamma>0 \,, \quad w=x,\hat{x} \quad (2.10a)$$

$$h_{\hat{u}} > \left(\lambda_M(G_{\hat{u}}) + \delta/\gamma\right)/\lambda_m(R_o') \quad (2.10b)$$

$$h_o > \left(\|g_o\|_M + \delta/\gamma\right)/\lambda_m(R_o') \quad (2.10c)$$

$$G_z = R_o K_z - C_z \,, \quad z=x,\hat{x},\hat{u} \quad (2.11a)$$

$$g_o = c_a + c_d \quad (2.11b)$$

$$C_x = B_o^{LM}(E - A(x,t)) \,, \quad C_{\hat{x}} = -K_e \,, \quad C_{\hat{u}} = B_o^{LM}\hat{B} \quad (2.12a)$$

$$c_a = B_o^{LM} a \,, \quad c_d = B_o^{LM} d \quad (2.12b)$$

the model-plant error ε remains uniformly bounded. Moreover, it tends to zero with a rate not less than an exponential one characterized by

$$1/T = 1/2\lambda_{max}(Q^{-1}P) \quad (2.13)$$

until region S_ρ, defined as

$$S_\rho = \{\varepsilon: \|v\| < \rho\} \ , \ \rho = \gamma \max\{\lambda_M(G_x), \lambda_M(G_{\hat{x}}), \lambda_M(G_{\hat{u}}), \|g_o\|_M\} \quad (2.14)$$

is reached; then it remains in this region and tends asymptotically to a ball of radius k_ρ, k being a suitable positive constant, with a rate not less than an exponential one characterized by the same time constant T as in (2.13). ΔΔ

Here $\lambda_m(F)$ ($\lambda_M(F)$ resp.) denotes the square root of the minimum (maximum resp.) eigenvalue of F^TF, $\|f\|_M$ the maximum value of the Euclidean norm of f, B_o^{LM} is the pseudoinverse of B_o given by $(B_o^T B_o)^{-1} B_o^T$, $\lambda_{max}(F)$ the maximum eigenvalue of square matrix F.

The proof of the above theorem follows immediately from some results given in [1].

The possibility of arbitrarily prescribing the time constant T is guaranteed by the following theorem.

Theorem 2. If the eigenvalues of E are real and distinct and we assume in (2.8)

$$Q = -2(Z^T)^{-1} \Lambda Z^{-1} \quad (2.15)$$

with Z such that

$$Z^{-1} E Z = \Lambda = \mathrm{diag}(\lambda_1, \ldots, \lambda_n) \quad (2.16)$$

then

$$P = (ZZ^T)^{-1} \quad (2.17)$$

$$T = -1/\lambda_{max}(E) \qquad \Delta\Delta \quad (2.18)$$

The proof of Theorem 2 is given in[1].

By using the above theorems, we can now provide a procedure for construction of a controller which satisfies the previously stated requirements. This is accomplished as follows:

i) choose a suitable reference model;

ii) choose an exponential rate of convergence $1/T$ and select a matrix E as in (2.9) having all its eigenvalues less than $-1/T$;

iii) determine P as in (2.17);

iv) compute $1/\lambda_m(R'_o)$;

v) choose suitable values for constants δ and γ;

vi) choose suitable linear feedback gains K_x, $K_{\hat{x}}$, $K_{\hat{u}}$ and select adaptation gains h_x, $h_{\hat{x}}$, $h_{\hat{u}}$ h_o satisfying (2.10).

3. ROBOT ARM MODEL AND STATEMENT OF THE PROBLEM

We refer to a ν degrees of freedom robot as a kinematical chain of $\nu+1$ rigid bodies (links) interconnected by ν joints. The relative displacement of each link is given by a joint variable q_i.

The equation of motion for the chain can be written as

$$J(q)\ddot{q}+c(q,\dot{q})+F(\dot{q})\dot{q}+g(q) = u \tag{3.1}$$

where

$$q = (q_1,\ldots,q_\nu)^T; \tag{3.2a}$$

$J(q)$ $\nu\times\nu$ positive definite symmetric inertia matrix; (3.2b)

$c(q,\dot{q}) = (\dot{q}^T C_1(q)\dot{q}, \ldots, \dot{q}^T C_\nu(q)\dot{q})^T$ $\nu\times 1$ vector defining Coriolis and centrifugal generalized forces; (3.2c)

$F(\dot{q}) = \mathrm{diag}(f_1(\dot{q}_1), \ldots, f_\nu(\dot{q}_\nu))$ $\nu\times\nu$ matrix defining the friction terms; (3.2d)

$g(q)$ $\nu\times 1$ vector defining the gravity terms; (3.2e)

u $\nu\times 1$ vector of input generalized forces. (3.2f)

Defining the state vector x as

$$x^T = (q^T\ \dot{q}^T) = (q^T\ w^T) \tag{3.3}$$

eqn (3.1) can be rewritten as

$$\dot{x} = a(x)+A(x)x+B_o R_o(x)u \tag{3.4}$$

where

$$a(x) = a(q) = \begin{pmatrix} 0 \\ -J^{-1}(q)g(q) \end{pmatrix} \tag{3.5a}$$

$$A(x) = \begin{pmatrix} 0 & I_\nu \\ 0 & -J^{-1}(q)\begin{pmatrix} w^T C_1(q) + f_1(1..0) \\ \vdots \\ w^T C_\nu(q) + f_\nu(0..1) \end{pmatrix} \end{pmatrix} \tag{3.5b}$$

$$B_o = \begin{pmatrix} 0 \\ I \end{pmatrix}; \quad R_o(x) = J^{-1}(q). \tag{3.5c}$$

In this paper we consider the so-called 'set-point regulation' problem. For the given initial state $x^T(0) = (q^T(0), \dot{q}^T(0))$ and the desired final state $x_d^T = (q_d^T, 0)$ find a feedback control $u(x)$ such that $x(t) \to x_d$.

To overcome the difficulties arising from the nonlinear and coupled nature of the problem and from variations in the loads, one can ask the manipulator to follow, as closely as possible, a linear, time invariant reference model as in (2.2); so doing, the problem becomes that of synthesizing a suitable adaptive control law guaranteeing accurate model tracking. In the next section we shall give a procedure to synthesize $u(x)$ based on the results given in Section 2.

4. CONTROLLER DESIGN

In accordance with the design procedure given in Section 2, the first step consists in the choice of a reference model. In this paper we propose a system made up of ν decoupled, linear, time invariant, second order subsystems of the form:

$$\ddot{\hat{q}}_i + \hat{a}_{i1}\dot{\hat{q}} + \hat{a}_{i2}\hat{q}_i = \hat{b}_i\hat{u}_i, \quad i=1,\dots,\nu \tag{4.1}$$

each one specifying the behaviour of a single joint. With such a choice, letting

$$\hat{x} = (\hat{q}_1, \dots, \hat{q}_\nu, \dot{\hat{q}}_1, \dots, \dot{\hat{q}}_\nu)^T, \quad \hat{u} = (\hat{u}_1, \dots, \hat{u}_\nu) \tag{4.2}$$

we have

$$\hat{A} = \begin{pmatrix} O & I_\nu \\ \hat{A}_{21} & \hat{A}_{22} \end{pmatrix}, \quad \hat{B} = \begin{pmatrix} O \\ \hat{B}_2 \end{pmatrix} \tag{4.3}$$

where

$$\hat{A}_{21} = -\text{diag}(\hat{a}_{12}, \ldots, \hat{a}_{\nu 2}), \quad \hat{A}_{22} = -\text{diag}(\hat{a}_{11}, \ldots, \hat{a}_{\nu 1}) \tag{4.4a}$$

$$\hat{B}_2 = \text{diag}(\hat{b}_1, \ldots, \hat{b}_\nu). \tag{4.4b}$$

Assuming that an exponential rate of convergence $1/T$ of the error is prescribed, the second step of the design is the choice of matrix E in (2.9). Taking into account the structures of $\hat{A}$ and B_o given by (4.3) and (3.5c) respectively, we can assume

$$E = \begin{pmatrix} O & I_\nu \\ -E_{21} & -E_{22} \end{pmatrix} \tag{4.5a}$$

$$E_{21} = \text{diag}(e_{12}, \ldots, e_{\nu 2}), \quad E_{22} = \text{diag}(e_{11}, \ldots, e_{\nu 1}) \tag{4.5b}$$

so that one has only to choose suitable values for e_{ij}, $i=1,..,\nu$, $j=1,2$. These, in accordance with Theorem 2, must be chosen so as to guarantee that all eigenvalues of matrix E will be real, distinct and less than $-1/T$. This can be accomplished in a very simple way noting that the eigenvalues of matrix E are the roots λ_{i1}, λ_{i2} of the equations

$$\lambda^2 + e_{i1}\lambda + e_{i2} = 0, \quad i=1,\ldots,\nu. \tag{4.6}$$

Indeed, according to (4.5b), the i-th and $(i+\nu)$-th column of matrix E in (4.5a) can be rewritten as

$$\begin{pmatrix} O \\ -r_i\lambda_{i1}\lambda_{i2} \end{pmatrix}, \quad \begin{pmatrix} r_i \\ r_i(\lambda_{i1} + \lambda_{i2}) \end{pmatrix} \tag{4.8}$$

where r_i is the i-th column of the identity matrix I_ν. From (4.5b) we have

$$(\lambda_{ij} I_2 - E) \begin{pmatrix} r_i \\ r_i\lambda_{ij} \end{pmatrix} = O, \quad i=1,\ldots,\nu, \; j=1,2 \tag{4.9}$$

so we can conclude that λ_{ij} are exactly the eigenvalues of matrix E and

$$u_{ij} = \begin{pmatrix} r_i \\ r_i \lambda_{ij} \end{pmatrix}, \quad i=1,\ldots,\nu, \quad j=1,2 \tag{4.10}$$

the corresponding eigenvectors.

Expression (4.10) for the eigenvectors of the matrix E turns out to be very useful in the third step of the design procedure, i.e. the determination of the matrix P.

If we assume, as suggested by Theorem 2, the matrix P given by (2.17), in virtue of (4.10) we can write

$$P = \begin{bmatrix} 2I_\nu & \mathrm{diag}\{\lambda_{i1}+\lambda_{i2}\}_{i=1}^{\nu} \\ \mathrm{diag}\{\lambda_{i1}+\lambda_{i2}\}_{i=1}^{\nu} & \mathrm{diag}\{\lambda_{i1}^2+\lambda_{i2}^2\}_{i=1}^{\nu} \end{bmatrix}^{-1} \tag{4.11}$$

and hence

$$P = \begin{bmatrix} \mathrm{diag}\left\{\dfrac{\lambda_{i1}^2+\lambda_{i2}^2}{(\lambda_{i1}-\lambda_{i2})^2}\right\}_{i=1}^{\nu} & -\,\mathrm{diag}\left\{\dfrac{\lambda_{i1}+\lambda_{i2}}{(\lambda_{i1}-\lambda_{i2})^2}\right\}_{i=1}^{\nu} \\ -\mathrm{diag}\left\{\dfrac{\lambda_{i1}+\lambda_{i2}}{(\lambda_{i1}-\lambda_{i2})^2}\right\}_{i=1}^{\nu} & \mathrm{diag}\left\{\dfrac{2}{(\lambda_{i1}-\lambda_{i2})^2}\right\}_{i=1}^{\nu} \end{bmatrix} \tag{4.12}$$

Eqn (4.12) gives the matrix P directly in terms of eigenvalues of the matrix E.

The fourth step of the procedure consists in the evaluation of $1/\lambda_m(R_o')$ which, in view of (3.5d), choosing $H=I_\nu$, becomes

$$1/\lambda_m(R_o') = 1/\lambda_m(J^{-1}(q)) = \lambda_{max}(J(q)). \tag{4.13}$$

The computation of $\lambda_{max}(J(q))$ can be performed by using any one of the search methods proposed in the literature; a good choice for the initial starting point could be a vector q maximizing the diagonal elements of the inertia matrix J(q); such a vector can often be obtained by means of simple mechanical considerations.

The choice of constants γ and δ (fifth step of the procedure) results from a compromise between tracking accuracy and chattering in the control input that can arise when the controller is implemented by means of a digital computer. Indeed, at a parity of γ, a low value for δ yields control signals of high value, also in the presence of small errors; therefore, if the sampling interval is not sufficiently small, changes in the error sign at the sampling frequency, and hence the chattering phenomenon, can occur. On the other hand, it can be easily noted that high values for δ can considerably deteriorate the tracking error.

To complete the design procedure, we give some criteria to choose the linear feedback matrices K_z, and to select the adaptation gains h_z.

A good choice for the matrices K_z is that minimizing the adaptation gains given by (2.10), (2.11), (2.12). To simplify the evaluation of such adaptation gains, we note that, since

$$B_O^{LM} = (O \quad I_\nu) \tag{4.14}$$

from (2.12), (3.5), (4.3), (4.4), (4.5) we have

$$C_x = \left(-\mathrm{diag}(e_{12},\dots,e_{\nu 2}) \quad -\mathrm{diag}(e_{11},\dots,e_{\nu 1}) \right.$$
$$\left. +J^{-1}(q) \begin{pmatrix} w^T C_1(q) + f_1 (1..0) \\ \vdots \\ w^T C_\nu(q) + f_\nu (0..1) \end{pmatrix} \right) \tag{4.15a}$$

$$C_{\hat{x}} = (\mathrm{diag}(e_{12}-\hat{a}_{12},\dots,e_{\nu 2}-\hat{a}_{\nu 2}) \quad \mathrm{diag}(e_{11}-\hat{a}_{11},\dots,e_{\nu 1}-\hat{a}_{\nu 1})) \tag{4.15b}$$

$$C_{\hat{u}} = \mathrm{diag}(b_1,\dots,b_\nu) \tag{4.15c}$$

$$c_a = -J^{-1}(q)g(q). \tag{4.15d}$$

Moreover, by simple manipulations, we have

$$(B_o^T P B_o)^{-1} B_o^T P E = (-\mathrm{diag}(e_{12},\dots,e_{\nu 2}) \quad -\tfrac{1}{2}\mathrm{diag}(e_{11},\dots,e_{\nu 1})) \tag{4.16}$$

so that the selection of the adaptation gains can be easily carried out only knowing the bounds of the uncertain parameters of the manipulator.

5. CONCLUSIONS

The paper has shown that a methodology of adaptive feedback control for nonlinear plants, recently proposed by the authors, can be successfully utilized for the positional control of robot arms. In this approach the design objectives are specified in terms of a linear reference model; the proposed controller guarantees accurate tracking between the state of the model and that of the plant also in presence of load variations.

It has been shown that the approach, though complicated nationally, can be considerably simplified by a suitable choice of the reference model and by a rearrangement of the dynamic equations of the manipulator; the consequence is that, by means of standard manipulations, the design problem boils down to the choice of some parameters satisfying certain inequalities.

6. REFERENCES

1. Ambrosino, G., Celentano, G., and Garofalo, F., "Robust Model Tracking Control for a Class of Nonlinear Plants", IEEE Trans. Automat. Contr., to appear.

2. Paul, R.P.C., (1981) "Robot Manipulators: Mathematics, Programming and Control", MIT Press.

3. Bejczy, A.K., (1974) "Robot Arm, Dynamics and Control", Tech. Memo 33-699, Jet Propulsion Laboratory.

4. Freund, E., (1977) "A Nonlinear Control Concept for Computer Controlled Manipulators", IFAC Symp. on Multivariable Control Systems, Toronto, Canada.

5. Luh, J.Y.S., Walker, M.W., and Paul, R.P.C. (1980) "On-Line Computational Scheme for Mechanical Manipulators", ASME J. of Dynamic Systems, Measurements, and Control, vol.102/69.

6. Utkin, V.I., (1977) "Sliding Modes and Their Application to Variable Structure Systems", MIR, Moscow.

7. Young, K.K.D., (1978) "Controller Design for a Manipulator Using Theory of Variable Structure Systems", IEEE Trans. on Syst., Man, and Cyb., vol.SMC-8, n.2.

8. Balestrino, A., De Maria, G., and Sciavicco, L., (1983) "An Adaptive Model Following Control for Robotic Manipulators", ASME J. of Dynamic Systems, Measurements, and Control, vol.105.

9. Corless, M., and Leitmann, G., (1981) "Continuous State Feedback Guaranteeing Uniform Ultimate Boundedness for Uncertain Dynamic Systems", IEEE Trans. Automat. Contr., vol.AC-26.

10. Gutman, S., and Palmor, Z., (1982) "Properties of Min-Max Controllers in Uncertain Dynamical Systems", SIAM J. Contr. Opt., vol.20.

IMPROVEMENT OF ROBUSTNESS VIA HYPERSTABLE VARIABLE STRUCTURE CONTROL

H.-P. Opitz

(Hochschule der Bundeswehr, Hamburg, FRG)

ABSTRACT

To improve robustness of a linear system, a dynamic variable structure controller is proposed. For the design of such a non-linear control system the hyperstability theory by V. M. Popov is applied. The implementation of this feedback control system is very simple because only knowledge of the measurement vector is needed.

The resulting closed loop system is robust against bounds on control and parameter uncertainties which belong to an admissible set of disturbances. By means of the hyperstability theory sufficient algebraic conditions are derived for this admissible set.

1. INTRODUCTION

Problems arise in the syntheses of control systems whenever there are bounds on control variables as well as large parameter variations, which change the dynamics of the whole system in a wide range.

This paper is concerned with stable linear, time-invariant plants with lumped parameters. For such systems some methods exist in order to design a linear state feedback law, which guarantees the stability in presence of bounds on control variables and parameter uncertainties[1,2]. However, in most cases this can only be achieved if one puts up with a very conservative dynamic behaviour of the entire system. One could imagine that a nonlinear dynamic controller is able to avoid this disadvantage. A method will be proposed which is characterized by adding a certain bilinear term to a constant output feedback, thus giving rise to variable structure control. The method turns out to be simply implementable.

2. PROBLEM FORMULATION

Let the stable linear, time-invariant plant to be controlled be given in the following form:

$$\dot{\underline{x}} = \underline{A}(\underline{\alpha})\underline{x} + \underline{B}(\underline{\alpha})\underline{u}, \quad \underline{x}(t_o) = \underline{x}_o, \tag{2.1}$$

$$\underline{y} = \underline{C}\,\underline{x}, \tag{2.2}$$

$$\underline{y}_M = \underline{C}_M\underline{x}, \tag{2.3}$$

with the n-dimensional state vector $\underline{x}$ of the plant, the m-dimensional input vector $\underline{u}$, the q-dimensional output vector $\underline{y}$ and the r-dimensional measurement vector $\underline{y}_M$. The disturbances of this plant are described by variations of a special fictitious parameter vector $\underline{\alpha}$ in the matrices $\underline{A}$ and $\underline{B}$. The vector $\underline{\alpha}$ should be an element of $\overline{M^*}\subset M$, M: all possible disturbances.

The problem now is to find a parameter vector $\underline{p}(t)$ in such a way that the nonlinear control law:

$$\underline{u} = -\underline{K}\,\underline{y} + \underline{G}\,\underline{w} + \sum_{i=1}^{m} p_i(t)\underline{N}_i\underline{y}, \tag{2.4}$$

$\underline{w}$: reference input,

$\underline{K}$: constant gain matrix,

$\underline{N}_i, \underline{G}$: known constant matrices,

guarantees the global asymptotic stability of the overall nonlinear control system. Moreover, it should be able to exhaust given bounds on control,

$$u_{i\,min} \leqq u_i \leqq u_{i\,max}, \quad i = 1, \ldots, m \tag{2.5}$$

aiming at an improvement of the dynamic behaviour, and parameter variations $\underline{\alpha}$ with $\underline{\alpha}\in\overline{M^*}$ of the plant should be allowable. Franke[3,4,5,6] has shown that it is possible to synthesize such a parameter adjustment law

$$\underline{p} = \underline{p}(\underline{x},\underline{y},\underline{w}) \tag{2.6}$$

by means of Ljapunov's direct method.

However, in order to realize this adjustment law it is necessary to measure the entire state vector. Normally this requirement cannot be satisfied. In order to avoid the use of an adaptive observer (remember the large parameter variations of the plant), here it should be ensured that only known quantities, such as $\underline{y}$, $\underline{y}_M$, and $\underline{w}$ enter into the parameter adjustment law:

$$\underline{p} = \underline{p}(\underline{y}_M, \underline{y}, \underline{w}). \tag{2.7}$$

In order to guarantee the global asymptotic stability of the nonlinear system (2.1) - (2.4) or (2.7) respectively, and to find a whole class of different parameter adjustment laws, use will be made of the hyperstability theory by V. M. Popov[7,8]. In the past this method has been applied for the design of adaptive control systems.

3. DESIGN OF THE NONLINEAR CONTROLLER

Let

$$\underline{p}_s = \underline{O},\ \underline{x}_s = -\left[\underline{A}(\underline{\alpha}_o) - \underline{B}(\underline{\alpha}_o)\underline{K}\right]^{-1}\underline{B}(\underline{\alpha}_o)\underline{G}\,\underline{w}_s$$

be the steady-state vectors of $\underline{p}$ and $\underline{x}$, if the plant is in the so-called nominal case $\alpha = \underline{\alpha}_o$ = const. To ensure global asymptotic stability of $\underline{x}_s$ in this nominal case it is helpful to rewrite equations (2.1) - (2.7) by introducing deviations from the pre-assigned steady state:

$$\underline{\Delta x} = \underline{x} - \underline{x}_s,\ \underline{\Delta p} = \underline{p} - \underline{p}_s,$$

$$\underline{\Delta y} = \underline{y} - \underline{y}_s,\ \underline{\Delta y}_M = \underline{y}_M - \underline{y}_{Ms},$$

$$\underline{\Delta u} = \underline{u} - \underline{u}_s,$$

$$\underline{y}_s = \underline{C}\,\underline{x}_s,\quad \underline{y}_{Ms} = \underline{C}_M\underline{x}_s,\quad \underline{u}_s = -\underline{K}\,\underline{C}\,\underline{x}_s + \underline{G}\,\underline{w}_s.$$

The system's equations in the nominal case can now be rewritten as

$$\underline{\Delta \dot{x}} = \underline{A}^*\underline{\Delta x} + \underline{B}(\underline{\alpha}_o)\underline{\Delta u}, \tag{3.1}$$

$$\underline{\Delta y} = \underline{C}\,\underline{\Delta x}, \tag{3.2}$$

$$\underline{\Delta y}_M = \underline{C}_M\underline{\Delta x}, \tag{3.3}$$

$$\underline{\Delta u} = \sum_{i=1}^{m} \Delta p_i \underline{N}_i(\underline{\Delta y} + \underline{y}_s), \tag{3.4}$$

$$\underline{\Delta p} = \underline{\Delta p}(\underline{\Delta y}_M, \underline{\Delta y}, \underline{w}), \tag{3.5}$$

$$\underline{A}^* := \underline{A}(\underline{\alpha}_o) - \underline{B}(\underline{\alpha}_o)\underline{K}\,\underline{C},\ \text{stable}. \tag{3.6}$$

The theory of hyperstability presumes that the linear part

of the system, i.e. equations (3.1) - (3.3), has an identical number of input and output components, thus defining an inner product of these two vectors. The dimensions of $\underline{\Delta u}$ and $\underline{\Delta y}_M$ are only then equal when the dimension r of the measurement vector $\underline{\Delta y}_M$ - in the sense of the hyperstability theory defined as output - is equal to m. This restriction can be avoided by introducing the new vector $\underline{\Delta \hat{y}}$ with the following properties:

a) $\underline{\Delta \hat{y}}$ has the same dimension as $\underline{\Delta u}$, i.e. m,

b) $\underline{\Delta \hat{y}}$ satisfies the equation

$$\underline{\Delta \hat{y}} = \underline{D}_K \underline{\Delta y}_M + \underline{D}\,\underline{\Delta u}, \tag{3.7}$$

with the constant matrices $\underline{D}_K$ and $\underline{D}$, which are not yet determined, $\underline{\Delta \hat{y}}$ now being interpreted as output vector of the linear part of the system. The resulting block diagram of the entire nonlinear system is shown in Fig. 1.

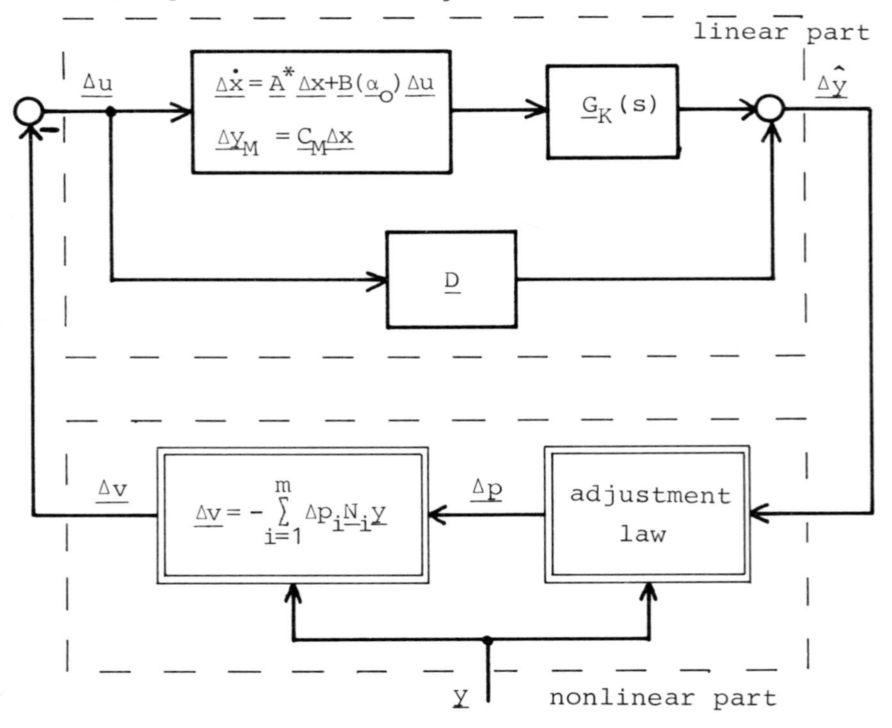

Fig. 1. Block diagram of the nonlinear control system

This nonlinear system is global asymptotically stable if the conditions of the hyperstability theory[7,8] hold:

a) $$\underline{G}^*(s) = \underline{D} + \underline{D}_K\underline{C}_M(s\underline{I} - \underline{A}^*)^{-1}\underline{B}(\underline{\alpha}_o) \quad (3.8)$$

is a strictly positive real transfer matrix,

b) the Popov-Integral inequality

$$\eta(0,T) = \int_0^T \underline{\Delta v}'\underline{\Delta \hat{y}}\, dt \geqq -\beta_o^2, \quad \forall T>0 \quad (3.9)$$

is satisfied.

The first demand is equivalent to the condition that there exists a solution of the equations of the Kalman-Yakubovich-Lemma[8]:

$$\underline{A}^{*\prime}\underline{P} + \underline{P}\,\underline{A}^* = -\underline{L}\,\underline{L}', \quad (3.10)$$

$$\underline{D}_K\underline{C}_M - \underline{B}'(\underline{\alpha}_o)\underline{P} = \underline{V}'\underline{L}', \quad (3.11)$$

$$\underline{D} + \underline{D}' = \underline{V}'\underline{V}, \quad (3.12)$$

namely matrices $\underline{L}$, $\underline{V}$ and $\underline{P}$, $\underline{P}$ positive definite. These equations can be utilized for the determination of the free matrices $\underline{D}_K$ and $\underline{D}$ by means of an additional performance index J.

$$J = \frac{1}{4}\,\mathrm{tr}(\underline{D} + \underline{D}')^2 \overset{!}{=} \min_{\underline{L},\underline{D}_K} .$$

This parameter optimization aims at very small absolute values of the elements of the matrix $\underline{D}$.

J has to be minimized by variation of $\underline{L}$ and $\underline{D}_K$ under the condition that the equations (3.10) - (3.12) are satisfied simultaneously. In any case a solution $\underline{D}_K$ and $\underline{D}$ can be found by means of numerical methods. The requirement (3.9) with (3.4), (3.7), $\underline{\Delta v} = -\underline{\Delta u}$ and the definitions

$$\gamma_i := (\underline{\Delta y} + \underline{y}_s)'\underline{N}_i'\underline{D}_K\underline{\Delta y}_M, \quad \underline{\gamma} = (\gamma_i),$$

$$g_{ij} := \frac{1}{2}(\underline{\Delta y} + \underline{y}_s)'\underline{N}_i'(\underline{D} + \underline{D}')\underline{N}_j(\underline{\Delta y} + \underline{y}_s), \quad \underline{G} = (g_{ij})$$

yield the new integral inequality:

$$\int_0^T \underline{\Delta p}'(\underline{\gamma} + \underline{G}\,\underline{\Delta p})\, dt \leqq \beta_o^2, \quad \forall T>0. \quad (3.13)$$

A sufficient criterion for a solution of (3.13) is the negative semidefinite choice of the whole left side of this inequality. By means of the Halanay-condition[9] for a positive semidefinite integral it can be shown that all (in general implicit) parameter adjustment laws with the principal structure (with $\underline{p}_s = \underline{0}$):

$$\underline{p} = -\int_0^t \underline{\tilde{K}}(t-\tau)[\underline{\gamma} + \underline{G}\,\underline{p} + \underline{R}(\underline{y}_M, \underline{y}, \underline{w})\underline{p}]d\tau, \qquad (3.14)$$

$\underline{R}$: any positive semidefinite square matrix,

$\underline{\tilde{K}}$: square transfer matrix with positive real Laplace-transform

guarantee (3.13). If it is possible to select $\underline{D} = \underline{0}$ in (3.10) - (3.12), (3.14) will be an explicit adjustment law for $\underline{p}$.

Therefore, in the nominal case $\underline{\alpha} = \underline{\alpha}_0$ the following class of nonlinear controllers guarantees global asymptotic stability when applied to system (2.1) - (2.3):

$$\underline{u} = -\underline{K}\,\underline{y} + \underline{G}\,\underline{w} + \sum_{i=1}^{m} p_i \underline{N}_i \underline{y}, \text{ control law}, \qquad (3.15)$$

$$\underline{p} = -\int_0^t \underline{\tilde{K}}(t-\tau)[\underline{\gamma} + \underline{G}\,\underline{p} + \underline{R}\,\underline{p}]\,d\tau, \text{ adjustment law}, \qquad (3.16)$$

$$\left.\begin{aligned} \gamma_i &= \underline{y}'\underline{N}_i'\underline{D}_K(\underline{y}_M - \underline{y}_{Ms}), && (3.17) \\ g_{ij} &= \frac{1}{2}\,\underline{y}'\underline{N}_i'(\underline{D} + \underline{D}')\underline{N}_j\underline{y}. && (3.18) \end{aligned}\right\} \text{input}$$

4. ROBUSTNESS PROPERTIES OF THIS CLASS OF NONLINEAR CONTROL SYSTEMS

The general form of this class of nonlinear control laws offers a large number of degrees of freedom, which can be utilized to satisfy robustness demands. In this paper robustness properties against bounds on control variables and parameter variations of the plant are of interest.

The positive semidefinite matrix $\underline{R}(\underline{y}_M, \underline{y}, \underline{p})$ in (3.14) can be specified in such a way as to guarantee and to exhaust bounds on control of the type (2.5) without endangering stability. For this purpose the matrices $\underline{N}_i$ are selected in the special form

$$\underline{N}_i = \begin{bmatrix} \underline{o}' \\ \vdots \\ \underline{n}_i' \\ \vdots \\ \underline{o}' \end{bmatrix}, \quad i = 1, \ldots, m$$

and $\underline{\tilde{K}}(t-\tau)$ is diagonal. This has been proved in [4,10] and therefore need to be repeated here in detail.

For robustness properties in case of parameter variations of the plant it is assumed that the dynamic matrix $\tilde{\underline{K}}$ of the adjustment law will contain an integral part. The system's equations with disturbances can be written as

$$\dot{\underline{x}} = \underline{A}(\underline{\alpha})\underline{x} + \underline{B}(\underline{\alpha})\underline{u},$$

$$\underline{y} = \underline{C}\,\underline{x},$$

$$\underline{y}_M = \underline{C}_M\underline{x},$$

$$\underline{A}(\underline{\alpha}) := \underline{A}(\underline{\alpha}_o) + \underline{\Delta A}(\underline{\alpha}),$$

$$\underline{B}(\underline{\alpha}) := \underline{B}(\underline{\alpha}_o) + \underline{\Delta B}(\underline{\alpha}).$$

The matrices $\underline{\Delta A}$ and $\underline{\Delta B}$ contain the disturbances in dependence of the fictitious parameter vector $\underline{\alpha}\varepsilon M$. If now control law (3.15) - (3.18) is used the question arises which set of disturbances will not endanger stability of the overall nonlinear system. Here an additional demand is that in admissible set of disturbances $(\underline{\Delta A},\underline{\Delta B})$, which is an element of $M^*\subset M$ should guarantee the same steady state as in the undisturbed case. A detailed analysis[10] shows that by means of the hyperstability theory one can derive a set of sufficient algebraic conditions, shown here:

a) Conditions to ensure the steady state

$$\det \underline{A}(\underline{\alpha}) \neq 0,$$

$$\det \left[\tfrac{1}{2}(\underline{D} + \underline{D}') - \underline{D}_K\underline{C}_M\underline{A}^{-1}(\underline{\alpha})\underline{B}(\underline{\alpha})\right] \neq 0,$$

$$\underline{C}\,\underline{A}^{-1}(\underline{\alpha})(\underline{I} + \underline{B}(\underline{\alpha})\underline{E})(\underline{\Delta A}\,\underline{x}_s + \underline{\Delta B}\,\underline{u}_s) \stackrel{!}{=} \underline{0}.$$

b) Conditions to ensure the global asymptotic stability

in the case $\underline{D} \neq \underline{0}$:

$$\underline{D}_K\underline{C}_M\underline{A}^{-1}(\underline{\alpha})(\underline{\Delta A}\,\underline{x}_s + \underline{\Delta B}\,\underline{u}_s) \stackrel{!}{=} \underline{0},$$

$\underline{G}(s) = \underline{D} + \underline{D}_K\underline{C}_M(s\underline{I} - \hat{\underline{A}}^*)^{-1}\underline{B}(\underline{\alpha})$ is strictly positive real,

$$\hat{\underline{A}}^* := \underline{A}(\underline{\alpha}) - \underline{B}(\underline{\alpha})\underline{K}\,\underline{C},$$

in the case $\underline{D} = \underline{0}$:

$\tilde{\underline{G}}(s) = \underline{D}_K\underline{C}_M(sI - \tilde{\underline{A}})^{-1}\underline{B}(\underline{\alpha})$ is strictly positive real,

$\tilde{\underline{A}} := \underline{A}(\underline{\alpha}) - \underline{B}(\underline{\alpha})\underline{K}\,\underline{C} + \sum_{i=1}^{m} p^*_{is}\underline{B}(\underline{\alpha})\underline{N}_i\underline{C}$, with

$$\underline{E} := [\tfrac{1}{2}(\underline{D} + \underline{D}') - \underline{D}_K \underline{C}_M \underline{A}^{-1}(\underline{\alpha})\underline{B}(\underline{\alpha})]^{-1} \underline{D}_K \underline{C}_M \underline{A}^{-1}(\underline{\alpha}),$$

$\underline{p}_s$, $\underline{x}_s$, $\underline{u}_s$: steady state without disturbances,

$\underline{p}_s^*$: steady state in the case of disturbances.

These conditions can be evaluated for each $\underline{\alpha}$ in a simple manner by numerical or analytical methods.

5. EXAMPLE

Let the stable linear, time-invariant disturbed plant to be controlled be given in the form

$$\dot{\underline{x}} = \begin{bmatrix} \dot{x}_1 \\ \dot{x}_2 \end{bmatrix} = \begin{bmatrix} -17 + \Delta a_{11} & -462 + \Delta a_{12} \\ 3.33 + \Delta a_{21} & 0 \end{bmatrix} \underline{x} + \begin{bmatrix} 479 + \Delta b_1 \\ 0 \end{bmatrix} u,$$

$$y = [0,1]\underline{x},$$

$$\underline{y}_M = \underline{x},$$

with disturbances $M = \{\Delta a_{11}, \Delta a_{12}, \Delta a_{21}, \Delta b_1\}$.

The design of a hyperstable variable structure control law yields:

$$u = 0.48 + p \cdot y,$$

$$p = -10\gamma - 100 \int_0^t \gamma d\tau,$$

$$\gamma = (0.1\, x_1 + 0.274\, x_2 - 0.137)x_2.$$

The above conditions now determine the set of admissible disturbances, not endangering stability and steady state. An analytic evaluation yields

$$M^* = \{\Delta a_{11}, \Delta a_{12}, \Delta a_{21}, \Delta b_1 : \Delta a_{11} < 17, \Delta a_{21} > -3.33, \Delta b_1 > -479,$$
$$0.1\, \Delta a_{11} + 0.274\, \Delta a_{21} < 0.78, |0.5\, \Delta a_{12} + 0.48\, \Delta b_1| < 0.48 \cdot$$
$$\cdot (479 + \Delta b_1)\}.$$

For the selection of five extreme parameter variations $\in M^*$:

	a)	b)	c)	d)	e)
Δa_{11}	- 3.78	4.25	0	-4.25	0
Δa_{12}	385	269.5	0	-631.25	- 231
Δa_{21}	2.917	- 1.663	0	2.99	- 1.66
Δb_1	53.22	-79.83	0	119.7	0 ,

the following two figures show the improvement of robustness via variable structure control. Fig. 2 shows the step response of the plant output y without any controller, Fig. 3 with variable structure control.

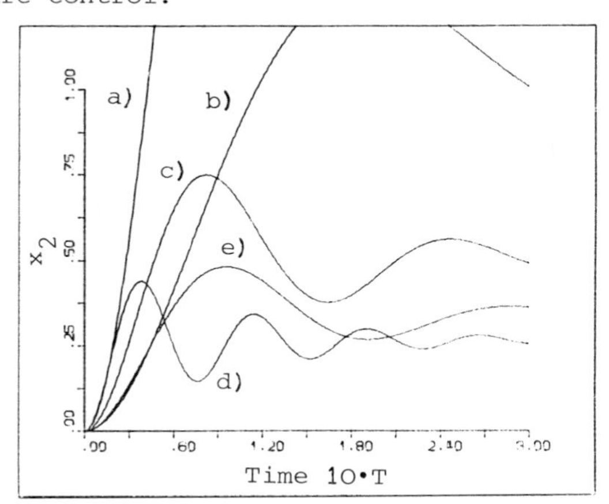

Fig. 2. Plant output without control for parameter variations a) - e)

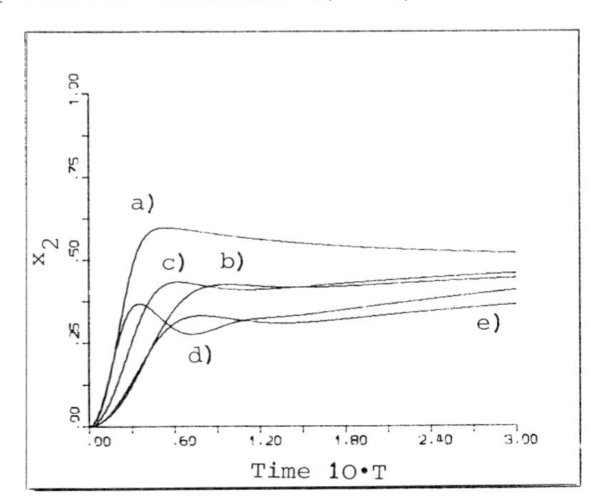

Fig. 3. Plant output with variable structure control for parameter variations a) - e)

The improvement of the dynamic behaviour is obvious and even better than results known before[6].

6. CONCLUSION

A non-linear dynamic controller has been proposed for variable structure feedback control of linear systems being subject to parameter disturbances and to bounds on control variables. An extension of these results for discrete-time systems is possible by using the discrete version of the hyperstability theory to synthesize discrete-time variable structure control systems[10]. In the discrete case all integral-inequalities have to be replaced by sum-inequalities. Especially the demand for a strictly discrete positive real transfer matrix $\underline{G}(z)$ in the discrete case complicates the design.

7. REFERENCES

1. Ackermann, J., (1981), Parameter Space Design of Robust Control Systems. IEEE Trans. Aut. Control, Vol. AC-26, pp. 1058-1072.

2. Kiendl, H., (1982), Calculation of the Maximum Absolute Value of Dynamical Variables in Linear Control Systems. IEEE Trans. Aut. Control, Vol. AC-27, pp. 86-89.

3. Franke, D., (1982), Variable structure control without sliding modes (in German). Regelungstechnik 30, pp. 271-276.

4. Franke, D., (1982), Exhausting bounds on control by means of soft variable structure control (in German). Regelungstechnik 30, pp. 348-355.

5. Franke, D., (1982), Soft Variable Structure Control on Non-Linear Distributed Parameter Systems. Fifth International Conference on Analysis and Optimization of Systems, Versaille/France.

6. Franke, D., (1983), A nonlinear dynamical controller with adaptive properties (in German). Regelungstechnik 31, pp. 369-374.

7. Popov, V.M., (1973), Hyperstability of Control Systems. Springer-Verlag, Berlin-Heidelberg-New York.

8. Landau, Y.D., (1979), Adaptive Control. Marcel Dekker Inc., New York.

9. Halanay, A., (1964), Noyaux positivement définis et stabilité des systéme automatique. C.R. Acad. Sc. Paris, t. 258.

10. Opitz, H.-P., (1984), Entwurf robuster, strukturvariabler Regelungssysteme mit der Hyperstabilitätstheorie, Fortschritt-Berichte der VDI-Zeitschriften, Reihe 8 Nr. 75, VDI-Verlag, Düsseldorf.

A ROBUSTNESS TEST FOR DISTRIBUTED FEEDBACK SYSTEMS

I. Postlethwaite and Y. K. Foo

(Department of Engineering Science, University of Oxford)

ABSTRACT

A robustness test is presented for a class of distributed feedback systems. The test does not require the plant and perturbed plant models to have the same number of unstable poles.

1. INTRODUCTION

This paper is concerned with the following problem: Given a stable feedback system with controller K(s) and nominal plant model G(s), under what conditions will the feedback system remain stable when G(s) is replaced by any plant model taken from a set $\{G_p(s)\}$? If the set $\{G_p(s)\}$ is finite and countable the problem is called the <u>simultaneous stabilization problem</u> [1][2], and when $\{G_p(s)\}$ is not countable we have the <u>robust stabilization problem</u>. A convenient way of characterizing $\{G_p(s)\}$ for the latter problem is to model it as a set of perturbed plant models around G(s). Much research has been carried out in this direction using singular-value-bounded perturbations, [3]-[6]. But with the exception of [5] all the results are for finite dimensional systems, and with the exception of [6] all require the plant and perturbed plant models to have the same number of unstable poles. The purpose of this paper is to extend the results of [6] to a class of distributed systems.

The paper is organised into six sections, of which this introduction is the first. Section 2 contains necessary preliminaries. In section 3, we present Nyquist-type stability criteria based on the work of Desoer et al., [5][7][12]. An inverse Nyquist-type criterion is given in section 4. The main aim of the paper is reached in section 5 where the robustness test of [6] is extended to a class of distributed systems.

Conclusions are given in section 6.

2. PRELIMINARIES

2.1 Notation

We will adopt the notation used in [5] and [7]. Let

$$f(t) = \begin{cases} 0 & t < 0 \\ f_a(t) + \sum_{i=0}^{\infty} f_i \delta(t - t_i) & t \geq 0 \end{cases} \tag{2.1}$$

where

$$\int_o^{\infty} |f_a(t)| e^{-\sigma t}\, dt < \infty,$$

$$t_o = 0,\ t_i > 0 \text{ for all } i > 0,$$

$$\sum_{i=0}^{\infty} |f_i| e^{-\sigma t_i} < \infty,$$

$\delta(.,t_i)$ is the Dirac delta distribution applied at t_i, and $\sigma \in R$, the field of real numbers.

$A(\sigma)$ is the set of all generalized functions (distributions) defined by (2.1).

$f \in A\_(\sigma)$ if, and only if, for some $\sigma_1 < \sigma$, $f \in A(\sigma_1)$.

$\hat{f}$ denotes the Laplace transform of f, and $\hat{A}(\sigma)$ denotes the set of all transfer functions $\hat{f}$ such that $f \in A(\sigma)$. Similarly, $\hat{A}\_(\sigma) \triangleq \{\hat{f} : f \in A\_(\sigma)\}$.

Note that $A(o)$ can be interpreted as a set of stable impulse responses, and $\hat{A}(o)$ a set of stable transfer functions.

Let $\hat{A}^{\infty}\_(\sigma)$ denote the subset of $\hat{A}\_(\sigma)$ consisting of those $\hat{f}$ that are bounded away from zero at infinity in $C_{\sigma+}$, the closed right-half plane $Re(s) \geq \sigma$. Then, $\hat{B}(\sigma) \triangleq [\hat{A}\_(\sigma)]\ [\hat{A}^{\infty}\_(\sigma)]^{-1}$ is the set of quotients $\hat{f} = \hat{n}/\hat{d}$ where $\hat{n} \in \hat{A}\_(\sigma)$ and $\hat{d} \in \hat{A}^{\infty}\_(\sigma)$. Note that all members of $\hat{B}(\sigma)$ are meromorphic in $C_{\sigma+}$.

$A(\sigma)^{mxn}$ denotes the set of mxn matrices of distributions in $A(\sigma)$. $A\_(\sigma)^{mxn}$ is similarly defined.

$\hat{B}(\sigma)^{mxn}$ denotes the set of mxn transfer function matrices with elements in $\hat{B}(\sigma)$. Any transfer function matrix M in $\hat{B}(\sigma)^{mxn}$ admits a right coprime factorization (RCF) i.e. $M = ND^{-1}$ where

$$N \varepsilon \hat{A}_{-}(\sigma)^{mxn}, \; D \varepsilon \hat{A}_{-}(\sigma)^{nxn},$$

$$\det D \varepsilon \hat{A}_{-}^{\infty}(\sigma), \text{ and } N, D \text{ are right coprime, [5].}$$

2.2 System description

We will consider the linear time-invariant feedback system shown in figure 1, consisting of a plant G and controller K, where

$$G(s) \; \varepsilon \; \hat{B}(\sigma)^{mxn} \text{ for some } \sigma \leq 0, \; m \geq n, \tag{2.2}$$

and

$$K(s) \; \varepsilon \; \hat{B}(\sigma)^{nxm} \tag{2.3}$$

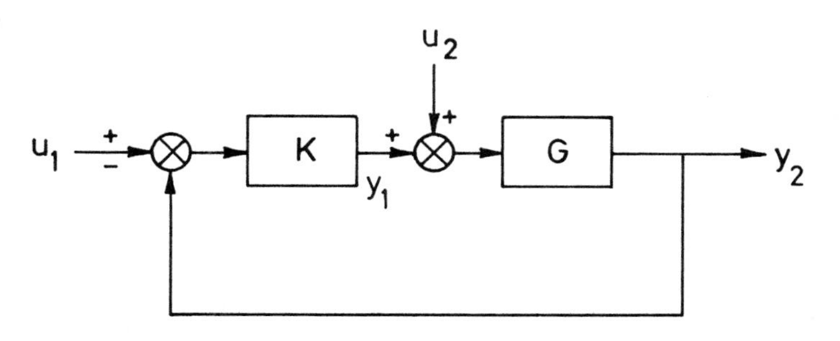

Fig.1. Feedback configuration

G(s), K(s) and G(s)K(s) have RCF's $N_G D_G^{-1}$, $N_K D_K^{-1}$ and $N_{GK} D_{GK}^{-1}$ respectively.

Let H_{yu} be the input-output map of the system, $u \mapsto y$, where $u = (u_1 u_2)^T$, $y = (y_1 y_2)^T$, and T denotes transposition; H_{yu} is a transfer function matrix.

2.3 Feedback stability

Definition [5]. The feedback system of figure 1 is $A_{-}(\sigma)$ - stable if, and only if $H_{yu} \; \varepsilon \; \hat{A}_{-}(\sigma)^{(n + 2m) \times (n + 2m)}$.

Normally σ would be taken equal to 0, but the extra flexibility in the definition allows a measure of relative stability to be included (if desired) by choosing σ negative.

2.4 Characteristic gain and inverse characteristic gain loci

The characteristic gain loci of a transfer function matrix $M(s) \in \hat{B}(\sigma)^{m \times m}$ are the image under the map $s \mapsto$ (eigenvalues of $M(s)$) of the clockwise contour $D(\sigma,r)$ consisting, with r arbitrarily large, of (i) a straight line segment from $(\sigma - jr)$ to $(\sigma + jr)$ except for the usual infinitesimal left indentations around any poles of $M(s)$ with real part equal to σ, and (ii) a semi-circle in $C_{\sigma+}$ joining $(\sigma - jr)$ to $(\sigma + jr)$.

The inverse characteristic gain loci of $M(s)$ are defined as the reciprocal of the nontrivial characteristic gain loci, except that along the straight line segment of $D(\sigma,r)$ there is no need to indent around poles since the inverse loci will then be finite, but we do construct infinitesimal left indentations around any zeros of the characteristic gain loci.

It has already been established in [7] that the characteristic gain loci form closed curves, and so the inverse characteristic gain loci must do likewise. Note that $M(s)$ may be singular in the definitions above.

2.5 Poles and zeros

The poles of G, K and GK are defined as the zeros of $X_G \triangleq \det D_G$, $X_K \triangleq \det D_K$, and $X_{GK} \triangleq \det D_{GK}$ respectively. The number of poles (counting multiplicities) of G in $C_{\sigma+}$ is denoted by P_G; P_K and P_{GK} are similarly defined. We say that there is a $C_{\sigma+}$ pole-zero cancellation in GK if $P_G + P_K \neq P_{GK}$.

As for lumped systems, [8], [9], we will define the characteristic poles (zeros) of GK as the poles (zeros, respectively) of the eigenvalue function $\lambda(s)$ defined by the characteristic equation $\det [\lambda I - GK] = 0$. The number of characteristic zeros (counting multiplicities) of GK in $C_{\sigma+}$ is denoted by Z.

A pole of GK which is not a characteristic pole is called a fixed pole; a detailed discussion of fixed poles for rational transfer functions is given in [9]. Note that a fixed pole of GK is also a pole of the feedback system of figure 1, and would be a fixed pole of $(I + GK)^{-1}GK$.

3. DIRECT NYQUIST STABILITY CRITERIA

In this section, we present four stability theorems derived from the work of Desoer et al., [5], [7]. The aim is to arrive at a stability test in terms of the characteristic gain loci as an aid to the development of the new robustness test in section 5.

Theorem 1. The feedback system of figure 1 is $A_-(\sigma)$ - stable if, and only if:

(i) det [I + GK] has no $C_{\sigma+}$ zeros, and

(ii) the $C_{\sigma+}$ poles of det [I + GK] are exactly the $C_{\sigma+}$ zeros of $X_G X_K$.

Proof. See the proof to theorem 1 in [5].

Theorem 2. The feedback system of figure 1 is $A_-(\sigma)$ - stable if, and only if:

(i) $\inf_{Res \geq \sigma}$ det [I + GK] > 0, and

(ii) there are no $C_{\sigma+}$ pole-zero cancellations in GK.

Proof. (i) <=> (i), Theorem 1, straightforwardly.

(ii) <=> (ii), Theorem 1, by definition.

(i) and (ii) <=> Theorem 1 <=> Stability.

Theorem 3. The feedback system of figure 1 is $A_-(\sigma)$ - stable if, and only if:

(i) the characteristic gain loci of GK do not go through -1,

(ii) the characteristic gain loci of GK encircle -1, P_{GK} times anticlockwise, and

(iii) there are no $C_{\sigma+}$ pole-zero cancellations in GK.

Proof. (i) and (ii) <=> (i), Theorem, 2, as shown in [7].

Theorem 4. The feedback system of figure 1 is $A_-(\sigma)$ - stable if, and only if, the characteristic gain loci of GK:

(i) do not go through -1, and

(ii) encircle -1, $P_G + P_K$ times anticlockwise.

Proof. (ii) <=> (ii) and (iii), Theorem 3, straightforwardly.

Theorems 3 and 4 provide tests which can be easily used to determine feedback stability and form the basis of the new robustness test to be presented later.

4. AN INVERSE NYQUIST STABILITY CRITERION

Consider the identity

$$(I + GK)^{-1} + (I + GK)^{-1} GK = I \tag{4.1}$$

We will denote the first term on the left (the sensitivity matrix) by S and the second term (the complementary sensitivity matrix) by T. From theorem 2, condition (i), a necessary condition for feedback stability is that S is analytic in $C_{\sigma+}$ if, and only, if, it has no characteristic poles and no fixed poles in $C_{\sigma+}$, and hence the following theorem.

Theorem 5. The feedback system of figure 1 will be $A_{-}(\sigma)$ - stable if, and only if:

(i) there are no characteristic poles of T in $C_{\sigma+}$,

(ii) there are no fixed poles of T in $C_{\sigma+}$, and

(iii) there are no $C_{\sigma+}$ pole-zero cancellations in GK.

Proof. (i) and (ii) <=> (i), Theorem 2, as argued above.

(iii) same as (iii), Theorem 2.

(i), (ii) and (iii) <=> Theorem 2 <=> Stability.

In the following two lemmas we show that there is a one-to-one correspondence between the inverse characteristic gain loci of T and those of GK. The results are analogous to the relationships between the characteristic gain loci of the return difference, I + GK, and those of GK.

Lemma 1. The characteristic zeros of T and GK are the same.

Proof. Consider the characteristic equation for T i.e.

$$\det [\lambda I - (I + GK)^{-1}GK] = 0. \tag{4.2}$$

Then, assuming as we must that $\det [(I + GK)^{-1}] \neq 0$, solutions of (4.2) with $\lambda = 0$ are equivalent to solutions of

$$\det [\lambda(I + GK) - GK] = 0, \ \lambda = 0$$

i.e. $\det [\lambda I - (1 - \lambda)GK] = 0, \ \lambda = 0$

Hence lemma 1.

Lemma 2. The number of anticlockwise encirclements of -1 by the inverse characteristic gain loci of GK is equal to the number of anticlockwise encirclements of the origin by the inverse characteristic gain loci of T.

Proof. Let $\lambda_i[.]$ denote the ith eigenvalue of [.].

Then

$$\lambda_i[T] = 1 - \lambda_i[S] = 1 - (1 + \lambda_i[GK])^{-1}, \quad \lambda_i[GK] \neq -1$$

$$= \lambda_i[GK](1 + \lambda_i[GK])^{-1}$$

Hence

$$\frac{1}{\lambda_i[T]} = 1 + \frac{1}{\lambda_i[GK]}, \quad \lambda_i[GK] \neq 0,$$

and lemma 2 follows naturally.

We can now state and prove an inverse Nyquist stability criterion.

Theorem 6. The feedback system of figure 1 will be $A_-(\sigma)$ - stable if, and only if:

(i) the inverse characteristic gain loci of GK do not go through -1,

(ii) the inverse characteristic gain loci of GK encircle -1, Z times anticlockwise.

(iii) there are no fixed poles of GK in $C_{\sigma+}$, and

(iv) there are no $C_{\sigma+}$ pole-zero cancellations in GK.

Proof. Consider the eigenvalue functions of T defined by the equation det $[\lambda I - T] = 0$. Because these functions are meromorphic in $C_{\sigma+}$ we can apply the Principle of the Argument as in [8] or [9] to show that the characteristic gain loci of T will encircle the origin Z times clockwise if, and only if, there are no characteristic poles of T in $C_{\sigma+}$, (condition (i) of theorem 5). Recall that Z is the number of characteristic zeros in $C_{\sigma+}$.

But the number of clockwise encirclements of the origin by the characteristic gain loci of T is equal to the number of anticlockwise encirclements of the origin by the inverse characteristic gain loci of T. Therefore, theorem 6 follows from theorem 5, using lemmas 1 and 2, and the fact that the fixed poles of T are the same as the fixed poles of GK.

Theorem 6 is not particularly useful as a test for determining feedback stability, but we will make use of it in the next section where we are interested in maintaining feedback stability under plant perturbations.

5. ROBUST STABILITY - A SWITCHING CRITERION

In this section we present a test, which if satisfied, is sufficient to guarantee that a stable feedback system will remain stable over a class of perturbed plant models. The class of perturbations may be looked upon as a region of uncertainty around the nominal plant G. We will consider a set $\{G_p(s)\}$ of possible plant models each of which is an element of $\hat{B}(\sigma)^{mxn}$ and assume that the set is arcwise connected, [6], [10], in a topology of unstable plants, [11]. By this we mean that for any $G_p(s)$ in the set, G(s) can be perturbed continuously along a path to $G_p(s)$ without any abrupt changes in the properties of the plant or GK. Therefore, as we move along any path in $\{G_p(s)\}$ the numbers $P_{\tilde{G}}$ and Z can only change as a result of a pole or zero moving across $D(\sigma,r)$. This leads to the following theorem which is an exact extension of the robustness test in [6] for lumped systems. Note that for simplicity we take $\sigma = 0$.

Theorem 7. Let $\{G_p(s)\}$ be an arcwise connected set of transfer function matrices in the topology of unstable plants.

Assume that every member $G_p(s)$ in $\{G_p(s)\}$ is an element of $\hat{B}(o)^{mxn}$ and that there exists $\Delta(j\omega)$ and $\tilde{\Delta}(j\omega)$ such that

$$G_p(j\omega) = [I + \Delta(j\omega)]G(j\omega) = [I + \tilde{\Delta}(j\omega)]^{-1}G(j\omega)$$

for all ω, where G(s) is a member of $\{G_p(s)\}$, known as the nominal plant model, such that the feedback system of figure 1 is $A_-(o)$ - stable.

Define:

$$\delta(\omega) = \sup_{G_p \varepsilon \{G_p(s)\}} \{\bar{\sigma}[\Delta(j\omega)]\}$$

$$\tilde{\delta}(\omega) = \sup_{G_p \varepsilon \{G_p(s)\}} \{\bar{\sigma}[\tilde{\Delta}(j\omega)]\}$$

where $\bar{\sigma}[.]$ denotes the maximum singular value of [.].

Then under these conditions the perturbed feedback system is $A_-(o)$ - stable for any member $G_p(s)$ from $\{G_p(s)\}$ if at each

$\omega \in [0,\infty]$, either

(i) $\bar{\sigma}[T(j\omega)] < \dfrac{1}{\delta(\omega)}$, or

(ii) $\bar{\sigma}[S(j\omega)] < \dfrac{1}{\tilde{\delta}(\omega)}$.

Proof. Since the proof is essentially the same as that for lumped systems in [6], we give only a brief outline here.

For any $G_p(s)$ in $\{G_p(s)\}$ we can find a path within the set connecting $G_p(s)$ to the nominal plant $G(s)$. The idea is to show that stability is maintained for any perturbed plant on the path, and hence any perturbed plant in the set.

By assumption the numbers P_G and Z can only change as a result of a pole or zero crossing $D(o,r)$. But if a pole crosses into or out of C_{o+} the characteristic gain loci of GK will automatically create or lose an extra encirclement to maintain stability. Similarly, if a characteristic zero crosses into or out of C_{o+} the inverse characteristic gain loci of GK will create or lose the extra encirclement required for stability. In fact, the only way the perturbed feedback system can become unstable is for the characteristic gain loci (or equivalently the inverse characteristic gain loci) to pass through the critical point, -1. This is prevented from happening by the switching criteria (i) and (ii) of the theorem.

Note that there is no need to test for unstable fixed poles or pole-zero cancellations. This is because if there is a $G_p(s)$ in $\{G_p(s)\}$ for which these occur then there will always be another model arbitrarily close to $G_p(s)$ for which they do not. And as the frequency responses of these neighbouring models will be arbitrarily close the switching criteria will (by necessity) not be satisfied.

6. CONCLUSIONS

A robustness test has been presented for a class of distributed feedback systems. The test involves a switching criterion which prevents the characteristic gain loci (or equivalently the inverse characteristic gain loci) from passing through -1 as the plant is perturbed continuously throughout a set of possible plant models. The test is an exact generalisation of that given for finite dimensional systems in [6].

7. ACKNOWLEDGEMENTS

The authors are grateful to the International Journal

of Control for permission to publish this work in these conference proceedings as well as in the IJC journal. I. Postlethwaite would like to thank the U.K. Science and Engineering Research Council for financial support.

8. REFERENCES

1. Vidyasagar, M., and Viswanadham, N., 1982 "Algebraic Design Techniques for Reliable Stabilization", IEEE Trans. Aut. Control, 27, 1085-1095.

2. Saeks, R., and Murray, J., 1982 "Fractional Representation Algebraic Geometry, and the Simultaneous Stabilization Problem", IEEE Trans. Aut. Control, 27, 895-903.

3. Foo, Y.K., and Postlethwaite, I., 1983 "Adequate/ inadequate representations of uncertainty and their implications for robustness analysis", Oxford University Enginnering Laboratory (O.U.E.L.), report no. 1498/83.

4. IEEE Trans. Aut. Control, Feb., 1981, 26, Special Issue.

5. Chen, M.J., and Desoer, C.A.,1982, "Necessary and sufficient condition for robust stability of linear distributed feedback systems", I.J.C., 35, 255-267.

6. Postlethwaite, I., and Foo, Y.K., 1984 "Robustness with simultaneous pole and zero movement across the $j\omega$-axis", O.U.E.L. report no. 1505/83, 1983; IFAC World Congress, Budapest, 1984; to appear in Automatica.

7. Desoer, C.A., and Wang, Y. -T., 1980 "On the Generalized Nyquist Stability Criterion", IEEE Trans. Aut. Control, 25, 187-196.

8. Postlethwaite, I., and MacFarlane, A.G.J., 1979 <u>A Complex Variable Approach to the Analysis of Linear Multivariable Feedback Systems</u>. Springer.

9. Smith, M.C., 1982, "A generalized Nyquist/root-locus theory for multi-loop feedback systems", Ph.D. thesis, Cambridge.

10. Foo, Y.K., and Postlethwaite, I.1984, "Extensions of the small-μ test for robust stability", O.U.E.L. report no. 1518/84.

11. Vidyasagar, M., Schneider, H., and Francis, B.A., 1982 "Algebraic and topological aspects of feedback stabilization" IEEE Trans. Aut. Control, 27, 880-894.

12. Callier, F.C., and Desoer, C.A., 1978 "An Algebra of Transfer Functions for Distributed Linear Time-Invariant Systems", IEEE Trans. Circuits and Systems, 25, 651-662.

IMPROVED STABILITY AND PERFORMANCE BOUNDS USING APPROXIMATE MODELS

D.H. Owens
(Department of Mathematics, University of Sheffield)

and

A. Chotai
(Department of Environmental Sciences, University of Lancaster)

ABSTRACT

Previous results on stability and performance assessment using approximate models are refined using filtering and exponential weighting techniques on the error in modelling plant step response characteristics.

1. INTRODUCTION

In this paper we study the problem introduced by the authors in ref [1] of undertaking multivariable feedback control design for a m-output/ℓ-input stable linear system G based upon the use of a simplified stable approximate or reduced-order model G_A. More precisely, if a forward path controller K is designed to produce acceptable stability and performance characteristics from the model G_A in the presence of linear measurement dynamics F in the configuration of Fig.1(a), we consider how information on the step response characteristics of the real plant G can be used to predict the stability and performance characteristics of the closed-loop system shown in Fig.1(b).

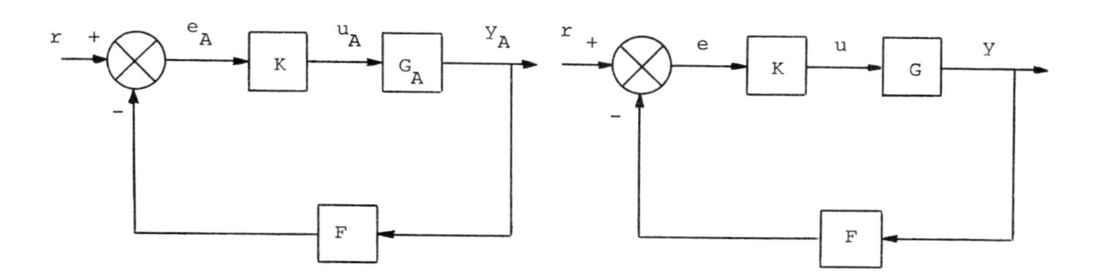

Fig.1. (a) Approximating and (b) implemented feedback schemes

The main results of this paper show that the procedures described in [1] can be improved in both their frequency domain and time-domain form by simulation-based data processing using filters and the technical trick of using exponentially weighted L_∞ spaces in the underlying fixed point problem. The results provide a substantial generalization of those described in ref [1].

2. BACKGROUND MATERIAL AND ASSUMPTIONS

In all that follows all elements G, G_A, K and F are assumed to be described by convolution operators in extended product L_∞ spaces. All impulse responses are assumed to be exponentially bounded and the step response matrices $Y(t)$ and $Y_A(t)$ of G and G_A respectively are assumed to be known. The modelling error will be characterized by the $m \times \ell$ matrix

$$E(t) = [E_1(t), \ldots, E_\ell(t)] \triangleq Y(t) - Y_A(t) \quad , \quad t \geq 0 \qquad (2.1)$$

If f is a scalar continuous function defined on $[0,\infty]$ and of bounded variation on any finite interval $[0,t]$, then $N_t(f)$ will denote [1] the norm of f on $[0,t]$ ie

$$N_t(f) \triangleq |f(0+)| + \sum_{j=1}^{k} |f(t_j) - f(t_{j-1})| + |f(t) - f(t_k)| \qquad (2.2)$$

where $0 = t_o < t_1 < t_2 < \ldots$ are the local minima and maxima of f on $[0,\infty]$ and k is the largest integer satisfying $t_k \leq t$. For $t = +\infty$ we define

$$N_\infty(f) = \sup_{t \geq o} N_t(f) = \lim_{t \to \infty} N_t(f) \qquad (2.3)$$

whenever the limit exists.

The development in the paper depends critically on the use of a vector form of the contraction mapping principle for fixed points and, in particular, on the use of vector norms on product Banach spaces and associated induced vector operator norms. The reader is referred to ref [1] for details which are omitted here for brevity.

3. FREQUENCY DOMAIN ANALYSIS

To motivate the general discussion we begin by considering the single-input/single-output case $m = \ell = 1$. If D represents the normal Nyquist contour in the complex plane generated by the imaginary axis $s = i\omega$, $-R \leq \omega \leq R$, and the semi-circle $|s| = R$,

Res $\geq$ 0 with R 'infinitely' large, then the basic sufficient condition ensuring the success of the approximation theorem can be stated as follows [1] in terms of the transfer functions of G, G_A, K and F:

Proposition 3.1: If K stabilizes G_A in the configuration of Fig.1(a), then it will stabilize the plant G in the configuration of Fig.1(b) if

$$\sup_{s \in D} \left| \frac{K(s)F(s)}{1+K(s)F(s)G_A(s)} \right| \Delta(s) < 1 \tag{3.1}$$

where $\Delta(s)$ is any available real-valued function on D with the property that

$$|G(s) - G_A(s)| \leq \Delta(s) \quad , \quad \forall \; s \in D \tag{3.2}$$

(Remark: stability is here, and in the following, interpreted as input/output stability. Asymptotic stability requires the addition of (generically satisfied) controllability and observability assumptions on GKF).

The result is a phase-independent result in the sense that it ignores the phase of the error $G-G_A$. Clearly, the best result is obtained by choosing

$$\Delta(s) = |G(s) - G_A(s)| \tag{3.3}$$

as it describes the gain characteristics of the error exactly. The 'best' frequency independent general choice of Δ is [1],

$$\Delta(s) = N_\infty(E) \tag{3.4}$$

obtained from the general result [1]:

Proposition 3.2: If L is a bounded convolution operator from $L_\infty(0,\infty)$ into itself with transfer function L(s) then, for all $s \in D$

$$|L(s)| \leq N_\infty(Y_L) \tag{3.5}$$

where $Y_L(t)$ is the unit step response of L.

It can be expected that, between these two extremes, there is an infinity of frequency dependent upper bounds on $|G(s)-G_A(s)|$ that can be used to refine the results of [1] and yet do not need detailed knowledge of G(s). The following result describes a class of bounds that can be obtained by filtering operations on the error data E(t):

Lemma 3.1: Let F_α be a filter with the properties that

(a) $E_\alpha \triangleq F_\alpha E \in L_\infty(0,\infty)$, and

(b) $F_\alpha^{-1}(s)$ is bounded and analytic in the open right-half plane.

Then, for all $s \in D$,

$$|G(s) - G_A(s)| \leq \Delta_\alpha(s) \triangleq |F_\alpha^{-1}(s)| N_\infty(E_\alpha) \tag{3.6}$$

Proof: Write $G-G_A = F_\alpha^{-1}(F_\alpha(G-G_A))$ and apply proposition 3.2 to $F_\alpha(G-G_A)$.

In the case of $F_\alpha = I$, the result reduces to the bound (3.4) used in previous studies. More generally, however, F_α produces a frequency-conscious bound capable of producing more refined results approaching that of (3.3). For example, choose $F_\alpha = (G-G_A)^{-1}$ and note that $E_\alpha(t) \equiv 1$ and hence that (3.6) reduces to (3.3). In practice this choice of F is not available but other, more simple choices, can intuitively be used to produce easily computed intermediate estimates.

Estimates obtained from a number of filters can also be used:

Theorem 3.1: Suppose that K stabilizes G_A and that $\{F_\alpha\}_{\alpha \in A}$ is a collection of filters satisfying the conditions of lemma 3.1, then K stabilizes G if (3.1) holds with

$$\Delta(s) = \Delta_A(s) \triangleq \inf_{\alpha \in A} \Delta_\alpha(s) \tag{3.7}$$

The result follows trivially from proposition 3.1 with

$$\Delta(s) \triangleq \inf_{\alpha \in A} \Delta_\alpha(s) \tag{3.8}$$

and is omitted for brevity. It is clear that suitable choice of $\{F_\alpha\}$ will enable a considerable refinement of the results of [1]. The result has a simple graphical interpretation described below:

Corollary 3.1: The conditions of theorem (3.1) are satisfied if the following conditions are satisfied

(i) $$\lim_{\substack{|s|\to\infty \\ \mathrm{Re}\, s\geq 0}} \left| \frac{K(s)F(s)}{1+K(s)F(s)G_A(s)} \right| \Delta_A(s) < 1 \qquad (3.9)$$

(ii) the inverse Nyquist plot of G_AKF with $s = i\omega$, $\omega\geq 0$, and superimposed 'confidence' circles of radius

$$r_A(\omega) \triangleq |G_A^{-1}(i\omega)|\ \Delta_A(i\omega) \qquad (3.10)$$

at each point generates a 'confidence band' that does not touch or contain the (-1,0) point of the complex plane.

Turning now to the multivariable case, it can be expected that the general structure and conclusions of the single-input/ single-output case will be retained. The main result is stated as follows:

Theorem 3.2: Suppose that K stabilizes G_A and that $\{F_\alpha^{(i,j)}\}_{\alpha\in A(i,j)}$, $1\leq i\leq m$, $1\leq j\leq \ell$, are filters satisfying the conditions of lemma 3.1 in the sense that

(a) $E_\alpha^{(i,j)} \triangleq F_\alpha^{(i,j)} E_{ij} \in L_\infty(0,\infty)$, and

(b) $(F_\alpha^{(i,j)}(s))^{-1}$ is bounded and analytic in the open right-half-plane for all $\alpha\in A(i,j)$, $1\leq i\leq m$, $1\leq j\leq \ell$.

Then K stabilizes the real plant G if

$$\sup_{s\in D} r(\, \| (I_\ell + K(s)F(s)G_A(s))^{-1} K(s)F(s) \|_P \Delta_A(s)) < 1 \qquad (3.11)$$

where $\Delta_A(s)$ is the $m\times\ell$ matrix with elements

$$(\Delta_A(s))_{ij} \triangleq \inf_{\alpha\in A(i,j)} (F_\alpha^{(i,j)}(s))^{-1} N_\infty(E_\alpha^{(i,j)}) \qquad (3.12)$$

Proof: Following the development of [1], a sufficient condition for stability is that

$$\sup_{s\in D} r((I_\ell+K(s)F(s)G_A(s))^{-1}K(s)F(s)(G(s)-G_A(s))) < 1 \quad (3.13)$$

and the result follows trivially from the inequality $\|G(s)-G_A(s)\|_P \leq \Delta_A(s)$.

The result has an interpretation identical to the results of [1] and, in the case of the choice of K, F and G_A diagonal, it can be realized in the form of an INA-type design process [1]-[3].

4. TIME DOMAIN ANALYSIS

The simulation philosophy implicit in the discussion of section 3 can also be used to bound the possible performance deterioration due to the approximation in a similar manner to that described in ref [1].

4.1. Input Assessment

The basic stability result underlying the performance assessment can be stated as follows [1]:

Proposition 4.1: If K stabilizes G_A, then it will also stabilize G if

$$r(N_\infty^P(W_A)) < 1 \quad (4.1)$$

where $W_A(t) = [W_A^{(1)}(t),\ldots,W_A^{(\ell)}(t)]$ and, $1\leq j\leq \ell$, $W_A^{(j)}(t)$ is the response from zero initial conditions of the system $(I+KFG_A)^{-1}KF$ to the error data $E^{(j)}(t)$.

The response of the input is described by [1]

$$u = W(u) \triangleq u_A - (I+KFG_A)^{-1}KF(G-G_A)u \quad (4.2)$$

which is here regarded as a fixed point equation in $X_\alpha^\ell(t_\alpha)$ with

$$X_\alpha(t_\alpha) \triangleq \{f : e^{\alpha t}f\in L_\infty(0,t_\alpha)\} \quad (4.3)$$

where t_α is to be determined from a contraction condition and the norm of f is taken to be the norm of $e^{\alpha t}f$ in $L_\infty(0,t_\alpha)$.

Proposition 4.2: Let L be a convolution operator from $L_\infty(0,\infty)$ into itself with impulse response satisfying $|h_L(t)| \leq h_o e^{-\lambda t}$, $t \geq 0$, for some $\lambda > 0$. If $t_\alpha < +\infty$, the induced operator norm of L restricted to $X_\alpha(t_\alpha)$ satisfies $||L|| \leq N_{t_\alpha}(Y_L;\alpha)$ for all α where

$$N_{t_\alpha}(Y_L;\alpha) \triangleq |Y_L(0+)| + \sum_{k=1}^{N} e^{\beta_k(\alpha)}(N_{t_k}(Y_L) - N_{t_{k-1}}(Y_L)) , \tag{4.4}$$

Y_L is the unit step response of L, $\beta_k(\alpha) = \alpha t_k$ if $\alpha \geq 0$ or αt_{k-1} if $\alpha < 0$ and $0 = t_o < t_1 < t_2 < \ldots < t_N = t_\alpha$ is any partition of $[0,t_\alpha]$ with the property that $0 < h_1 \leq t_k - t_{k-1} \leq h_2 < +\infty$, $k \geq 1$. Moreover, if $t_\alpha = +\infty$, L maps $X_\alpha(t_\alpha)$ into itself for $\alpha < \lambda$ with $||L|| \leq N_\infty(Y_L;\alpha) < +\infty$.

Proof: The induced norm of L in $X_\alpha(t_\alpha)$ is

$$||L|| = |Y_L(0+)| + \int_0^{t_\alpha} e^{\alpha t}|h_L(t)|dt$$

$$\leq |Y_L(0+)| + \sum_{k=1}^{N} e^{\beta_k(\alpha)} \int_{t_{k-1}}^{t_k} |h_L(t)|dt \tag{4.5}$$

which is just (4.4). It is clearly finite if t_α is finite. If $t_\alpha = +\infty$ then $N = +\infty$. If $\alpha \leq 0$ then $e^{\beta_k(\alpha)} \leq 1$ so that $N_\infty(Y_L;\alpha) \leq N_\infty(Y_L) < +\infty$. If $0 < \alpha < \lambda$ then

$$e^{\beta_k(\alpha)} \int_{t_{k-1}}^{t_k} |h_L(t)|dt \leq h_o(t_k - t_{k-1})e^{\alpha t_k - \lambda t_{k-1}}$$

$$\leq h_o(t_k - t_{k-1})e^{(\alpha-\lambda)t_k + \lambda(t_k - t_{k-1})}$$

$$\leq h_o h_2 e^{(\alpha-\lambda)kh_1} e^{\lambda h_2} \tag{4.6}$$

and the (infinite) series in (4.4) converges. This completes the proof.

It is easily verified that $N_{t_\alpha}(Y_L;\alpha)$ is continuous and monotonically increasing in both α and t_α with $N_o(Y_L;\alpha) = 0$ and $N_t(Y_L;0) = N_t(Y_L)$. Also

---

Proposition 4.3: If condition (4.1) holds, then

(i) for each choice of α, there exists $t_\alpha^* > 0$ such that $t_\alpha \leq t^*$,

$$r(N_{t_\alpha}^P(W_A;\alpha)) < 1 \tag{4.7}$$

where $N_{t_\alpha}^P(W_A;\alpha)$ is the $\ell x \ell$ matrix with $(i,j)^{th}$ entry $N_{t_\alpha}((W_A)_{ij};\alpha)$,

(ii) there exists $\alpha^* \geq 0$ such that we can choose $t_\alpha^* = +\infty$ for $\alpha \leq \alpha^*$, and

(iii) if $L = (I+KFG_A)^{-1}KF(G-G_A)$ satisfies the conditions of proposition 4.2 then we can choose $\alpha^* > 0$.

Proof: Follows simply from the previous discussion and is omitted for brevity.

We now state the main result of this section.

Theorem 4.1: Suppose that the conditions of proposition 4.1 hold and define

$$\eta(t) = -(\int_0^t W_A(t-t')H_o(t')dt')\beta \tag{4.8}$$

where $H_o(t)$ is the impulse response matrix of $(I+KFG_A)^{-1}K$ and $\beta \in R^\ell$. If $u_A(t)$ (resp $u(t)$) is the input response of the approximating (resp real) feedback scheme of Fig.1(a) (resp Fig.1(b)) to the step set point signal $r(t) \equiv \beta$, $t \geq 0$, then, for any choice of α, $1 \leq j \leq \ell$,

$$|u_j(t) - u_j^{(1)}(t)| \leq \varepsilon_j^\alpha(t) \quad , \quad 0 \leq t \leq t_\alpha \tag{4.9}$$

where

$$\text{(a)} \quad u^{(1)}(t) \triangleq u_A(t) + \eta(t) \quad , \quad t \geq 0 \tag{4.10}$$

(b) $$\varepsilon^{\alpha}(t) \triangleq e^{-\alpha t}(I_{\ell}-N_t^{P}(W_A;\alpha))^{-1}N_t^{P}(W_A;\alpha) \sup_{0\le t'\le t} e^{\alpha t'}\|\eta(t')\|_P \tag{4.11}$$

(the supremum being interpreted with respect to the partial ordering in R^{ℓ} [1]), and

(c) t_{α} is any time choice such that (4.7) holds true.

Proof: Condition (c) ensures that the unique solution to (4.2) can be obtained by successive approximation [1] as W is a P-contraction in $X_{\alpha}^{\ell}(t)$ for any finite $t\le t_{\alpha}$ with

$$\|W(x) - W(y)\|_P \le N_t^{P}(W_A;\alpha)\,\|x-y\|_P \tag{4.12}$$

for all $x,y\in X_{\alpha}^{\ell}(t)$. Let the initial guess be $u^{(o)} = u_A$ to yield the first iterate $u^{(1)}$ and the norm estimate in $X_{\alpha}^{\ell}(t)$

$$\|u-u^{(1)}\|_P \le (I-N_t^{P}(W_A;\alpha))^{-1}N_t^{P}(W_A;\alpha)\,\|\eta\|_P \tag{4.13}$$

Equation (4.9) then follows from the definition of the vector norm.

In the case of $\alpha = 0$, the result reduces to previously published work [1]. The exponential weighting adds a refinement that can in principle be used to tighten the bounds. In practice, it is envisaged that ε_j^{α} will be computed for a variety of choices of α in a selection set A and extending (4.9) to the form

$$|u_j(t) - u_j^{(1)}(t)| \le \inf_{\substack{\alpha\in A\\ t\le t_{\alpha}}} \varepsilon_j^{\alpha}(t) \tag{4.14}$$

by noting that both u and $u^{(1)}$ are independent of α. The possibilities inherent here can be illustrated by recalling that previous studies suffered from the problem that the bound $\varepsilon^{o}(t)$ is monotonically increasing and hence that the uncertainty at infinity $\varepsilon^{o}(\infty)>0$. In contrast, the bound (4.9) may have the property that $\varepsilon^{\alpha}(t)\to 0$ $(t\to+\infty)$ provided that we choose $\alpha>0$ such that (i) $t_{\alpha} = +\infty$ (ii) condition (4.7) holds and (iii) $e^{\alpha t}\eta(t)$ is uniformly bounded. The technical background to the existence of such choices is omitted for brevity but we note that it is necessary that $\eta(t)\to 0$ $(t\to+\infty)$ which can only be ensured if $E(t)\to 0$ $(t\to+\infty)$ ie the plant and model must have identical steady state characteristics!

4.2. Output Assessment: The Single-input/single-output Case

In the case of $m = \ell = 1$, output performance deterioration can be assessed in a similar manner to input as described in section 4.1. The output response is described by [1]

$$y = W_o(y) \triangleq y_A + (I+KFG_A)^{-1}K(G-G_A)(r-Fy) \tag{4.15}$$

which is regarded as a fixed-point equation in $X_\alpha(t_\alpha)$. Input-output stability is guaranteed if Proposition 4.1 holds which, in the case of $m = \ell = 1$, reduces to

$$N_\infty(W_A) < 1 \tag{4.16}$$

where W_A is the response of $(I+KFG_A)^{-1}KF$ to the error data E. A similar technique to that used in the proof of theorem 4.1 then yields the result:

Theorem 4.2: Let (4.16) hold and define $\eta(t)$ to be the response of the linear system $(I+KFG_A)^{-1}K(I+KFG_A)^{-1}$ to the error data $E(t)$. If $y_A(t)$ (resp $y(t)$) is the output response of the approximating (resp real) feedback scheme of Fig.1(a) (resp Fig.1(b)) to a unit step demand signal r, then, for any choice of α,

$$|y(t) - y^{(1)}(t)| \leq \varepsilon^\alpha(t) \quad , \quad 0 \leq t \leq t_\alpha \tag{4.17}$$

where $y^{(1)}(t) \triangleq y_A(t)+\eta(t)$, $t\geq 0$,

$$\varepsilon^\alpha(t) \triangleq e^{-\alpha t}(1-N_t(W_A;\alpha))^{-1}N_t(W_A;\alpha) \sup_{0\leq t'\leq t} e^{\alpha t'}|\eta(t')| \tag{4.18}$$

and t_α is any time choice such that $N_{t_\alpha}(W_A;\alpha) < 1$.

Proof: The result follows in a similar manner to Theorem 4.1 noting that W_o is a contraction on $X_\alpha(t)$ for any finite $t \leq t_\alpha$ with contraction constant $N_t(W_A;\alpha)$ and using successive approximation with initial iterate $y(o) = y_A$ and consequent second iterate $y^{(1)} = y_A+\eta$. The details are omitted for brevity.

As in the discussion following theorem 4.1 the choice of several α in an index set A can be used to refine the result to the form

$$|y(t) - y^{(1)}(t)| \leq \inf_{\substack{\alpha\in A \\ t\leq t_\alpha}} \varepsilon^\alpha(t) \tag{4.19}$$

and by choosing $\alpha>0$, it is possible to ensure that $\varepsilon^{\alpha}(t)\to 0$ $(t\to\infty)$ by careful choice of α providing that the modelling error $E(t)\to 0$ $(t\to\infty)$.

Similar results hold for the multivariable case but are omitted for brevity.

5. REFERENCES

1. Owens, D.H., and Chotai, A., (1983). "Robust controller design for linear dynamic systems using approximate models", Proc.IEE, 130, Pt.D, 45-56.

2. Rosenbrock, H.H., (1974). "Computer-aided-design of control systems", Academic Press.

3. Owens, D.H., (1978). "Feedback and multivariable systems", Peter Peregrinus.

DESIGN OF LOW ORDER COMPENSATORS FOR HIGH PERFORMANCE CONTROL SYSTEMS

J.O. Gray and D. Valsamis

(Department of Electronic and Electrical Engineering, University of Salford)

ABSTRACT

On the design of controllers which meet specified closed loop linear system performance in the time domain, it is shown that the application of simple functional analytic concepts on input - output system descriptions provides an easily computable assessment of system performance deterioration that is envisaged when a simplified low order equivalent system is implemented to approximate some ideal, often high order, controller.

1. INTRODUCTION

Recent years have seen the increasing application of design techniques that ensure acceptable or optimum system performance in the time domain. The popularity of this approach is due to the fact that many practical engineering control problems are often specified in terms of time domain performance functionals such as rise time, settling time, etc. Design procedures based on time domain performance functional inequalities and their solution via optimisation algorithms have been developed by Zakian[1] and Mayne[2]. Although optimisation-based methods are capable of deriving optimum compensators for specific design problems, current trends seem to aim towards the development of simpler approximate design methods which require little computation and yet are capable of arriving at high performance control system designs. This approach has been enhanced by the application of functional analysis which provides easily computable worst-case bounds on approximation-induced performance errors. The application of functional bounds has been demonstrated by Owens and Chotai[3] on the design of controllers for approximate plant representations.

In this work we concentrate on the design of low order feedback controllers which ensure specified system performance in the time domain. This extends previous work by Gray and Valsamis[4,6] where compensator synthesis has been achieved by transforming some ideal system input-output specification into an open loop input-output relation for the ideal compensator and subsequently deriving the parameters of a suitable compensator model which induce the given input-output map. Although it is possible to obtain an accurate input-output fit by some compensator of arbitrarily large order, it is often preferable to examine whether a simpler compensator is sufficient in satisfying the given system's performance specifications. In this context, elementary functional analytic concepts are employed for the derivation of simple inequalities which relate the deterioration in closed-loop system performance to the synthesis (identification) error that is associated with the open-loop synthesis of the ideal compensator.

2. MATHEMATICAL PRELIMINARIES

The assumed system configuration is shown in Fig. 1. The following assumptions are made with respect to the system's parameters:

(i) Q is a well defined scalar linear non-anticipative and minimum phase system mapping input variables $e(t)$ into output variables $y(t)$.

(ii) F is a scalar linear convolution system mapping input variables $y(t)$ into output variables $z(t)$. F is initially unknown and is the design objective.

(iii) Explicit system variables are initially zero (zero initial conditions) and belong to the space of finite, scalar, real-valued functions of time $L\,[0\;,\;\infty]$ with a suitable norm defined on it.

In the context of system relations in a functional analytic setting we consider system variables belonging to a Banach space X and system components that represent operators mapping X to itself. With reference to this work, suitable spaces X are the popular $L_\infty[0,\infty]$ and $L_1[0,\infty]$ spaces.

The norm of a truncated scalar function of time $x(t) \in X$, in the L_1 and L_∞ spaces is defined by the following relations, Desoer and Vidyasagar[5]:

$$||P_T x||_1 = \int_0^T |x(t)| \, dt \tag{1}$$

and

$$||P_T x||_\infty = \underset{0 \leq t \leq T}{\text{ess sup}} |x(t)| \tag{2}$$

The truncated operator norm of a linear convolution operator A in X is denoted by $||A_T||$ and is induced from the norm in X. The norm and the induced operator norm in X satisfy the well known axioms:

$$||P_T x|| \geq 0 \tag{3}$$

$$||P_T (x+w)|| \leq ||P_T x|| + ||P_T w|| \tag{4}$$

$$||P_T Ax|| \leq ||A_T|| \; ||P_T x|| \tag{5}$$

$\forall$ x, w $\in$ X, $\forall$ A mapping X to itself.

The induced operator norm in L_∞ of a linear non-anticipative convolution operator A, with impulse response function $h_A(t)$, is given by Desoer and Vidyasagar[5]:

$$||A_T||_\infty = \int_0^T |h_A(t)| \, dt = ||P_T h_A||_1 \tag{6}$$

Alternatively, it has been shown by Owens and Chotai[3] that, if $x_A(t)$ is the unit step response of A, from zero initial conditions, then an equivalent definition of $||A_T||_\infty$ is given by:

$$||A_T||_\infty = N_T \{x_A\} \tag{7}$$

where $N_T \{x_A\}$ is the total variation of $x_A(t)$ in the interval $t \in [0,T]$ and is defined as:

$$N_T \{x_A\} = |x_A(0+)| + \sum_{k=1}^{n} |x_A(t_k) - x_A(t_{k-1})| + |x_A(T) - x_A(t_n)| \tag{8}$$

where the instants $t_0, t_1, .. t_n$ denote consecutive local maxima and minima of the function x_A for $t < T$. $t_0 = 0$ and t_n denotes the last local maximum or minimum of x_A in the interval $[0,T]$. The total variation method for the evaluation of the operator norm $||A_T||_\infty$ offers the practical advantage that an accurate assessment of the operator norm can be deduced by mere inspection of the unit step response of A.

3. THE EFFECT OF SYNTHESIS ERROR ON THE INPUT-OUTPUT BASED DESIGN OF FEEDBACK COMPENSATORS

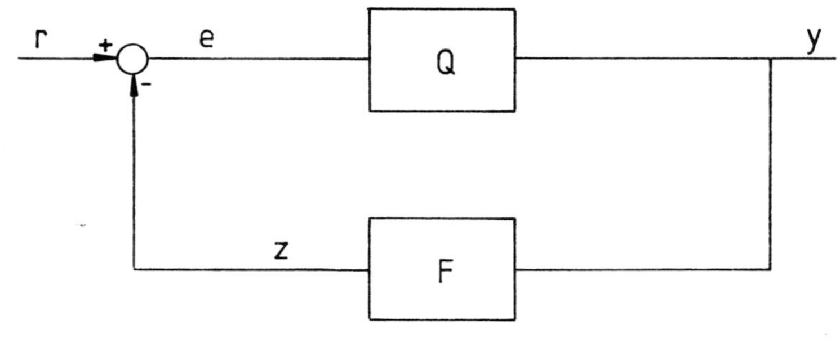

Fig. 1

The input-output synthesis of system compensation in the time domain implies the determination of a controller such that the system's response $y(t)$ to an input $r(t)$ (usually chosen as the unit step function $u(t)$) satisfies certain performance constraints.

Here, we shall be more specific by translating such performance constraints into an ideal response function $y^*(t)$ which in turn defines the idealised system operator S^* : $y^*=S^*r$. Consequently, with reference to feedback compensation, the design problem is simply to synthesise a suitable feedback element F such that the system of Fig. 1 realises the ideal form S^* or an adequate approximation to S^*. In the remainder of this work we assumed that $r(t) = u(t)$.

Provided that $y^*(t)$ is a realistic response function, it is reasonable to assume that there exists a linear feedback compensator F^* which realises S^* and, as such, satisfies the input-output mapping:

$$z^* = F^* y^* \tag{9}$$

where

$$z^* = r - Q^{-1} y^* \tag{10}$$

The function $Q^{-1}y^*(t)$ denotes the inverse of the response of the forward path operator Q and represents the control variable which induces the ideally specified response $y^*(t)$.

A practical problem which is often associated with system synthesis is that F^* may be unrealisable and/or infinite order. Even if F^* is a realisable and finite order dynamical system it may be a relatively high order linear system and therefore its implementation could be unrealistic. On the other hand, as is often the case, there could exist a simple low order approximate feedback compensator F_a which nevertheless achieves the specified system performance in an adequate manner. This is justified on the basis that a certain degree of tolerance with respect to the idealised response $y^*(t)$ is invariably acceptable for most practical purposes.

It would be useful to see how the approximation (F_a) of the ideal feedback F^* affects the overall system's response. In particular if $y(t)$ denotes the response of the system with feedback F_a, we will show the absolute maximum performance error $||P_T(y-y^*)||_\infty$ can be bounded by a simple and easily computable functional.

Let z_a denote the response of F_a to input $y^*(t)$:

$$z_a = F_a y^* , \tag{11}$$

and hence define the identification error function $\delta(t)$,

$$\delta(t) = z^*(t) - z_a(t) \tag{12}$$

It is clear that $\delta(t)$ reflects the difference between the ideal and approximate feedback compensators in that, if H is the linear operator defined as $H = F^*-F_a$, then $\delta = H y^*$.

Theorem 1

Provided that:

(i) $N_T\{\delta\} < 1$, where $N_T\{\delta\}$ denotes the total variation of $\delta(t)$ up to time T, defined by equation (8), and

(ii) S^* is a stable system,

the following inequality bounds the absolute maximum performance error:

$$||P_T(y-y^*)||_\infty \leqslant \frac{N_T\{\delta\}}{1 - N_T\{\delta\}} ||P_T y^*||_\infty = \phi_T \tag{13}$$

Furthermore, provided that $\lim_{T\to\infty} N_T\{\delta\} < 1$, inequality (13) implies that the approximate feedback compensator F_a ensures the input-output stability of the closed-loop system.

Proof

Observe that from the definition of the identification error operator $H = F^* - F_a$, the system's loop equation can be written as:

$$y = Q\, r - Q\, F^*y + Q\, H\, y, \qquad (14)$$

therefore

$$y = (1+Q\, F^*)^{-1} Q\, r + (1+Q\, F^*)^{-1}\, Q\, H\, y \qquad (15)$$

Notice that $(1+Q\, F^*)^{-1}Q$ represents the idealised closed-loop system operator S^* ; hence,

$$y - y^* = S^*\, H\, y \qquad (16)$$

or from the commutativity of linear scalar convolution operators,

$$y - y^* = H\, S^*\, y \qquad (17)$$

Truncating (17) and taking norms in L_∞, we can deduce the following inequality from the fundamental axiom given by (5),

$$||P_T(y-y^*)||_\infty \leq ||H\, S_T^*||_\infty\; ||P_T y||_\infty \qquad (18)$$

The truncated operator norm $||HS_T^*||_\infty$ is defined as the total variation of the response of $H\, S^*$ to a unit step input. That is:

$$||HS_T^*||_\infty = N_T\{H\, S^*u(t)\} = N_T\{Hy^*(t)\} = N_T\{\delta(t)\} \qquad (19)$$

Substituting (19) in (18), and observing that by right of the triangle inequality (4) : $||P_T y||_\infty \leq ||\, P_T(y-y^*)\, ||_\infty + ||P_T y^*||_\infty$ the following inequality can be obtained.

$$(1-N_T\{\delta\})||P_T(y-y^*)||_\infty \leq N_T\{\delta\}\;||P_T y^*||_\infty \qquad (20)$$

Inequality (13) can be immediately deduced noting that condition (i) of the theorem ensures that $1 - N_T\{\delta\}$ is non-negative.

Inequality (13) has a useful graphical interpretation as an uncertainty band, equidistant from the ideal response y*, which bounds the possible location of the approximately compensated system's response y(t). That is:

$$Y_L(T) \leq y(T) \leq y_U(T), \tag{21}$$

where

$$Y_L(T) = y^*(T) - \phi_T \; ; \text{ and } y_U(T) = y^*(T) + \phi_T \tag{22}$$

The fact that $N_T\{\delta\}$ is a non-decreasing function of T, implies that the condition $N_\tau\{\delta\} < 1$ ensures the existence of the bound ϕ_T for all $T \in [0,\tau]$. For practical purposes, however, the condition $\lim_{T\to\infty} N_T\{\delta\} < 1$ is a more realistic requirement as it ensures that the performance bound exists over the entire time interval $T \in [0,\infty]$ and thus, provided that S* is stable, guarantees the stability of the approximately compensated system.

The practical importance of (13) and the associated uncertainty band, lies in that these do not explicitly depend on the parameters of the approximate compensator. In fact, the r.h.s. of (13) depends merely on the norm of the idealised system's unit step response function y* and the total variation of the synthesis error δ. In relation to the time domain input-output synthesis of feedback compensators, the synthesis error function $\delta(t)$ is readily determined as part of the computation for the derivation of F_a. As a result the evaluation of the absolute bound ϕ_T is straight-forward.

It is thus possible to utilise the performance bound ϕ_T in an efficient procedure for the time domain synthesis of feedback compensators. This approach is enhanced by the fact that the performance bound depends on the norm of the idealised response y* which is normally known a priori from the desired system performance specification. Consequently from the tolerated performance deterioration the designer can immediately determine the maximum acceptable total variation of the identification error which ensures that the closed loop system's response lies within the appropriate uncertainty (tolerance) band. In this manner the designer can exploit the possibility of deriving relatively simple and low order feedback compensators F_a which are sufficient in meeting stringent system performance specifications and at the same time enable a more efficient and economical realisation than the idealised compensator F*.

Although a certain degree of uncertainty is normally tolerated during the transient part of the system's response, it is often imperative to ensure that the system attains the exact specified steady state level to a unit step input. Consequently it is important to examine under what conditions does the approximate compensator F_a ensure that the closed loop system's response converges to the idealised response as $t \to \infty$. The following theorem is concerned with this practical requirement.

Theorem 2

Subject to the conditions:

(i) $\lim_{T\to\infty} N_T\{\delta\} < 1$,

(ii) $||\delta||_1 = \int_0^\infty |\delta(t)| \, dt < \infty$, and

(iii) the idealised system S* is stable,

the approximate system's response y(t) converges to the steady state level of the idealised response y*(t): $\lim_{t\to\infty} y(t) = \lim_{t\to\infty} y^*(t)$.

Proof

Truncate and take the norm in L_1 of equation (16). Subsequently, observing that $S^*H\,y = S^*H\,(y-y^*) + S^*H\,y^*$ we can invoke the triangle inequality (4) to obtain:

$$||P_T(y-y^*)||_1 \leq ||P_T S^*H(y-y^*)||_1 + ||P_T S^*Hy^*||_1 . \quad (23)$$

Recalling a well-known property of the L_1 norm which states that for a linear convolution operator A, the relation $x = A\,w$ can be bounded by:

$$||x||_1 \leq ||A||_\infty \, ||w||_1 \quad (24)$$

(see Desoer and Vidyasagar [5]), and noting that $H\,y^* = \delta$, inequality (23) can be replaced by the stronger:

$$||P_T(y-y^*)||_1 \leq N_T\{\delta\}\ ||P_T(y-y^*)||_1 + ||S_T^*||_\infty\ ||P_T\delta||_1 \quad (25)$$

Condition (i) ensures that $1 - N_T\{\delta\} > 0$ and therefore (25) can be written as

$$||P_T(y-y^*)||_1 \leq \frac{||S_T^*||_\infty\ ||\delta||_1}{1 - N_T\{\delta\}} \quad (26)$$

In the limit as $T \to \infty$ we deduce:

$$||y-y^*||_1 \leq \frac{||S^*||_\infty\ ||\delta||_1}{1 - N_\infty\{\delta\}} = \Theta \quad (27)$$

Conditions (ii) and (iii) ensure that $||\delta||_1 < \infty$ and that $||S^*||_\infty < \infty$, respectively. Consequently Θ is a positive finite number. As such, inequality (27) implies that:

$$||y-y^*||_1 \triangleq \int_0^\infty |y(t)-y^*(t)|\ dt < \infty \quad (28)$$

and thus proves the theorem.

Theorem 2 simply confirms an intuitively obvious result, but is interesting in the sense of strengthening the methodology of the input-output approach to the synthesis of compensators. In fact, theorem 2 complements theorem 1 by providing sufficient conditions which ensure that the approximately compensated system's response $y(t)$ will not only lie close to the idealised response $y^*(t)$ but it will attain the same steady state value as $y^*(t)$. The only real additional requirement is condition (ii) which, in practical terms, implies that the synthesis (identification) error $\delta(t)$ must converge asymptotically to zero.

Consequently, the requirement $\delta(t) \to 0$, as $t \to \infty$ can be incorporated in the actual synthesis procedure to ensure that the approximate feedback compensator F_a satisfies the exact specified steady state gain for the system. This is particularly useful in relation to the design of feedback controllers for tracking systems.

The application of simple functional analytic concepts can also be useful in the assessment of other important properties of the performance error $y(t) - y^*(t)$. For example, the rate of change of the unit step response can be a major practical requirement. In such cases it would be useful to assess the worst possible rate of change of the performance error from the properties of the synthesis error. In fact, it can be shown that the conditions of theorem 1 ensure the validity of the following first derivative bound:

$$||P_T(\dot{y}-\dot{y}^*)||_\infty \leq \frac{N_T\{\delta\}}{1 - N_T\{\delta\}} ||P_T\dot{y}^*||_\infty , \tag{29}$$

where

$$||P_T\dot{x}||_\infty = \operatorname*{ess\,sup}_{0\leq t \leq T} |\dot{x}(t)|. \tag{30}$$

The proof of (29) is similar to the proof of inequality (13). Simply note that, since we have assumed zero initial conditions, equation (17) implies the derivative equation: $\dot{y} - \dot{y}^* = H\ S^*\ \dot{y}$.

It is not difficult to realise that higher derivative bounds on the performance error can similarly be derived. Furthermore as both absolute and derivative bounds depend on the very same parameter function $N_T\{\delta\}$, the determination of such bounds is computationally efficient.

The previous results provide useful assessments on the worst possible performance deterioration in both an absolute and a derivative sense. Consequently it is possible to adopt classical system synthesis or identification methods for the determination of suitable, possibly low order, feedback compensators on the basis of a specified upper bound on the total variation of the synthesis error. The latter can simply be chosen from a knowledge of the idealised closed-loop system performance and the tolerable performance deterioration with respect to the system's unit step response. The main advantage of the results lies in their simple forms and in that both absolute and derivative bounds depend on the very same parameter function $N_T\{\delta\}$. Although any value of $N_T\{\delta\}$ less than one, satisfies the validity of the functional inequalities, much smaller values are normally necessary in order to ensure that the performance bounds are practically meaningful. A disadvantage of the absolute and derivative bounds is that these represent mere sufficient conditions. As a result, it is important to realise that the assessments can be rather conservative and that the condition $N_\infty\{\delta\} \geq 1$ does not necessarily imply that the approximately compensated system is unstable.

4. EVALUATION

This paper has described certain functional analytic results that have useful application in the time domain synthesis of approximate (low order) feedback compensators for specified system performance. The approach involves the determination of an idealised input-output description for the compensator which reflects some specified closed loop system's unit step response performance. Subsequently, a compensator F_a can be synthesised to generate the given input-output map in an approximate manner. Based on simple descriptions of the synthesis error (identification error) $\delta(t)$, the results of this work provide easily computable assessments of the approximation-induced deterioration in the closed-loop system's performance.

In particular, theorem 1 generates an uncertainty band which contains the approximately compensated system's response and is also capable of bounding the worst possible rate of change of the performance error. Furthermore, provided $N_\infty\{\delta\} < 1$, the theorem ensures the stability of the approximately compensated system. Theorem 2 imposes an additional condition and thus guarantees that the idealised and approximate system's unit step responses converge to the same steady state value. The results have simple and easily computable forms and can be combined with classical time domain synthesis techniques for the derivation of low order feedback controllers that meet tight system performance specifications.

The simple functional inequalities that have been presented in this work can be extended to the synthesis of compensators in relation to other system configurations. For example, similar results can easily be derived for the input-output synthesis of approximate (low order) error-actuated compensators.

5. ACKNOWLEDGEMENT

The authors thank Dr. D.H. Owens of the University of Sheffield for his useful suggestions.

6. REFERENCES

1. Zakian, V. and Al Naib, U. (1973) Design of dynamical and control systems by the method of inequalities, Proc. IEE, Vol. 120, pp 1421-1427.

2. Becker, R.G., Heunis, A.J. and Mayne D.Q. (1979) Computer-aided design of control systems via optimisation, Proc. IEE, Vol. 126, pp 573-584.

3. Owens, D.H. and Chotai, A. (1983) Robust controller design for linear dynamic systems using approximate models, Proc. IEE, part D, Vol. 130, pp 45-56.

4. Gray, J.O. and Valsamis, D. (1982) The direct synthesis of compensators for nonlinear feedback systems, IFAC Symp. Purdue.

5. Desoer, C.A. and Vidyasagar, M. (1975) Feedback systems: Input-output properties, Academic Press.

6. Gray, J.O. and Valsamis, D. (1984) Multivariable nonlinear system design in IFAC Congress, Budapest.

TWO PROBLEMS IN THE DESIGN OF DISCRETE DECENTRALISED CONTROL SYSTEMS SUBJECT TO FINITE PERTURBATIONS

A. Locatelli, R. Scattolini and N. Schiavoni

(Dipartimento di Elettronica, Politecnico di Milano, Italy)

ABSTRACT

The problem of synthesising a decentralised regulator which supplies the control system with asymptotic stability and zero steady-state error is considered in this paper under the assumption that the instrumentation or the plant parameters are subject to finite perturbations.

1. INTRODUCTION

This paper is concerned with the design of discrete-time feedback control systems. It addresses the general problem of synthesising a decentralised regulator which guarantees stability as well as zero steady-state error despite the presence of constant exogenous signals. It is moreover desired that these properties hold not only when the closed-loop system is in nominal conditions but also, to the maximum possible extent, when the system parameters undergo finite perturbations. Two specific kinds of perturbations are separately considered, relevant to the instrumentation and the plant under control, respectively. Consistently, two different qualitative problems are discussed.

We first present the Reliable Robust Decentralised Regulator Problem (RRDRP)[1] which basically aims at synthesising a regulator capable of a satisfactory behaviour also when one of the feedback loops opens because of a failure of the corresponding sensor and/or actuator.

Then we focus the attention on the Simultaneous Robust Decentralised Regulator Problem (SRDRP)[2]. It consists of synthesising a regulator which is required to behave in a satisfactory way when connected to any element of a given finite set of systems.

This is apparently a significant goal to be pursued whenever the unique plant under control is strongly nonlinear or may undergo so large parameter variations as to motivate a description of it in terms of a finite bunch of linear systems rather than by means of a single one.

Constructive sufficient solvability conditions for both problems are presented in Sections 2 and 3, respectively. They almost naturally lead to defining regulators which, unfortunately, are likely to produce very unsatisfactory transients of the resulting control systems. Therefore, it seems advisable to introduce some quantitative aspects into the problem formulation. This is done in Section 4, where two optimal control problems are stated with reference to RRDRP and SRDRP.

2. THE RELIABLE ROBUST DECENTRALISED REGULATOR PROBLEM

In order to state RRDRP, first assume that the plant under control is described by

$$x(t+1) = A^{o}x(t) + B^{o}u(t) + M^{o}d \tag{2.1.a}$$

$$y(t) = C^{o}x(t) + N^{o}d \quad , \tag{2.1.b}$$

where $x(t) \in R^{n}$, $u(t) \in R^{m}$, $y(t) \in R^{m}$ and $d \in R^{r}$ are the state, control, output and disturbance vectors, respectively. The disturbance d is constant.

The behaviour of the i-th sensor and actuator, $i \in M \overset{\Delta}{=} \{1, 2, \ldots, m\}$, is specified by (see Fig. 1)

$$\tilde{y}_i(t) = \alpha_i \, y_i(t) \quad , \tag{2.2.a}$$

$$u_i(t) = \beta_i \, \tilde{u}_i(t) \quad , \tag{2.2.b}$$

respectively, where $\alpha_i = 1$ ($\beta_i = 1$) if the sensor (actuator) is properly working and $\alpha_i = 0$ ($\beta_i = 0$) if a failure occurs.

The regulator as a whole consists of m noninteracting single-input-single-output local regulators. The output of the i-th one is $\tilde{u}_i(t)$ which coincides with the i-th control variable $u_i(t)$ whenever the i-th actuator is properly working; the input is

$$\tilde{e}_i(t) \overset{\Delta}{=} \bar{y}_i - \tilde{y}_i(t) \quad , \tag{2.3}$$

where $\bar{y}_i$ is the constant set-point for $y_i(t)$. The scalar $\tilde{e}_i(t)$ equals the i-th system error $e_i \overset{\Delta}{=} \bar{y}_i - y_i(t)$ whenever the i-th sensor is properly working. Furthermore, the i-th local regulator possesses perfect information on the operating condition of

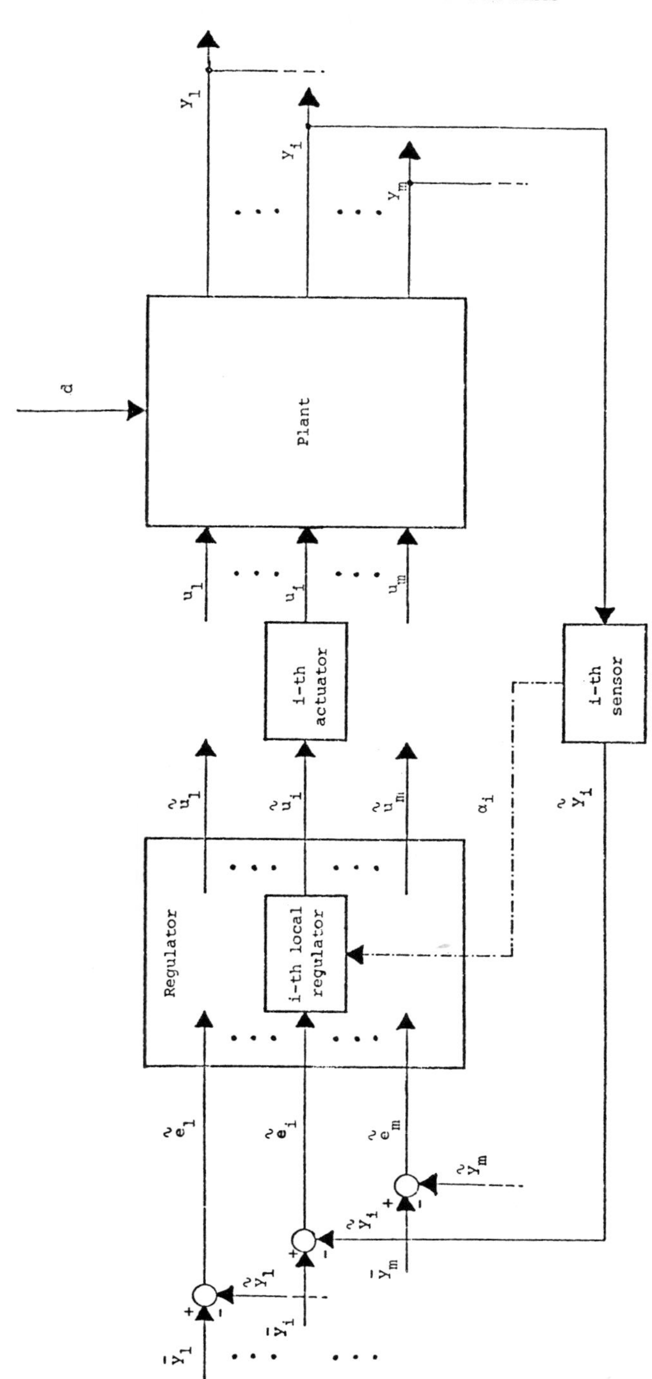

Fig. 1 The control system considered in the RRDRP

the corresponding sensor, namely it knows the value taken on by α_i. In particular, the i-th local regulator is assumed to be described by

$$z^i(t+1) = F^i z^i(t) + g^i \tilde{e}_i(t) \qquad (2.4.a)$$

$$\tilde{u}_i(t) = \gamma_i(h^{iT} z^i(t) + k^i \tilde{e}_i(t), \alpha_i) \quad , \qquad (2.4.b)$$

where $z^i(t) \in R^{\nu_i}$ and the function $\gamma_i(\cdot,\cdot): R\times\{0,1\} \rightarrow R$ is defined by

$$\gamma_i(\xi, \alpha_i) = \begin{cases} \gamma_{i0}\,\xi & , \text{ if } \alpha_i = 0 \\ \gamma_{i1}\,\xi & , \text{ if } \alpha_i = 1 \end{cases} . \qquad (2.4.c)$$

Therefore, for given α_i's, the overall regulator is linear.

If all sensors and actuators behave properly, i.e., $\alpha_j \beta_j = 1$, $j \in M$, we shall say that the regulator is fully connected to the plant and denote the dynamical matrix of the overall system (2.1) - (2.4) by $\hat{A}^o_\ell$. If a failure occurs in the ℓ-th loop, i.e., $\alpha_\ell \beta_\ell = 0$ and $\alpha_j \beta_j = 1$, $j \in M^\ell \overset{\Delta}{=} \{1, 2, \ldots, \ell-1, \ell+1, \ldots, m\}$, the overall control system will apparently be constituted by the cascade connection of two subsystems. Their eigenvalues are those of F^ℓ together with those of a matrix henceforth denoted by $\hat{A}^\ell$.

If we desire to design a regulator such as to guarantee asymptotic stability and robust zero regulation of all error variables when it is fully connected to the plant and to preserve stability as well as robust zero regulation of the m-1 error variables not involved in the failure when a failure occurs, we must solve the following problem.

RRDRP

For all $i \in M$ find ν_i, F^i, g^i, h^i, k^i, γ_{i0} and γ_{i1} such that:

(a) If the regulator is fully connected to the plant:

(i) the eigenvalues of matrix $\hat{A}^o$ all lie within the unit circle,

(ii) $\lim_{t\rightarrow\infty} e_j(t) = 0$, $j \in M$, for all (constant) $\bar{y}$ and d,

(iii) condition (ii) holds in a robust way, that is for all perturbations of the parameters of system (2.1) such that condition (i) still holds;

(b) If a failure occurs in the ℓ-th loop, $\ell \in M$:
(iv) the eigenvalues of matrix $\hat{A}^{\ell}$ all lie within the unit circle,
(v) $\lim_{t\to\infty} e_j(t) = 0$, $j \in M^{\ell}$, for all (constant) $\bar{y}$ and d,
(vi) condition (v) holds in a robust way, that is for all perturbations of the parameters of system (2.1) such that condition (iv) still holds. □

Sufficient solvability conditions are presented for RRDRP in the case of asymptotically stable plants. In spite of this assumption, the problem is however not trivial at all since the requirement of robust zero error regulation entails, as well known, the use of integral-type regulators, so that the resulting open-loop system is anyhow not asymptotically stable. These conditions can be proved by exploiting multivariable root-loci techniques along the same lines already successfully adopted in the past[3,4].

For any matrix Σ, let $\lambda_h(\Sigma)$ denote its h-th eigenvalue and $\Sigma(h)$ the matrix resulting after the h-th row and column have been deleted. Moreover, let $G = C^o(I - A^o)^{-1} B^o$ be the static gain matrix of system (2.1).

Theorem 2.1

If (i) $|\lambda_h(A^o)| < 1$, $h = 1, 2, \ldots, n$; (ii) there exists a diagonal matrix $V \triangleq \mathrm{diag}(v_1, v_2, \ldots, v_m)$ such that $\mathrm{Re}(\lambda_j(GV)) < 0$, $j \in M$, and $\mathrm{Re}(\lambda_j(G(i)V(i))) < 0$, $j \in M$, $i \in M$, then there exists $\bar{\delta} > 0$ such that, for any $\delta \in (0, \bar{\delta})$, the local regulators (2.4) with

$$\nu_i = F^i = g^i = \gamma_{i1} = 1, \quad k^i = \gamma_{i0} = 0, \quad h^{iT} = -\delta v_i \qquad (2.5)$$

provide a solution to RRDRP. □

Unfortunately, condition (ii) of this theorem is not given in terms of the problem data and can in general be checked in an easy way for very small values of m only. However, there are special and significant cases where it is possible to test this condition in terms of matrix G and to give an explicit expression of V. The following corollary makes reference to the simplest of such cases.

Corollary 2.1

If (i) $|\lambda_h(A^o)| < 1$, h = 1, 2, ..., n; (ii) matrix $G \stackrel{\Delta}{=} \{\mu_{ij}\}$ is diagonally dominant or triangular and nonsingular, then there exists $\bar{\delta} > 0$, such that for any $\delta \in (0, \bar{\delta})$, the local regulators (2.4), (2.5) with $v_i = \mu_{ii}/|\mu_{ii}|$ provide a solution to RRDRP. □

The range of applicability of this corollary can significantly be extended by exploiting the fact that under specified circumstances (see,e.g., Theorem A2 in Šiljak[5]) it is possible to find a diagonal matrix V* such that GV* is diagonally dominant even though G is not. This amounts to introducing a first nondynamic regulator into the loop which modifies the system equations so that Corollary 1 can subsequently be applied.

3. THE SIMULTANEOUS ROBUST DECENTRALISED REGULATOR PROBLEM

In order to state SRDRP, assume that the plant under control is described by the q systems

$$x^j(t + 1) = A^j x^j(t) + B^j u^j(t) + M^j d^j \tag{3.1.a}$$

$$y^j(t) = C^j x^j(t) + N^j d^j \quad , \tag{3.1.b}$$

where, for any $j \in \mathcal{Q} \stackrel{\Delta}{=} \{1, 2, ..., q\}$, $x^j(t) \in R^{n_j}$, $u^j(t) \in R^m$, $y^j(t) \in R^m$ and $d^j \in R^{r_j}$ are the state, control, output and disturbance vectors, respectively. The disturbance is constant.

The overall regulator for the q systems (3.1) consists again of m noninteracting single-input-single-output local regulators the i-th of which makes the i-th component $u_i^j(t)$ of $u^j(t)$ to depend only on the i-th component $e_i^j(t)$ of the error vector

$$e^j(t) \stackrel{\Delta}{=} \bar{y}^j - y^j(t) \quad , \tag{3.2}$$

where $\bar{y}^j$ is the constant set-point for $y^j(t)$. In particular, the i-th local regulator, $i \in \mathcal{M}$, is assumed to be described by

$$z^i(t + 1) = F^i z^i(t) + g^i e_i^j(t) \tag{3.3.a}$$

$$u_i^j(t) = h^{iT} z^i(t) + k^i e_i^j(t) \quad , \tag{3.3.b}$$

where $z^i(t) \in R^{\nu_i}$.

If we desire to design a unique m-tuple of regulators (3.3) which guarantee closed-loop asymptotic stability and robust zero regulation of all the error vectors (3.2), we must solve the following problem.

SRDRP

For all $i \in M$, find ν_i, F^i, g^i, h^i and k^i such that, for all $j \in Q$,

(i) system (3.1) - (3.3) is asymptotically stable,

(ii) $\lim_{t\to\infty} e^j(t) = 0$,
for all (constant) $\bar{y}^j$ and d^j.

(iii) condition (ii) holds in a robust way, that is for all perturbations of the parameters of systems (3.1) such that condition (i) still holds. □

For this problem, constructive sufficient conditions can be given when systems (3.1) are asymptotically stable. Two simple results are here reported, which parallel Theorem 2.1 and Corollary 2.1. Let $G^j \triangleq C^j(I - A^j)^{-1} B^j$ be the static gain of the j-th system (3.1).

Theorem 3.1

If, for $j \in Q$, (i) $|\lambda_h(A^j)| < 1$, $h = 1, 2, \ldots, n_j$; (ii) there exists a diagonal matrix $V \triangleq \mathrm{diag}(v_1, v_2, \ldots, v_m)$ such that $\mathrm{Re}(\lambda_k(G^j V)) < 0$, $k \in M$, then there exists $\bar{\delta} > 0$ such that, for any $\delta \in (0, \bar{\delta})$, the local regulators (3.3) with

$$\nu_i = F^i = g^i = 1, \quad k^i = 0, \quad h^{iT} = -\delta v_i \tag{3.4}$$

provide a solution to SRDRP. □

4. REGULAR DESIGN VIA PARAMETER OPTIMISATION

It should be noted that only qualitative aspects of the control systems behaviour are taken into consideration by the two previously stated problems. This might considerably reduce their significance mostly in view of the fact that the derived sufficient solvability conditions directly suggest the use of purely integral low-gain regulators which in turn are likely to produce sluggish transients of the resulting closed-loop systems. Therefore, in order to supply the control systems with satisfactory dynamic behaviour, it is advisable to use more complex local regulators and to suitably select their parameters, for instance, by solving optimal control problems. This can be done for both RRDRP and SRDRP.

Let us consider first RRDRP. With reference to the notation of Section 2, let $\nu_i \geq 1$, $i \in M$, be given and assume that y_i, $i \in M$, d, x(o) and $\bar{z}^i(o)$, $i \in M$, are zero-mean independent random variables with known covariances.

Moreover, assume that, when the regulator is fully connected to the plant, the transient performance of the control system is evaluated through the index

$$J_o \stackrel{\Delta}{=} E\{\sum_{t=0}^{\infty} \|\xi^o(t) - \xi^o_\infty\|^2_{Q^o} + \|e^o(t)\|^2_{R^o} + \|u(t) - u_\infty\|^2_{S^o}\},$$

where Q^o, R^o and S^o are given positive semidefinite matrices,

$$\xi^o \stackrel{\Delta}{=} \left[x^T \; z^{1T} \; z^{2T} \; \ldots \; z^{mT}\right]^T, \quad e^o \stackrel{\Delta}{=} \left[e_1 \; e_2 \; \ldots \; e_m\right]^T,$$

$$\xi^o_\infty = \lim_{t\to\infty} \xi^o(t), \quad u_\infty \stackrel{\Delta}{=} \lim_{t\to\infty} u(t)$$

and $E\{\cdot\}$ is the expectation operator. Then, when a failure occurs in the ℓ-th loop, $\ell \in M$, the transient performance of the damaged control system is evaluated through the index

$$J_\ell \stackrel{\Delta}{=} E\{\sum_{t=0}^{\infty} \|\xi^\ell(t) - \xi^\ell_\infty\|^2_{Q^\ell} + \|e^\ell(t)\|^2_{R^\ell} + \|u^\ell(t) - u^\ell_\infty\|^2_{S^\ell}\},$$

where Q^ℓ, R^ℓ and S^ℓ are given positive semidefinite matrices,

$$\xi^\ell \stackrel{\Delta}{=} \left[x^T \; z^{1T} \; \ldots \; z^{\ell-1T} \; z^{\ell+1T} \; \ldots \; z^{mT}\right]^T,$$

$$e^\ell \stackrel{\Delta}{=} \left[e_i \; \ldots \; e_{\ell-1} \; e_{\ell+1} \; \ldots \; e_m\right]^T, \quad u^\ell \stackrel{\Delta}{=} \left[u_1 \; \ldots \; u_{\ell-1} \; u_{\ell+1} \; \ldots u_m\right]^T,$$

$$\xi^\ell_\infty \stackrel{\Delta}{=} \lim_{t\to\infty} \xi^\ell(t), \quad u^\ell_\infty \stackrel{\Delta}{=} \lim_{t\to\infty} u^\ell(t).$$

Finally, by assuming that a set of positive scalars σ_j, $j \in \{0\} \cup \in \cup M$, is given, the Optimal Reliable Robust Decentralised Regulator Problem (ORRDRP) can be stated.

ORRDRP

Find F^i, g^i, h^i, k^i and γ_{il} which, together with $\gamma_{iO} = 0$, solve RRDRP and minimise

$$J \stackrel{\Delta}{=} \sum_{\ell=0}^{m} \sigma_\ell J_\ell . \qquad \square$$

Consider now SRDRP. With reference to the notation of Section 3, let $\nu_i \geq 1$, $i \in M$, be given and assume that $\bar{y}^j$, d^j, $x^j(o)$, $j \in \mathcal{Q}^i$, and $z^i(o)$, $i \in M$, are independent random variables with known covariances. Then, the transient behaviour of the j-th, $j \in \mathcal{Q}$, control system is evaluated through the index

$$J_j \stackrel{\Delta}{=} E\{\sum_{t=0}^{\infty} \|\chi^j(t) - \chi^j_\infty\|^2_{Q^j} + \|e^j(t)\|^2_{R^j} + \|u^j(t) - u^j_\infty\|^2_{S^j}\},$$

where Q^j, R^j and S^j are given positive semidefinite matrices,

$\chi^j \stackrel{\Delta}{=} [x^{jT}\ z^{1T}\ z^{2T}\ \dots\ z^{mT}]^T$, $\quad \chi^j_\infty = \lim_{t\to\infty} \chi^j(t)$,

$u^j_\infty = \lim_{t\to\infty} u^j(t)$.

Finally, by assuming that a set of positive scalars σ^j, $j \in \mathcal{Q}$, is given, the Optimal Simultaneous Robust Decentralised Regulator Problem (OSRDRP) can be stated.

OSRDRP

Find F^i, g^i, h^i and k^i, $i \in M$, which solve SRDRP and minimise

$$J \stackrel{\Delta}{=} \sum_{j=1}^{q} \sigma_j J_j \ . \qquad \square$$

Following a line of reasoning already exploited in the past [1,2,6,7], it can be shown that the above stated ORRDRP and OSRDRP can be recast into two constrained nonlinear programming problems endowed with a number of peculiar features.

First, the parametrisation of the local regulators can be selected in such a way that the robust asymptotic zero error constraints ((ii), (iii), (v), (vi) as for RRDRP and (ii), (iii) as for SRDRP) are always satisfied under certain stability conditions ((i), (iv) as for RRDRP and (i) as for SRDRP). In particular, it is necessary and sufficient that every local regulator incorporates an integral action.

Second, it can easily be recognised that, whenever the relevant control systems are asymptotically stable, the evaluation of both the performance indices and their gradients simply calls for solving a number of Lyapunov equations. Moreover, the meaning itself of the performance indices actually guarantees that, apart from very critical cases, the stability region in the parameter space is not abandoned during the optimisation.

Finally, the problem of initialising the optimisation algorithms, that is the preliminary finding of a stabilising solution, can often be overcome by exploiting the constructive solvability conditions given in the theorems and corollaries above.

Hence, ORRDRP and OSRDRP can be considered to be fairly easy to solve.

5. CONCLUDING REMARKS

This paper has considered two specific problems arising in the design of decentralised control systems subject to finite perturbations. The results presented here can be generalised along different lines: significant are the cases of polynomial exogenous signals and multivariable local regulators.

Among the problems still open special attention might be deserved by the finding of more detailed models of the perturbations undergone by the control systems.

ACKNOWLEDGEMENTS

This paper has partially been supported by Centro di Teoria dei Sistemi, CNR, and MPI.

6. REFERENCES

1. Locatelli, A., Romeo, F., Scattolini, R., and Schiavoni, N., (1983) "A parameter optimization approach to the design of reliable robust decentralized regulators", A. Straszak (Ed.): Proc. IFAC/IFORS Symposium on Large Scale Systems: Theory and Applications, Warsaw: Scientific Printers, pp. 291-296.

2. Schiavoni, N., (1983) "On the simultaneous design of decentralized linear control systems", Proc. 22nd IEEE Conference on Decision and Control, pp. 529-530.

3. Guardabassi, G., Locatelli, A., and Schiavoni, N., (1981) "On the initialization problem in the parameter optimization of structurally constrained industrial regulators", Large Scale Systems, Vol. 3, pp. 267-277.

4. Locatelli, A., and Schiavoni, N., (1983) "On the initialization problem in the design of structurally constrained regulators", Int. J. Contr., Vol. 38, pp. 717-733.

5. Šiljak, D.D., (1978) "Large scale dynamic systems: stability and structure", North Holland, New York.

6. Davison, E.J., and Ferguson, I., (1981) "The design of controllers for the multivariable robust servomechanism problem using parameter optimization methods", IEEE Trans. Automat. Contr., Vol. AC-26, pp. 93-110.

7. Guardabassi, G., Locatelli, A., Maffezzoni, C., and Schiavoni, N., (1983) "Computer-aided design of structurally constrained multivariable regulators. Part 1: problem statement, analysis and solution. Part 2: applications". IEE Proc. - Part D, Vol. 130, pp. 155-172.

MODIFIED OLD AND NEW ADAPTIVE LAWS FOR APPLICATIONS WITH A NEW MRAC SCHEME

J. Chandrasekhar
(Aeronautical Engineering Department, I.I.T., Bombay, India)

and

M.P.R. Vittal Rao
(Department of Electrical and Electronic Engineering, University of Ulster at Jordanstown)

ABSTRACT

For Lyapunov MRAC, two families of modified adaptive laws having different forms (I, P+I, P+I+D, I+DI, P+I+DI) are presented. These laws are derived using modified Lyapunov functions with a new differentiator-free MRAC scheme that has a simple structure even when the relative degree of the plant $n^* \geq 2$.

1. INTRODUCTION

In recent years, the theory of Lyapunov Model Reference Adaptive Control (MRAC) has made promising advances[1-12]. The important forms of adaptive laws[1-6] for Lyapunov MRAC are: Integral (I), Proportional+Integral (P+I) and Proportional +Integral+ Derivative (P+I+D) - referred to as old laws in this paper. Colburn and Boland III[6] have classified these old laws into two families, based on a system error function. Similarly, new Integral+Double Integral (I+DI) and Proportional+Integral +Double Integral (P+I+DI) forms of adaptive laws were developed[7,8]. These works[1-8] assumed that: 1) all the plant output derivatives are available; and 2) all the plant parameters are directly incremented by the controller parameters. But, in reality, since these assumptions can not be met, these old and new laws, as such, have not been useful for applications.

However, there has been research effort[9-11] to design differentiator-free MRAC schemes with suitable structure, since Parks[9] had suggested a way, by using the Kalman-Yacubovich Lemma (KYL). Later, Monopoli's scheme[10], using an augmented error signal and the KYL, has been the major work. Recently, Narendra and Valavani[11], pointing out a stability problem with Monopoli's scheme, proposed a stable MRAC scheme. But, this scheme has a complicated structure with many auxiliary signals and filters, when (n-m), defined as the relative degree n^* of the plant with n poles and m zeros, is $\geqslant 2$. These schemes[9-11] use only 'I' form of adaptive laws. Of late, Chandrasekhar and Vittal Rao[12] have suggested a differentiator-free Lyapunov MRAC scheme, which does not need such auxiliary signals and so many filters when $n^* \geqq 2$, and which has a flexibility to use different forms of adaptive laws.

In this context, with the new MRAC scheme[12], this paper presents modified old and new adaptive laws that are suitable for applications.

2. MODIFIED ADAPTIVE LAWS

2.1 The MRAC Scheme[12]

For ease of exposition, consider the MRAC system shown in Fig. 1 for the plant with n = 3 and m = 2 referred with a model without zeros.

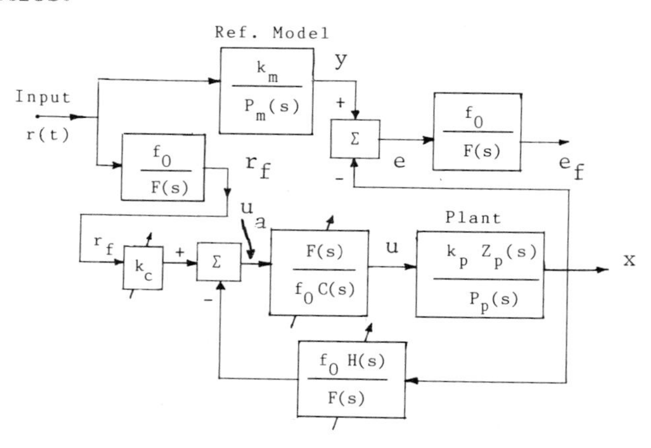

Fig.1. Structure of the MRAC

Plant*:

$$P_p(s)x = k_p Z_p(s)u; \; P_p(s) = s^3 + g_2 s^2 + g_1 s + g_o;$$

$$Z_p(s) = s^2 + b_1 s + b_o \tag{2.1}$$

The assumptions on the plant are: 1) k_p is a positive unknown constant; 2) the number of zeros m and the number of poles n are known; 3) the parameters g's and b's are unknown constants; 4) the zeros are restricted to only L.H.S. of s-plane; and 5) the plant output derivatives are not available.

Reference Model:

$$P_m(s)y = k_m r; \; P_m(s) = s^3 + a_2 s^2 + a_1 s + a_o \tag{2.2}$$

This reference model is chosen to represent the desired behaviour of the plant with its controller.

Plant with the Controller: From Fig.1, it can be seen that the controller essentially generates the plant input signal u, as

$$u = [F(s)/f_o C(s)]u_a \tag{2.3}$$

where $u_a = k_c r_f - H(s)x_f; \; r_f = [f_o/F(s)]r$

and

$$x_f = [f_o/F(s)]x \tag{2.4}$$

with $C(s) = s^2 + c_1 s + c_o$ (mth order);

$$H(s) = h_2 s^2 + h_1 s + h_o \text{ (n-1th order)} \tag{2.5}$$

and $F(s) = s^3 + f_2 s^2 + f_1 s + f_o$ (nth order, prechosen stable polynomial). (2.6)

Substituting Eqns. (2.3-6) in Eqn. (2.1) gives

$$[P_p(s) + k_p H(s)]x = k_c k_p r - \delta_1 \dot{u} - \delta_o u;$$

$$\delta_j = k_p(c_j - b_j) \quad (j = 0,1) \tag{2.7}$$

*In this paper, 's' is used as the differential operator or the Laplace transform variable, depending on the context.

Filtered Error Equation: Defining the system error as $e = y - x$; subtracting Eqn. (2.7) from Eqn. (2.2); and adding suitable terms lead to the system error equation as

$$P_m(s)\ e = \phi_c r + \delta_1 \dot{u} + \delta_O u + (\sum_{i=O}^{2} \psi_i s^i)\ x = f(\theta), \qquad (2.8)$$

where $f(\theta)$ is the parameter misalignment function, with $\theta^T = (\phi_c, \delta_1, \delta_O; \psi_i)$ and

$$\phi_c = k_m - k_c k_p; \ \psi_i = g_i + k_p h_i - a_i. \qquad (2.9)$$

Now, multiplying both sides of Eqn. (2.8) by $f_O/F(s)$ gives the filtered error equation as

$$P_m(s) e_f = \phi_c r_f + \delta_1 \dot{u}_f + \delta_O u_f + (\sum_{i=O}^{2} \psi_i s^i) x_f = f_f(\theta) \qquad (2.10)$$

where $$e_f = [f_O/F(s)]e; \quad u_f = [f_O/F(s)]u \qquad (2.11)$$

and $f_f(\theta)$ is the filtered parameter misalignment function. In state-space notation, Eqn. (2.10) becomes,

$$\dot{\underline{e}}_f = A_m \underline{e}_f + \underline{f} \text{ with } \underline{e}_f^T = (\underline{e}_f, \dot{\underline{e}}_f, \ddot{\underline{e}}_f); \ \underline{f}^T = (O, O, f_f(\theta)) \qquad (2.12)$$

and A_m is the stable reference model matrix in the phase variable canonical form.

2.2 Modified New Laws

Consider the Lyapunov function V_n, given by

$$V_n = \underline{e}_f^T P \underline{e}_f + (V_{c1} + V_{r1} + V_{x1}) + (V_{c2} + V_{r2} + V_{x2}) \qquad (2.13)$$

with

$$V_{c1} = \sum_{j=O}^{1} (1/\eta_j)(\delta_j + \rho_j Y_{uj} + \lambda_j \int Y_{uj}\, dt)^2 \qquad (2.14)$$

$$V_{r1} = (1/\eta_2)(\phi_c + \rho_2 Y_r + \lambda_2 \int Y_r\, dt)^2 \qquad (2.15)$$

$$V_{x1} = \sum_{i=O}^{2} (1/\alpha_i)(\psi_i + \sigma_i Y_{xi} + \beta_i \int Y_{xi}\, dt)^2 \qquad (2.16)$$

$$V_{c2} = \sum_{j=0}^{1} (1/\lambda_j)(\delta_j + \rho_j\, Y_{uj})^2 + \rho_j(1 + \eta_j/\lambda_j)(\int Y_{uj}\, dt)^2 \qquad (2.17)$$

$$V_{r2} = (1/\lambda_2)(\phi_c + \rho_2\, Y_r)^2 + \rho_2(1 + \eta_2/\lambda_2)(\int Y_r\, dt)^2 \qquad (2.18)$$

$$V_{x2} = \sum_{i=0}^{2} (1/\beta_i)(\psi_i + \sigma_i\, Y_{xi})^2 + \sigma_i(1 + \alpha_i/\beta_i)(\int Y_{xi} dt)^2 \qquad (2.19)$$

where

$$Y_{uj} = Z_f\, u_f^{(j)};\quad Y_r = Z_f\, r_f;\quad Y_{xi} = Z_f\, x_f^{(i)} \qquad (2.20)$$

$$Z_f = W_f + \gamma\, f_f(\theta);\quad W_f = \sum_{k=0}^{2} p_{(k+1)3}\, e_f^{(k)};\quad (\gamma \geqq 0) \qquad (2.21)$$

and $\underline{e}_f$, ϕ_c, δ_j, ψ_i are as defined above; $x_f^{(.)}$, $u_f^{(.)}$ and $e_f^{(.)}$ are (.)th derivatives of the x_f, u_f and e_f respectively; (ρ's, η's, λ's, σ's, β's, α's) are arbitrary positive constants; Z_f is a filtered system error function, with p's as the elements of the positive definite symmetric matrix P determined from the Lyapunov equation

$$A_m^T\, P + P\, A_m = -\, Q \qquad (2.22)$$

for the choice of Q as any positive definite symmetric matrix.

Now, differentiating V_n in Eqn. (2.13) with respect to time, substituting $\dot{\underline{e}}_f$ from Eqn. (2.12) and choosing

$$\dot{\delta}_j = -\rho_j \frac{d}{dt}\, Y_{uj} - \lambda_j\, Y_{uj} - \eta_j \int Y_{uj}\, dt \qquad (j = 0,1) \qquad (2.23)$$

$$\dot{\phi}_c = -\rho_2 \frac{d}{dt}\, Y_r - \lambda_2\, Y_r - \eta_2 \int Y_r\, dt \qquad (2.24)$$

$$\dot{\psi}_i = -\, \sigma_i \frac{d}{dt}\, Y_{xi} - \beta_i\, Y_{xi} - \alpha_i \int Y_{xi}\, dt \qquad (i = 0,1,2) \qquad (2.25)$$

lead to the 'Proportional + Integral + Double Integral' (P+I+DI) form of adaptive laws, as

$$c_j(t) = -\, \rho_j Y_{uj} - \lambda_j \int Y_{uj}\, dt - \eta_j \iint Y_{uj}\, dt\, dt + c_j(0) \qquad (2.26)$$

$$k_c(t) = \rho_2 Y_r + \lambda_2 \int Y_r \, dt + \eta_2 \iint Y_r \, dt\, dt + k_c(0) \tag{2.27}$$

$$h_i(t) = -\sigma_i Y_{xi} - \beta_i \int Y_{xi} \, dt - \alpha_i \iint Y_{xi} \, dt\, dt + h_i(0) \tag{2.28}$$

and
$$\dot{V}_n = (\dot{V}_{n1} + \dot{V}_{n2} + \dot{V}_{n3} + \dot{V}_{n4}) + \dot{V}_{n5} \tag{2.29}$$

with
$$\dot{V}_{n1} = - \underline{e}_f^T Q \underline{e}_f; \quad \dot{V}_{n2} = -2 \, [f_f(\theta)]^2 \tag{2.30}$$

$$\dot{V}_{n3} = - 2 \, [\rho_2 Y_r^2 + \sum_{j=0}^{1} \rho_j Y_{uj}^2 + \sum_{i=0}^{2} \sigma_i Y_{xi}^2] \tag{2.31}$$

$$\dot{V}_{n4} = - 2\lambda_2 [\int Y_r \, dt]^2 - \sum_{j=0}^{1} 2\lambda_j [\int Y_{uj} \, dt]^2 - \sum_{j=0}^{2} 2 \beta_i [\int Y_{xi} \, dt]^2 \tag{2.32}$$

$$\dot{V}_{n5} = - 2(1 + \eta_2/\lambda_2) \, \phi_c \int Y_r \, dt - \sum_{j=0}^{1} 2(1 + \eta_j/\lambda_j) \, \delta_j \int Y_{uj} dt$$
$$- \sum_{i=0}^{2} 2(1 + \alpha_i/\beta_i) \, \psi_i \int Y_{xi} dt \tag{2.33}$$

In the R.H.S. of the Eqn. (2.29), $\dot{V}_{n1}$ is negative definite, since the Q matrix has been chosen as positive definite; $\dot{V}_{n2}$, $\dot{V}_{n3}$ and $\dot{V}_{n4}$ are negative definite; and $\dot{V}_{n5}$ is sign-indefinite. However, it can be seen that for $\lambda_j >> \eta_j (j=0,1,2)$, $\beta_i >> \alpha_i$ $(i=0,1,2)$, the sum of the first four terms dominates in magnitude over the last $\dot{V}_{n5}$ term.

Thus, for $\lambda_j >> \eta_j (j=0,1,2)$ and $\beta_i >> \alpha_i (i=0,1,2)$, the filtered system error equation, Eqn. (2.10), is asymptotically stable, i.e., $\underline{e}_f \to 0$ as $t \to \infty$. This condition, in turn, implies $\underline{e} \to 0$ as $t \to \infty$, since $e_f = [f_0/F(s)]e$ with $F(s)$ as the prechosen stable polynomial.

It is important to note that these laws require the n derivatives of x_f and e_f for generating Y_{xi} and Z_f; and the (m - 1) derivatives of u_f for generating the signals Y_{uj} (see Eqns. (2.20 - 21)). All these filtered derivative signals are available (see Figs.1 and 2).

2.3 Modified Old Laws

Consider the Lyapunov function V_{mo}, given by

$$V_{mo} = \underline{e}_f^T P \underline{e}_f + V_1 + V_2 + V_3 \tag{2.34}$$

with
$$V_1 = \sum_{j=0}^{1} (1/\lambda_j)[\delta_j + \rho_j Y_{uj} + \eta_j \frac{d}{dt} Y_{uj}]^2 + \sum_{j=0}^{1} \eta_j Y_{uj}^2 \tag{2.35}$$

$$V_2 = (1/\lambda_2)[\phi_c + \rho_2 Y_r + \eta_2 \frac{d}{dt} Y_r]^2 + \eta_2 Y_r^2 \tag{2.36}$$

$$V_3 = \sum_{i=0}^{2} (1/\beta_i)[\psi_i + \sigma_i Y_{xi} + \alpha_i \frac{d}{dt} Y_{xi}]^2 + \sum_{i=0}^{2} \alpha_i Y_{xi}^2 \tag{2.37}$$

Now following the steps of Sec. 2.2 and choosing[5]

$$\dot{\delta}_j = - \rho_j \frac{d}{dt} Y_{uj} - \lambda_j Y_{uj} - \eta_j \frac{d^2}{dt^2} Y_{uj} \quad (j=0,1) \tag{2.38}$$

$$\dot{\phi}_c = -\rho_2 \frac{d}{dt} Y_r - \lambda_2 Y_r - \eta_2 \frac{d^2}{dt^2} Y_r \tag{2.39}$$

$$\dot{\psi}_i = - \sigma_i \frac{d}{dt} Y_{xi} - \beta_i Y_{xi} - \alpha_i \frac{d^2}{dt^2} Y_{xi} \quad (i = 0,1,2) \tag{2.40}$$

lead to the 'P+I+D' form of adaptive laws,

$$c_j(t) = - \rho_j Y_{uj} - \lambda_j \int Y_{uj}\, dt - \eta_j \frac{d}{dt} Y_{uj} + c_j(0) \tag{2.41}$$

$$k_c(t) = \rho_2 Y_r + \lambda_2 \int Y_r\, dt + \eta_2 \frac{d}{dt} Y_r + k_c(0) \tag{2.42}$$

$$h_i(t) = - \sigma_i Y_{xi} - \beta_i \int Y_{xi}\, dt - \alpha_i \frac{d}{dt} Y_{xi} + h_i(0) \tag{2.43}$$

and
$$\dot{V}_{mO} = \dot{V}_{n1} + \dot{V}_{n2} + \dot{V}_{n3} \tag{2.44}$$

The $\dot{V}_{n1}$, $\dot{V}_{n2}$ and $\dot{V}_{n3}$ are as given in Eqns. (2.30 - 31). Thus, the $\dot{V}_{mO}$ is negative definite, since the Q matrix has been chosen as positive definite and the $\dot{V}_{n2}$ and $\dot{V}_{n3}$ are either negative or zero. Therefore, the filtered system error equation, Eqn. (2.10), is asymptotically stable,[1-8] with the adaptive laws of Eqns. (2.41 - 43) resulting in $\underline{e} \to 0$ as $t \to \infty$.

2.4 Families of Adaptive Laws

Two families of adaptive laws[6] of different forms can be obtained from the above development. This feature is summarized in Table I.

Table I

OLD LAWS Eqns.(2.41-43)			
Form	I	P+I	P+I+D
Conditions	$\rho_j = 0$; $\eta_j = 0$ $\sigma_i = 0$; $\alpha_i = 0$	$\eta_j = 0$ $\alpha_i = 0$	$\lambda_j \neq 0$; $\rho_j \neq 0$; $\eta_j \neq 0$ $\beta_i \neq 0$; $\sigma_i \neq 0$; $\alpha_i \neq 0$
NEW LAWS Eqns.(2.26-28)			
Form	I+DI	P+I+DI	
Conditions	$\rho_j = 0$ $\sigma_i = 0$	$\lambda_j \neq 0$; $\rho_j \neq 0$; $\eta_j \neq 0$ $\beta_i \neq 0$; $\sigma_i \neq 0$; $\alpha_i \neq 0$	
$(j = 0,\ldots m;\ i = 0,\ldots n-1)$			
See Eqn.(2.21); Family - I : $\gamma = 0, Z_f = W_f$; Family - II : $\gamma \neq 0, Z_f = W_f + \gamma P_m(s)\, e_f$			

2.5 Realization Aspects of the MRAC

Consider the scheme shown in Fig. 1. The block before the plant poses a realization problem as the order of F(s) is n, while the order of C(s) is m ($\leq$ n-1).

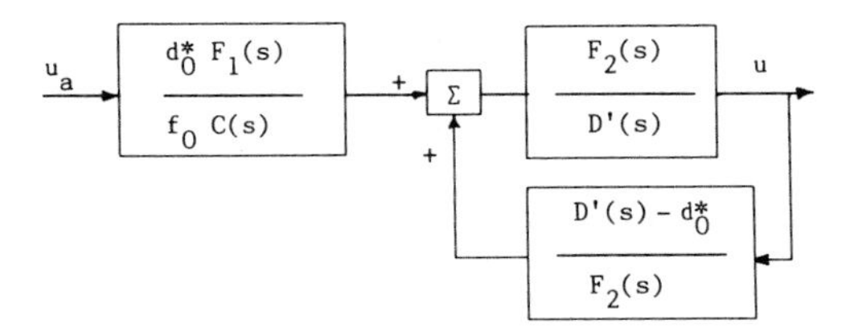

Fig.2. Realization of $[F(s)/f_0.C(s)]$

This difficulty can be circumvented by choosing F(s) as a product of two stable polynomials $F_1(s)$ and $F_2(s)$ of orders m and n* (=n - m) respectively; and, realizing the $F_2(s)$ separately, using positive feedback as shown in Fig.2, where

$$D'(s) = s^{n^*}+\ldots + d_1's + d_0';\ D^*(s) = D'(s) - d_0^* \text{ with } d_0^* > 0 \tag{2.45}$$

are suitably chosen stable polynomials. Since u_a consists of two signals, which have undergone filtering through the inverse of F(s), this realization would work in this context[12]. Next, since

$$u_f = [f_0/F(s)]\ u = [f_0/F(s)]\ [F(s)/f_0\ C(s)]\ u_a = u_a/C(s) \tag{2.46}$$

the required signals u_f and its (m-1) derivatives can be obtained without separately passing u through $[f_0/F(s)]$.

2.6 Modification When the Reference Model has Zeros

When the reference model is chosen to have zeros, the only modification is the redefinition of the signal r_f as

$$r_f = [f_0\ Z_m(s)/F(s)]r \tag{2.47}$$

instead of $r_f = [f_0/F(s)]r$ as in Eqn. (2.4). Here, the $Z_m(s)$ is the prechosen numerator polynomial of the reference model transfer function.

3. SIMULATION RESULTS

Digital simulation for the following examples has been carried out on VAX 11/780.

Example 1: The reference model without zeros having $k_m = a_o = 3$ is chosen. The plant is of first order with the parameters set (for simulation) as $k_p = 1$, $g_o = a_o$ and the controller gain $k_p(0) = 1.0$. The $F(s) = s + 10$. The input is a step of magnitude 3 units. The P+I form of adaptive law of family - II is used, with the adaptive gains $\rho_2 = 0.05$, $\lambda_2 = 1.0$. The system error response is shown in Fig. 3 a.

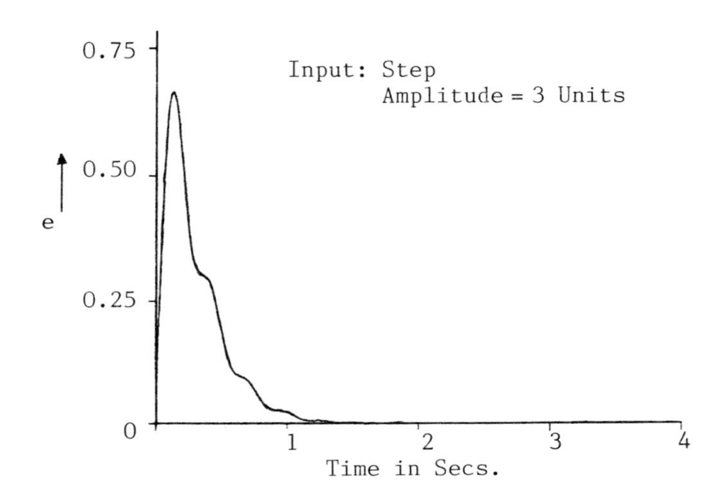

Fig.3a. System error response for Example 1

Example 2: A third order plant without zeros is considered so that $n^* = 3$. The reference model is chosen without zeros, $k_m = a_o = a_2 = 6$; $a_1 = 11$. For simulation, the plant parameters are set as $g_2 = a_2$; $g_1 = a_1$; $g_o = 4$, $k_p = 2$. The controller parameters are initially set as $k_c = 1.0$ and $h_o = 0$. The input is a square wave of amplitude unity and frequency 0.1 Hz. The P+I+DI form of adaptive law of family-I is used, with the adaptive gains $\rho_2 = \sigma_o = 2$; $\lambda_2 = \beta_o = 4$; $\eta_2 = \alpha_o = 0.1$. The system error response is shown in Fig. 3b. In this example, the $F(s) = (s+2.5)(s+1.5)$ is chosen; it is only of (n-1) th order (unlike n th order, as stated earlier) polynomial, which is sufficient for family-I laws.

In both the examples, the system error converged to zero in a satisfactory manner, while other signals in the controller remained bounded.

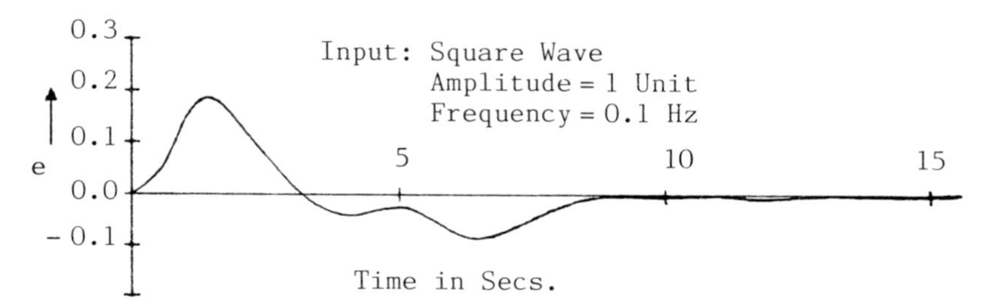

Fig.3b. System error response for Example 2

4. CONCLUSION

Lyapunov model reference adaptive control problem is addressed in this paper. Two families of modified adaptive laws of different forms (I, P+I, P+I+D, I+DI, P+I+DI) are developed, using modified Lyapunov functions with a new differentiator-free MRAC scheme. These laws do not assume that the plant parameters are directly incremented by the controller parameters and do not require plant output derivatives. Even when the plant relative degree $n^* \geq 2$, the MRAC scheme and the laws have simple structure, with considerable flexibility; do not require many auxiliary signals and filters; hence, appear to be suitable for applications. The reference model can be without or with zeros. Simulation results on simple examples show satisfactory working of the laws and the scheme. However, more research needs to be done towards resolving the issues of robustness of the scheme for plant modelling errors, relative speeds of error convergence for different forms of laws, and bounds on the choice of the adaptive gains.

ACKNOWLEDGEMENTS

The second author gratefully thanks the authorities of the Indian Institute of Technology, Bombay, India; Prof. K.S. Narendra of Yale University, U.S.A.; Prof. I.J. Nagrath of B.I.T.S., Pilani, India; and Dr. G.W. Irwin of the Queen's University of Belfast, U.K., for their encouragement.

5. REFERENCES

1. Winsor, C.A. and Roy, R.J. (1968) "Design of model reference adaptive control systems by Liapunov second method," IEEE Trans. Automat. Contr., vol.AC-13, p.204.

2. Gilbert, J.W., Monopoli, R.V. and Price, C.F. (1970)

"Improved convergence and increased flexibility in the design of model reference adaptive control systems,' Proc. IEEE Symp. Adaptive Processes, Decision and Control.

3. Sutherlin, D.W. and Boland III, J.S. (1973) "Model reference adaptive control system design technique," ASME J. Dynamic Systems, Measurement and Control, vol. 95, pp.374-379.

4. Cate, A.J.U.T. (1975) "Improved convergence of Lyapunov model reference adaptive systems by a parameter misalignment function," IEEE Trans. Automat. Contr., vol.AC-20, pp.132-134.

5. Colburn, B.K. and Boland III, J.S. (1976), "Extended Lyapunov model reference adaption law for model reference adaptive systems," IEEE Trans. Automat. Contr., vol.AC-21, pp.879-880.

6. Colburn, B.K. and Boland III, J.S. (1978) "A unified approach to model reference adaptive systems, Part-I: theory; and Part-II: application of conventional design techniques," IEEE Trans. Aerosp. and Electron. Syst., vol.AES-14, pp.501-523.

7. Chandrasekhar, J. and Vittal Rao, M.P.R. (1980) "A new form of adaptive law for Liapunov MRAC, " Proc. Joint Automatic Control Conference, Part-II, paper no. FP-2D.

8. Chandrasekhar, J. and Vittal Rao, M.P.R. (1982) "Proportional + integral + double integral adaptive laws for Liapunov MRAC," Proc. American Control Conference, S.no.TA-5, pp.557-561.

9. Parks, P.C. (1966) "Liapunov redesign of model reference adaptive control systems," IEEE Trans. Automat. Contr., vol.AC-11, pp.362-367.

10. Monopoli, R.V. (1974) "Model reference adaptive control with an augmented error signal," IEEE Trans. Automat. Contr., vol.AC-19, pp.474-484.

11. Narendra, K.S. and Valavani, L.S. (1978) "Stable adaptive controller design-direct control," IEEE Trans. Automat. Contr., vol.AC-23, pp.570-583.

12. Chandrasekhar, J. and Vittal Rao, M.P.R. (1984) "A new differentiator-free Lyapunov MRAC with increased flexibility and simple structure," Proc. IX IFAC World Congress at Budapest, Hungary, S.no.14.4, paper no. 39.

KEYWORD INDEX